Palgrave Studies in Animals and Literature

Series Editors
Susan McHugh
Department of English
University of New England
Auburn, ME, USA

Robert McKay
School of English
University of Sheffield
Sheffield, UK

John Miller
School of English
University of Sheffield
Sheffield, UK

Various academic disciplines can now be found in the process of executing an 'animal turn', questioning the ethical and philosophical grounds of human exceptionalism by taking seriously the nonhuman animal presences that haunt the margins of history, anthropology, philosophy, sociology and literary studies. Such work is characterised by a series of broad, cross-disciplinary questions. How might we rethink and problematise the separation of the human from other animals? What are the ethical and political stakes of our relationships with other species? How might we locate and understand the agency of animals in human cultures?

This series publishes work that looks, specifically, at the implications of the 'animal turn' for the field of English Studies. Language is often thought of as the key marker of humanity's difference from other species; animals may have codes, calls or songs, but humans have a mode of communication of a wholly other order. The primary motivation is to muddy this assumption and to animalise the canons of English Literature by rethinking representations of animals and interspecies encounter. Whereas animals are conventionally read as objects of fable, allegory or metaphor (and as signs of specifically human concerns), this series significantly extends the new insights of interdisciplinary animal studies by tracing the engagement of such figuration with the material lives of animals. It examines textual cultures as variously embodying a debt to or an intimacy with animals and advances understanding of how the aesthetic engagements of literary arts have always done more than simply illustrate natural history. We publish studies of the representation of animals in literary texts from the Middle Ages to the present and with reference to the discipline's key thematic concerns, genres and critical methods. The series focuses on literary prose and poetry, while also accommodating related discussion of the full range of materials and texts and contexts (from theatre and film to fine art, journalism, the law, popular writing and other cultural ephemera) with which English studies now engages.

Series Board
Karl Steel (Brooklyn College)
Erica Fudge (Strathclyde)
Kevin Hutchings (UNBC)
Philip Armstrong (Canterbury)
Carrie Rohman (Lafayette)
Wendy Woodward (Western Cape)

More information about this series at
http://www.palgrave.com/gp/series/14649

Laurence Talairach

Animals, Museum Culture and Children's Literature in Nineteenth-Century Britain

Curious Beasties

palgrave
macmillan

Laurence Talairach
Alexandre Koyré Center for the
History of Science and Technology
Paris, France

ISSN 2634-6338   ISSN 2634-6346   (electronic)
Palgrave Studies in Animals and Literature
ISBN 978-3-030-72526-6   ISBN 978-3-030-72527-3   (eBook)
https://doi.org/10.1007/978-3-030-72527-3

© The Editor(s) (if applicable) and The Author(s), under exclusive licence to Springer
Nature Switzerland AG 2021
This work is subject to copyright. All rights are solely and exclusively licensed by the
Publisher, whether the whole or part of the material is concerned, specifically the rights of
translation, reprinting, reuse of illustrations, recitation, broadcasting, reproduction on
microfilms or in any other physical way, and transmission or information storage and retrieval,
electronic adaptation, computer software, or by similar or dissimilar methodology now
known or hereafter developed.
The use of general descriptive names, registered names, trademarks, service marks, etc. in this
publication does not imply, even in the absence of a specific statement, that such names are
exempt from the relevant protective laws and regulations and therefore free for general use.
The publisher, the authors and the editors are safe to assume that the advice and information
in this book are believed to be true and accurate at the date of publication. Neither the
publisher nor the authors or the editors give a warranty, expressed or implied, with respect
to the material contained herein or for any errors or omissions that may have been made.
The publisher remains neutral with regard to jurisdictional claims in published maps and
institutional affiliations.

Cover illustration: © AF Fotografie / Alamy Stock Photo

This Palgrave Macmillan imprint is published by the registered company Springer Nature
Switzerland AG.
The registered company address is: Gewerbestrasse 11, 6330 Cham, Switzerland

To my curious beasties, with love

Acknowledgements

Animals have been part of my academic life for more than a decade. Whether living, stuffed, skeletonised, dried or pickled in preserving fluid, they have inspired me to probe into their secret histories. This project originated in 2008, when I started working with museums of natural history and medical museums, as head of a research programme developed in collaboration with the Toulouse Natural History Museum. The many conferences, study days and workshops that were organised over several years enabled me to meet many scholars from different disciplinary fields who helped me look at animals through various lenses.

Animals, Museum Culture and Children's Literature in Nineteenth-Century Britain: Curious Beasties would not have been possible without the help of the many people I have met and worked with in the years it took me to complete this book. I wish to thank the scholars and students who came to Toulouse to talk about insects, marine creatures or extinct species, as well as the many colleagues and students from my department who helped organise the events and contributed to the discussions. Many thanks are due, in particular, to Nathalie Dessens, who introduced me in 2011 to Randy Malamud, then visiting professor at the University of Toulouse Jean Jaurès. Meeting Randy opened my eyes to what zoos used to be like and still are. I was also fortunate to receive invitations to present my work from scholars from other fields, such as geologists and palaeontologists, as in May 2011 before the French Geological Society, where I talked about Edith Nesbit's 'dinosaurus'. This book would also have been different had I not been invited by Mariaconcetta Costantini to participate in the eighth international CUVSE (Centro Universitario di Studi Vittoriani e Edoardiani) conference in October 2015 at D'Annunzio University of Chieti-Pescara, Italy. There I met Raffaella Antinucci, who shared with me her extensive knowledge of Edward Lear and knew how much Lear would inspire my own research. I am also grateful to my colleagues from the 'Children's Literature and Science' group at Edinburgh Napier University, who make research on juvenile literature so exciting.

This book is also based on extensive research on two mid-Victorian children's periodicals which was made possible by generous fellowships from two libraries holding children's literature collections: the Ezra Jack Keats/Janina Domanska Research Fellowship Programme at the University of Southern Mississippi and the Armstrong Browning Library Fellowship at Baylor University respectively awarded in 2015 and 2016. The staff of both institutions did much to assist my research and help me locate materials, making my time in the archives stimulating and productive. I would especially like to thank Ellen Ruffin, curator of de Grummond collection at the University of Southern Mississippi, and Rita S. Patteson, then director of the Armstrong Browning Library, as well as Jennifer Borderud, Cynthia Burgess, Christi Klempnauer and Glenda Ross for their help, kindness and hospitality during my stay. I am very grateful as well for the assistance of Jennifer Borderud, now director of the Armstrong Browning Library, in tracing some of the images reproduced in *Curious Beasties*. Special thanks as well to Fabienne Moine, without whom I would never have experienced four magical weeks of research at Baylor.

Curious Beasties has also benefited from fieldwork performed at London Natural History Museum (Library and Archives Collection) and the British library, funded by my research centre, the Alexandre Koyre Centre for the History of Science and Technology (UMR 8560). I wish to thank the staff at both institutions for enabling me access to sometimes long-forgotten letters, as well as my research centre's executive committee and administrative staff for enabling me to cross the Channel as often as possible to access primary sources.

Since 2017, I have had the pleasure of working closely with colleagues from many natural history museums, both in France and around the world, who have provided me with information on animal collections and collectors. Having myself gone through the looking-glass of children's literature and become a children's writer to tell stories of animals in museums, I much benefitted from their experience regarding the history of museology and the mediation of knowledge. Parts of Chap. 5, on Lewis Carroll, owe a lot to a visiting fellowship at Oxford University Museum of Natural History in July 2018. I wish to thank Eliza Howlett, Chris Jarvis, Jim Kennedy, Hilary Ketchum, Darren Mann, Philip Powell, Wendy Shepherd, Zoe Simmons, Paul Smith, director of Oxford University Museum of Natural History, and Janet Stott for making my stay in Oxford so enjoyable whilst discovering thousands of objects whose stories I had often read about but never seen nor touched, as well as Lee Macdonald from the Museum of the History of Science, for helping me trace some of Charles Dodgson's photographs.

While completing this project, the combined use of the scholarship and theory of children's literature, animal studies, history of science, and literature and science studies has often been difficult, and I would like to thank the staff at Palgrave Macmillan, as well as series editors and three anonymous reviewers who provided many suggestions for improvement and enabled me to find a more engaging narrative.

A number of people have read part or all of the manuscript as it has developed in its various forms, and I would like to express my most sincere thanks for their patience, suggestions and advice. Leslie Albiston read the very first chapters of the book as this project was in its initial stages and encouraged me to keep on writing. Richard Somerset read an entire first draft of the book, and his intellectual rigour, astute comments and expertise in nineteenth-century popular science were more than useful. At a time when all frontiers were shut down and libraries closed, Neil Davie agreed to meticulously read the third draft of *Curious Beasties* while I had lost all hope of ever completing the book. I owe him more than I can say for his enthusiasm, insight and scholarly generosity. I do hope that one day I can repay their kindness.

Finally, thank you to my loving and inspiring curious beasties. Although they will here remain nameless, like many of the animals displayed in museums and zoos, they have taught me over the years that the best beasties roar and snore, and know about animals galore. This book is dedicated to them.

Several chapters of *Curious Beasties* include portions of papers delivered at various conferences, such as the International IBBY Conference (International Board of Books for Young People/NCRCL [National Centre for Research in Children's Literature]), *Deep into Nature: Ecology, Environment and Children's Literature*, held at Roehampton University in

November 2008 ('Victorian Children's Literature and the Natural World: Parables, Fairy Tales and the Construction of "Moral Ecology"'); the International Conference *Dinosaurs–Their Kith and Kin: A Historical Perspective*, organised by the French Geological Society in Paris in May 2011 ('"Are we not brothers, we and the dinosaurus?": The Cultural Impact of the Crystal Palace Park') and the conference on *Animal Bodies in Science, Art and Commerce in Early Modern Europe*, held at Humboldt University, Berlin, in December 2015 ('"We were obliged to take Harry to see the wild beasts at Exeter Exchange": Exotic animals, children's education and polite culture in Georgian Britain'). Sections from former papers delivered in 2009 at the Toulouse Natural History Museum have also been developed ('Victorian Women Popularisers of Science and Ecology' and 'From Margaret Gatty to Arabella Buckley: Victorian Women Popularisers of Science and Evolutionary Theory'). A paper on 'Extinct Creatures in Nesbit's Fiction', originally delivered in March 2011 at the EXPLORA International Conference *Lost and Found: In Search of Extinct Species*, as well as sections from a paper given at the 2012 EXPLORA International Conference *Into the Deep: Monstrous Creatures, Alien Worlds* ('Bringing the Sea to the City: The Craze for Aquaria and the Popularisation of Marine Life in Victorian England') inform Chaps. 4 and 6.

CONTENTS

1	Introduction	1
2	Wild and Exotic 'Beasties' in Early Children's Literature	23
3	Victorian Menageries	67
4	Young Collectors	125
5	Nonsense 'Beasties'	169
6	Prehistoric 'Beasties'	219
7	Epilogue	273
	Select Bibliography	277
	Index	301

LIST OF FIGURES

Fig. 2.1	'The Dowager Lady Toucan first cut it', *The Peacock 'At Home'*, *by a Lady and The Butterfly's Ball: An Original Poem, by Mr. Roscoe* (London: John Harris, 1834)	62
Fig. 3.1	Charles Camden, 'The Travelling Menagerie', *Good Words for the Young* (1871). (Courtesy of Armstrong Browning Library. Baylor University, Waco, Texas)	103
Fig. 3.2	Charles Camden, 'The Travelling Menagerie', *Good Words for the Young* (1872). (Courtesy of Armstrong Browning Library. Baylor University, Waco, Texas)	107
Fig. 4.1	M. B. S., 'Children on the Shore', *Aunt Judy's May-Day Volume for Young People* (London: Bell and Daldy, 1867)	136
Fig. 5.1	Frontispiece, Tom Hood, *From Nowhere to the North Pole. A Noah's Ark-Æcological Narrative* (London: Chatto & Windus, 1875)	172
Fig. 6.1	Henry Kingsley, 'The Boy in Grey', *Good Words for the Young* (1869). (Courtesy of Armstrong Browning Library. Baylor University, Waco, Texas)	239

CHAPTER 1

Introduction

CATALOGUE

Of Pretty Little Books for Children, sold by Mrs. Newbery, at the Corner of St. Paul's Churchyard, Ludgate Street.

...

A Description of the Tower of London, the Wild Beasts and Birds, and other Curiosities contained therein. A Description of the Custom House, the Monument, the Royal Exchange, London Bridge and Bethlem Hospital. Embellished with six Copper-Plates. Price Six-pence.

A Description of Guildhall, with the History of the Giants, Gog, and Magog, an Account of the Lord Maoyr's [sic] Shew, a Description of the Bank of England, the Mansion-House, the India-House, and the Foundling and St. Thomas's Hospitals. Embellished with Copper-Plates. Price 6 d.

An Accurate and Historical Account of St. Paul's Cathedral, with the various Changes the first Church underwent by Lightning, Fire, and other Accidents. A Description of the Cathedral as it now stands, St. Bartholomew's Hospital, Black Friars Bridge, the British Museum, Temple Bar, Northumberland House and the Statue of Charles I. at Charing Cross, adorned with six Copper-Plates. Price Six-Pence.

A Description of Westminster Abbey, with a particular Account of the Monuments and Curiosities contained therein, a Description of Westminster-Hall, the House of Lords, the House of Commons, the Queen's Palace, the Banquetting House, the Horse Guards, and the Admiralty Office, adorned with cuts. Price Six-Pence.

© The Author(s), under exclusive license to Springer Nature Switzerland AG 2021
L. Talairach, *Animals, Museum Culture and Children's Literature in Nineteenth-Century Britain*, Palgrave Studies in Animals and Literature, https://doi.org/10.1007/978-3-030-72527-3_1

N.B. The four last Articles, will be found an agreeable Pocket Companion, in viewing those Public Edifices; and may be had in Two Volumes neatly bound. Price 2 s.[1]

How we understand the Victorians today is shaped to a large extent by what they left us—in particular the innumerable collections of objects and specimens they assembled and displayed in museums throughout the nineteenth century. Indeed, public museums proliferated in nineteenth-century Britain, rising from fewer than 60 in 1850 to over 240 in 1887. By the first decades of the twentieth century, their numbers had reached 500.[2] Moreover, if museums were the 'hallmarks of Victorian advancement',[3] they were also considered places of entertainment, including for children. As illustrated above, as early as the end of the eighteenth century, Francis Newbery and his wife were promoting museum visits for children in their pocket companions for visitors to London. Over a century later, Edith Nesbit argued in *The Story of the Amulet* (1906) that children could, 'for instance, visit the Tower of London, the Houses of Parliament, the National Gallery, the Zoological Gardens, the various Parks, the Museums at South Kensington, Madame Tussaud's Exhibition of Waxworks, or the Botanical Gardens at Kew'.[4] The collections accumulated throughout the eighteenth and nineteenth centuries and presented to the general public in the second half of the nineteenth century both illuminated the Victorians' passion for collecting and symbolised the vast British empire. In other words, the objects on display in the numerous venues open to the public illustrated the expansion of knowledge throughout the long nineteenth century just as they embodied British power and hegemony.

Children were a key audience for these shows of knowledge. In particular, this book will show that zoological and palaeontological specimens brought back from around the world, whether living, stuffed or fossilised,

---

[1] Advertising at the back of *The Natural History of Birds by T. Telltruth. Embellished with Curious Cuts* (London: F. Newbery, 1778).

[2] Susan Pearce, *Museums, Objects, and Collections: A Cultural Story* (Washington, DC: Smithsonian Institution, 1992), p. 107, qtd. in Ruth Hoberman, *Museum Trouble. Edwardian Fiction and the Emergence of Modernism* (Charlottesville and London: University of Virginia Press, 2011), p. 10.

[3] Barbara J. Black, *On Exhibit: Victorians and their Museums* (Charlottesville and London: University Press of Virginia, 2000), p. 4.

[4] Edith Nesbit, *The Story of the Amulet* (s.l.: Amazon, s.d.), p. 94.

were of particular interest to late Georgian, Victorian and even Edwardian pedagogues writing for juvenile audiences. The general public's interest in European exploration of the globe, and the new knowledge those voyages revealed about the natural world, was strikingly reflected in the growing number of natural history titles published from the 1730s onwards. The range of titles available exemplifies the extent to which natural history had by that period 'emerged as a prestigious Enlightenment discipline'.[5] According to Harriet Ritvo, two natural history books were published every year between 1730 and 1750, rising to eight in the last decades of the eighteenth century—an exponential rise that continued during the following century.[6] At the same time, the number of animal species known to science increased dramatically. As Glynis Ridley notes, from 150 species identified in John Ray's *Catalogue of Quadrupeds* (1693), the number grew to 300 in Carl Linnaeus's *Systema naturæ* (1735), and by 1900, 1000 new species were being discovered and named each year.[7]

Quite logically, therefore, natural history found its way into children's fiction, and animals were given a prominent place in the narratives, and not just because animals were seen as an appropriate subject for children. As Ritvo contends, zoological discourse enabled popularisers to present 'a moral hierarchy in the animal kingdom based on the hierarchy of orders in human society'.[8] Thus, the popularisation of zoology often revealed an 'anxiety about the maintenance of social discipline' which was crucial to the British imperialist project.[9] In addition, natural history's emphasis upon classification—especially perhaps after the publication of Charles Darwin's *On the Origin of Species by Means of Natural Selection, or the Preservation of Favoured Races in the Struggle for Life* in 1859—offered a way of knowing the world, of exploring it and of *acting* in it. In particular, as this study will highlight, hunting, buying, selling, exchanging, preserving, classifying, exhibiting and reconstructing animals—in other words, *practising* natural history—were activities very likely to be represented or even encouraged in children's literature in the hope of moulding young minds into future British and imperial citizens.

[5] Harriet Ritvo, *The Animal Estate: The English and Other Creatures in the Victorian Age* (Cambridge, Mass.: Harvard University Press, 1987), p. 8.

[6] Ritvo, *The Animal Estate*, p. 8.

[7] Glynis Ridley, 'Introduction: Representing Animals', *Journal for Eighteenth-Century Studies*, 34.4 (2010): 431–6, p. 431.

[8] Ritvo, *The Animal Estate*, p. 8.

[9] Ritvo, *The Animal Estate*, p. 8.

Animals, Museum Culture and Children's Literature in Nineteenth-Century Britain: Curious Beasties will argue that the popularisation of natural history for young audiences cannot be separated from the growth of both zoological and palaeontological displays presenting amazing 'beasties' to children. Although it is often unclear who precisely the intended audience was for zoological shows and exhibits (as well as that of the contemporary literature popularising the new creatures discovered in the world or brought to Britain),[10] studying how children's literature dealt with the new ordering of the world from the time of its emergence to its heyday in the second half of the nineteenth century and the early years of the following one provides us with key insights into the construction of science in the long nineteenth century. The many natural history museums that opened during the period offered visitors an experience of the natural world under control. Whether presenting stuffed or live specimens (such as in zoological gardens), these institutions suggested a particular reading of the natural world and provided a sense of order to an ever-growing body of scientific knowledge. The systematic attempt at classifying the animal kingdom was reflected both in the galleries of the newly opened sites of knowledge and in scientific and popular publications, whether these were aimed at adult or young audiences.

Furthermore, even if '[n]atures of all kinds existed before European colonialism', Alan Bewell argues that over the course of the eighteenth and nineteenth centuries 'the natures that materially mattered most to the British were those that existed at a distance from England'.[11] In addition, imperial expansion, the appropriation of and mastery over an ever-expanding natural world and the vast enterprise of classification were inseparable, as naturalists regularly returned to England with collections of thousands of unnamed and undescribed species—species which had to be labelled and stored until the long process could begin of deciding how they fitted into the existing taxonomic system. The challenge that naturalists faced, as Harriet Ritvo contends, inevitably pointed to the weakness of that taxonomic system: '[A] system that was too rigid to make places for new discoveries, or too limited by the state of knowledge existing when it

---

[10] Ritvo explains that the size of the volume only could indicate whether works were intended for young or adult readers, the books sometimes having identical contents. She gives Thomas Bewick's *History of Quadrupeds* as a significant example. Ritvo, *The Animal Estate*, p. 9.

[11] Alan Bewell, *Natures in Translation: Romanticism and Colonial Natural History* (Baltimore: Johns Hopkins University Press, 2017), pp. 8–9.

had been devised to tolerate occasional realignments, was a system whose time had come and gone'.[12] The taxonomic anomalies 'undermined previous taxonomic assumptions and structures, [and] created a vacuum of zoological authority, if not of power'.[13] Moreover, as they increasingly pointed to a more chaotic Darwinian world, the 'curious beasties' newly imported to Britain or dug out of the earth prompted wider scientific debate. The vexed question of their rightful place in the animal kingdom led to frequent disagreements and controversy. Thus, although this book will not dwell specifically on the connections between these intriguing new creatures and the emergence of evolutionary theory, the issue of evolution will percolate into parts of this study, especially when dealing with such issues as the classification or extinction of plant and animal species.

Animals, Museum Culture and Children's Literature in Nineteenth-Century Britain: Curious Beasties examines the trajectory of children's literature in Britain from the second half of the eighteenth century to the end of the Edwardian period in relation to the rising presence of 'curious' animals, described, collected and displayed in a variety of venues, ranging from zoological gardens and natural history museums to shops. In so doing, it aims to highlight the centrality of animals to the construction of knowledge. The exploration of children's literature in the long nineteenth century as seen through the prism of popular zoology and palaeontology proposed in *Curious Beasties* is informed by Harriet Ritvo's illuminating study of taxonomic practices in eighteenth- and nineteenth-century Britain. It also aims to further Tess Cosslett's seminal study of animals in British children's literature, by connecting some of the nonhuman creatures she examines in canonical children's texts of the long nineteenth century to the fast-developing museum culture of this era.[14] Children's literature emerged, indeed, at a time when dizzying numbers of new species were flooding into England with scientific expeditions—from giraffes and hippopotami to kangaroos, wombats, platypuses and sloths. These amazing creatures inspired wonder in all who saw them, but they also confronted naturalists with the daunting task of classifying such new species, especially New World creatures. Consequently, the representation of these often strange and curious 'beasties', suddenly joining a menagerie of

[12] Ritvo, *The Animal Estate*, p. 11.
[13] Ritvo, *The Animal Estate*, pp. 13–14.
[14] Tess Cosslett, *Talking Animals in British Children's Fiction, 1786–1914* (Aldershot: Ashgate, 2006).

more familiar animals, sometimes created odd associations which this book will examine. The display cases of museums, with their sometimes scientifically dubious readings of the natural world and surprising connections between species,[15] along with the era's 'classificatory conundrums'[16] such as platypuses or kangaroos, inspired some strange encounters in children's literature. The stable system that museums presented through meticulously arranged displays imposed a classification and reading of the world which children's literature (at times humorously) reflected or even questioned. At the same time, the puzzles faced by scientists when examining those odd creatures (frequently debating their genuine nature) pervaded books for children in sometimes very surprising ways, and now and then even the shops which capitalised upon the commerce in stuffed or prepared beasts went through the looking-glass of children's literature.

That literature also emerged at a time when technological innovations made possible changes in reading patterns. Not only did literacy rates rise during the period, but the advent of the steam press and the increased speed of distribution facilitated by the expansion of the railways meant that books and periodicals were cheaper to produce, sell and circulate. According to Helen Cowie, two key phases marked the expansion of the book trade: the 'distribution revolution', between 1830 and 1850, and the 'mass production revolution' of the 1870s.[17] Moreover, as Jessica Straley explains, the boom in the number of schools established in the early nineteenth century (many of them created by religious societies) opened education to a new generation of pupils for whom new subjects had to be designed. Hence, '[e]ducators sought lessons based on everyday objects and practical skills, and the natural sciences suggested a profitable supplement to the traditional curriculum'.[18]

Animals, Museum Culture and Children's Literature in Nineteenth-Century Britain: Curious Beasties examines how children's writers took part in the Victorian appetite for mass education and presented the world and its curious creatures to children, often borrowing from museum

---

[15] Ritvo gives the example of Bullock's Museum where kangaroos were presented with monkeys; Harriet Ritvo, *The Platypus and the Mermaid and Other Figments of the Classifying Imagination* (Cambridge, Mass., and London: Harvard University Press, 1997), p. 7.

[16] Ritvo, *The Animal Estate*, p. 7.

[17] Helen Cowie, *Exhibiting Animals in Nineteenth-Century Britain: Empathy, Education, Entertainment* (Basingstoke: Palgrave Macmillan, 2014), p. 104.

[18] Jessica Straley, *Evolution and Imagination in Victorian Children's Literature* (Cambridge: Cambridge University Press, 2016), p. 59.

culture and its objects to map out that world. Looking at the way in which those 'curious beasties' symbolised British imperialism and colonial ideologies, the book traces changes in the representation and interpretation of such animals, in order to understand the role played by children's literature in reflecting, laying bare or subverting imperial discourses and ideologies of totality. To do so, it focuses in particular on some of the emblematic zoological specimens which marked the long nineteenth century and fascinated the Victorians, both professionals and the lay public. It shows how collections and displays of animals removed in time and space from their natural habitat raised in striking form the question of the human/animal divide; it highlights as well how children's literature embraced the era's material culture, and how museums, especially natural history museums, served as reference models in many children's books and stories. Indeed, *Curious Beasties* seeks to demonstrate that even in children's fantasies and fairy tales, nineteenth-century reality was shaped by the 'things'—be they stuffed objects or live specimens—that typically belonged to museum culture. Oscillating as they did between the realm of fantasy or myth and that of reality (particularly when they were unknown to science and/or appeared composite in nature), curious creatures could simultaneously transport and instruct, and stimulate wonder while remaining firmly anchored in reality.

Nineteenth-century experiments with collections of stuffed animals often placed familiar creatures within fantastic realms. For instance, the German Hermann Ploucquet's collection, exhibited at the 1851 Great Exhibition and described by Queen Victoria as 'really marvellous', displayed a frog shaving another frog whilst another carried an umbrella.[19] Some of the anthropomorphised animals presented to the public by the taxidermist of the Royal Museum in Stuttgart illustrated fables as much as they pointed to the practice of natural history. Likewise, Walter Potter's most celebrated taxidermy display, *The Death and Burial of Cock Robin* (1861), used stuffed animals to reproduce a nursery rhyme, binding the world of childhood fantasy to that of the natural world, and thus reality. As these examples show, animals were very much part of Victorian material culture and many of them, interestingly, were used as much to instruct children as to offer them journeys into fantastic lands. By miniaturising the world and bringing it to the child's level, taxidermied displays also presented a tamed and controlled picture of the world. But many of the

[19] Asa Briggs, *Victorian Things* (Stroud: Sutton Publishing [1988] 2003), p. 50.

curious creatures we will encounter in the following chapters were not simply exotic animals from far-away lands that children could only read about in books. They could also be creatures that were hard to define or classify, that seemed to be an odd mix of several animals or that had become extinct in spite of their ostensible size and strength. Through their strangeness, they illustrated the world's infinite wonders just as they emphasised the era's obsession with classification—and sometimes even the elusiveness of all attempts at ordering the world.

As Thomas Richards has argued, the new imperialism of the 1860s and the British empire that was constructed throughout the Victorian period was essentially 'a fiction'.[20] Control over the British colonies was often little more than an attempt at 'collect[ing] and collat[ing] information', and the construction of an 'archive'.[21] For Richards, therefore, the British empire 'was more productive of knowledge than any previous empire in history', explaining why 'the literature of the late nineteenth century was so obsessed with the control of knowledge'.[22] *Animals, Museum Culture and Children's Literature in Nineteenth-Century Britain: Curious Beasties* contends that the way in which the children's literature of the period presented and popularised natural history and its practice (from the collecting and preserving of specimens to their exhibition) offers a striking illustration of how the Victorians attempted to capture the elusive reality of imperial Britain. The fragments of knowledge scattered throughout the narratives, both in children's fiction and non-fiction, bridged the gap between text and object, recurrently pointing to the reality that lay outside the text, from the collection in the cabinet or the exhibit in the museum to the beast in the cage. Curious creatures encapsulated the wider world from which they had been extracted, for children to visualise and grasp, to master and possess. *Curious Beasties* will deal, of course, with the master narratives that underpinned the popularisation of knowledge throughout the long nineteenth century, such as the imperial myth. Such narratives, as we shall see, informed many a book for children. However, this study will examine above all the 'objects' themselves that became emblematic of Victorian museum culture; in particular, the creatures that children were invited to discover in fictionalised form in books, living and breathing in

---

[20] Thomas Richards, *The Imperial Archive: Knowledge and the Fantasy of Empire* (London and New York: Verso, 1993), p. 1.

[21] Richards, *The Imperial Archive*, p. 3.

[22] Richards, *The Imperial Archive*, pp. 4–5.

zoological gardens or stuffed in museums. The museum will thus be construed in myriad forms, as Barbara J. Black puts it, 'as emblem ... as institution, ... as image, [or] as practice ... that illuminates the ideological workings of Victorian society and literature'.[23] The places that we will call 'museums' will be the venues driven by collecting enterprises, including zoos and menageries of all sorts—places which, through exhibiting 'curious beasties' and inviting audiences to look at them, inevitably exposed the human/animal divide. By 'museum culture', I therefore mean the material spaces where 'curious beasties' were displayed and articulated through an epistemological framework—what Elizabeth Weiser terms 'highly crafted spaces, offering symbolic narratives via words, objects, and architecture that together shape the actions of their nations'.[24]

As Eilean Hooper-Greenhill has shown, from the beginning, the functions of museums were twofold:

> [F]irstly, to bring objects together within a setting and a discourse where the material things (made meaningful) could act to represent all the different parts of the existent; and secondly, having assembled a representative collection of meaningful objects, to display, or present, this assemblage in such a way that the ordering of the material both represented and demonstrated the knowing of the world.[25]

It will be seen throughout this study that both museums and children's literature played a key part in popular education. Both participated alike in the construction of a national identity, and both used curious specimens to attract audiences and prick their readers' interest. As the 'curious beasties' displayed in Britain found their way into children's literature, however, they helped capture systems of representations, whether the texts were allegorical, realistic, nonsensical or fantastic—sometimes disentangling the threads of representation and laying bare the links between the materiality of scientific specimens and the narratives of empire.

Consequently, because, as Barbara J. Black puts it, '[n]o other age collected with such a vengeance and to such spectacular proportions [and

---

[23] Black, *On Exhibit*, p. 4.

[24] Elizabeth Weiser, *Museum Rhetoric: Building Civic Identity in National Spaces* (University Park: Penn State University Press, 2017), p. 7.

[25] Eilean Hooper-Greenhill, *Museums and the Shaping of Knowledge* (London and New York: Routledge, 1992), p. 82.

10 L. TALAIRACH

because] [n]o other age treated the museum as an enterprise',[26] *Curious Beasties* will probe the myriad significances of the 'curious beasties' child audiences grew up with in the long nineteenth century—whether realistically described in books, fictionalised or sometimes even mythologised. It will underline the extent to which such creatures resonated with the Georgians', Victorians' and Edwardians' collections of live animals and with the chimaeras that were created, sold and purchased, and circulated in museums and private collections. It will analyse how the sense of wonderment produced by museum culture was rendered in books for children and presented to them, especially at a time when the Victorians were striving to master the world, its rules, laws and organisation, like Lewis Carroll's Alice once she had tumbled into Wonderland. Above all, this study will trace the way in which 'curious beasties', be they exotic, unknown or extinct, participated in the *making* of natural history through the practices that naturalists pursued, developed and promoted. To this end, *Curious Beasties* will explore how the main activities involved in the practice of natural history, namely the exhibition of live 'curious beasties' (Chaps. 2 and 3), the making of collections, whether living or dead (Chaps. 3 and 4), the classification of intriguing creatures (Chap. 5) and the reconstruction of extinct animals (Chap. 6), were represented in children's literature, as well as how young audiences were encouraged to participate in a practical way in the making of natural history. The chapters will follow a (generally) chronological structure, starting with early (Georgian) children's literature (Chap. 2) and ending with Edwardian children's literature (Chap. 6), although the central chapters will mainly focus on mid-Victorian children's texts. The book will also adopt in part a generic approach, since a chapter will be devoted to nonsense literature for children (Chap. 5).

As Chap. 2 highlights, the Georgian scientific enterprise permeated early children's literature. The development of children's literature from the second half of the eighteenth century, climaxing in the highly entertaining works published by John Harris at the turn of the nineteenth century (especially works by Sarah Catherine Martin, William Roscoe and Catherine Dorset), coincided with the discovery, analysis and description of new and sometimes surprising creatures, many of them originating in the New World. Unknown animals such as platypuses, koalas, wombats or echidna, mainly discovered in the 1790s, arrived in Britain and were

[26] Black, *On Exhibit*, p. 17.

described in the first decade of the nineteenth century. George Shaw had described the platypus from a dried specimen in 1799, and Everard Home dissected one in 1802.[27] The first echidna was spotted by Captain William Bligh of the *Bounty* in Tasmania in 1790 and a drawing was made of it. In 1792, George Shaw described the species from a pickled specimen sent from Australia on HMS *Gorgon* and Home classified it a decade later, using Shaw's description and William Bligh's drawing. The first koala was spotted in 1798 and a live specimen caught in Australia in 1803, whilst the very first wombats arrived in France in Nicolas Baudin's *Geographe* in 1804.[28] As for kangaroos and wallabies, the creatures had been spotted and described as early as in the mid-seventeenth century,[29] but the first kangaroos did not reach Britain until the 1790s.[30] Interestingly, although several of these animals were popularised in Thomas Bewick's *General History of Quadrupeds* (such as the wombat which appeared in the fifth edition of 1807, amongst other exotic animals[31]), hardly any of these curious Australasian 'beasties' appeared in children's literature at the turn of the nineteenth century.

Animals in mid- to late-eighteenth-century children's books were frequently used merely as metaphors for aspects of the human—mirroring, for instance, the ways in which 'the categorization of animals reflected the rankings of people both figuratively and literally'.[32] Chapter 2 provides a survey of eighteenth- and early nineteenth-century children's literature in order to show how the children's books and stories of the period reflected Enlightenment Britain, its growing scientific knowledge about the natural world and the non-European species regularly brought back to Britain and kept in menageries. As Chap. 2 shows, many wild and exotic beasts brought to Britain, collected and exhibited, were found in the children's literature of the era. Literary animals thus participated in the period's questioning of the relationship between humans and animals: they were often involved in discourses on morality and sensibility, as well as on what

[27] Ritvo, *The Platypus and the Mermaid*, pp. 14–15. The animal's mammary glands remained however undiscovered until 1824, and the fact that it laid eggs was unknown until William Hay Caldwell (1859–1941) caught an egg-bearing female in 1884.

[28] John Simons, *Rossetti's Wombat: Pre-Raphaelites and Australian Animals in Victorian London* (S.l: Middlesex University Press, 2008), p. 25.

[29] Simons, *Rossetti's Wombat*, p. 30.

[30] Simons, *Rossetti's Wombat*, p. 34.

[31] Simons, *Rossetti's Wombat*, p. 18.

[32] Ritvo, *The Platypus and the Mermaid*, p. xii.

it meant to be human, and in this way helped Georgian educationalists model ideals about gentility. However, Chap. 2 argues that whilst serving early pedagogues' didactic lessons and helping them shape ideal class relations, the curious and less curious creatures which permeated Georgian children's literature, from fable literature to moral stories, offered, at times at least, an insight into the animal mind which evoked the era's growing consciousness of animal thoughts and feelings. Moreover, anthropomorphised creatures and talking animals were not only useful to teach children about social mores; they could also help Georgian pedagogues question forms of enslavement.

The study of the appropriation of animals in early children's literature is pursued in Chap. 3, which focuses on representations of Victorian menageries. The first decades of the nineteenth century were, indeed, marked by the founding of the Zoological Society of London in 1826 and the opening of the London Zoological Gardens in 1828. In the nineteenth century, as Teresa Mangum has shown, animals—'living, working, preening, suffering, dying, and dead'—were ubiquitous, the Victorian city often becoming a 'veritable animal sensorium'.[33] The various contexts in which they were encountered explain why, as Martin Danahay puts it, the nineteenth century 'witnessed a profound shift in both the perception and treatment of animals'.[34] This was illustrated, for instance, by the founding of the Society for the Prevention of Cruelty to Animals (1824) and the rise of anti-vivisectionism in the last decades of the century. Anti-cruelty measures throughout the century reflected, as Diana Donald has shown, how the 'predominance of enlightened attitudes to animals' exemplified 'moral progress'.[35] For this reason, Chap. 3 chooses to focus on changing discourses on caged creatures in an era of imperial development. Several types of publications for juvenile audiences are examined, ranging from abecedaria and popular science articles and books to children's fiction. Chapter 3 shows that the display of animals, both living creatures and stuffed or 'skeletonized' ones, many of them 'curious', became reflected in the children's literature of the period. However, many articles and stories

---

[33] Teresa Mangum, 'Animal Angst: Victorians Memorialize their Pets', in Deborah Denenholz Morse and Martin A. Danahay (eds), *Victorian Animal Dreams: Representations of Animals in Victorian Literature and Culture* (Aldershot: Ashgate, 2007), pp. 15–34, p. 15.

[34] Martin A. Danahay, 'Nature Red in Hoof and Paw: Domestic Animals and Violence in Victorian Art', in Morse and Danahay (eds), *Victorian Animal Dreams*, pp. 97–119, p. 97.

[35] Diana Donald, '"Beastly Sights": The Treatment of Animals as a Moral Theme in Representations of London, c.1820–1850', *Art History*, 22.4 (Nov. 1999): 514–44, p. 516.

published in Victorian children's periodicals such as *Good Words for the Young* (first edited by Norman Macleod and later George MacDonald) and *Aunt Judy's Magazine* (founded by Margaret Gatty), both aimed at middle-class boys and girls, upheld, rather than undermined the divisions between the human and the animal. Throughout the second half of the nineteenth century, such articles and stories played a vital role in establishing 'popular understandings of the way in which the British empire as a whole incorporate[d] and inform[ed] Englishness'.[36] Magnifying as they did modes of representation of the British empire characteristic of the period, the articles and stories published in these periodicals often used menageries and scenes of caged (or hunted) animals as a way of teaching children how to master imperial culture. Thus, Chap. 3 looks at representations of animals caught in the wild and put in cages in order to highlight how 'beasties' were both captured by Victorian children's literature and child characters, increasingly recording and writing the empire in the course of the nineteenth century.

As will be seen in the following chapter, moreover, the boom in children's periodicals in the second half of the nineteenth century coincided with the expansion of public museums. This explains, perhaps, why Victorian children's literature sought to transmit the 'curatorial imperative'[37] to its young audiences and why the collecting enterprise of the nineteenth century reverberated so much in the children's literature of the period and in the various discourses which shaped narratives dealing with collecting. Often intertwined with expectations related to the gendered pursuit of knowledge, collecting was however not simply an illustration or aspect of the practice of natural history. In the second half of the nineteenth century, collecting was described as essential to both the practice of natural history and to the making of science. Writers often illuminated how inseparable knowledge of natural history was from collecting. For Barbara Black, 'representations of the museum … promoted the fantasy that collecting brought the world home and thus domesticated the world into home'.[38] Inviting therefore children to collect the world in their turn so as to observe and possess it at home, as a miniature museal activity, was

---

[36] Kurt Koenigsberger, *The Novel and the Menagerie: Totality, Englishness, and Empire* (Columbus: Ohio State University Press, 2007), p. 4.

[37] Black, *On Exhibit*, p. 149.

[38] Black, *On Exhibit*, p. 150.

a frequent means of acculturation which was interestingly often represented and even encouraged in children's literature.

In the second half of the nineteenth century, children were offered manifold activities that would allow them to practise collecting. Barbara Black mentions, for example, games which involved collecting, such as the 'Panorama of Europe: A New Game and a Day at the Zoo', whose objective was to paste bars over various exotic animals.[39] Adventure literature was also 'the most obvious avenue by which the museum came home to children':[40] indeed, adventure stories fictionalised 'the museum enterprise while ordaining innumerable Victorian boys into a community of collectors'.[41] Building on Chap. 3, Chap. 4 moves away from modes of writing the wild and the unknown to focus instead on the various meanings informing the word 'museum' and the significance of the collection. As Laura White has argued, the Victorian craze for natural history spurred 'shifts in thinking about the human/nature divide', especially amongst 'the adherents of conservative orthodoxy'. In addition to natural history books, the publication of which expanded enormously in this period, the 'fad of natural history collecting … also made evident a widespread cultural anxiety about what nature actually means and is'.[42] The question of how mid-Victorian children's magazines illuminated the relationship between humans and more or less familiar creatures by inviting children to collect the natural world around them lies at the heart of this chapter, which focuses on *Aunt Judy's Magazine*, edited by the naturalists and children's writers Margaret Gatty and her daughters, Juliana Horatia Ewing and Horatia Katherine Frances Gatty. Chapter 4 examines, therefore, the pedagogical strategies in a Victorian children's periodical published during the new imperialism of the 1860s, pointing out how these writers familiarised children with the practice of natural history, such as collecting, exchanging, preserving and exhibiting specimens, hence reframing the world for young audiences. By encouraging children to collect the world,

---

[39] Black, *On Exhibit*, p. 151.

[40] Black, *On Exhibit*, p. 151.

[41] Black, *On Exhibit*, p. 152. Black focuses on late-Victorian novels, however, such as Rudyard Kipling's *Kim* (1901) and Edith Nesbit's *The Story of the Amulet* (1906), to explore 'the significance of acquisition', as well as children's magazines, such as *Boys of the Empire* (1888–93) or *The Boy's Own Paper* (1879–1967)—the titles of which clearly indicated a young male readership.

[42] Laura White, *The* Alice *Books and the Contested Ground of the Natural World* (Abingdon, New York: Routledge, 2017), p. 2.

Aunt Judy's Magazine, just like *Good Words for the Young*, as discussed in Chap. 3, shaped the vast and growing British empire into a miniature world capable of being readily mastered. With their emphasis on collecting, these magazines reflected the imperial desire to contain the wild and the unknown, illustrating how constructions of knowledge were dependent upon the construction of collections. As these examples of Victorian children's literature illustrate, nineteenth-century museum culture did not so much reflect the growth of knowledge of the era as evidence the pivotal role of museums in the making of knowledge and in its diffusion: the vaster the British empire became, the greater became the emphasis on the desire to contain it in the children's literature of the period. Controlling and possessing a world transformed into collectible, purchasable or exchangeable objects, as some of these periodicals suggested, lay at the heart of the Victorian child's education. Using Susan Stewart's analysis of collections as a means of 'replac[ing] history with classification, with order beyond the realm of temporality',[43] Chap. 4 thus looks at the collecting enterprise which permeates *Aunt Judy's Magazine*, and analyses the way in which children's collections (whether real or represented) 'refram[ed] objects within a world of attention and manipulation of context'.[44]

Jim Endersby has observed that the mid-Victorian natural history sciences 'were pre-eminently concerned with collecting and classifying, activities that some practitioners of the physical sciences regarded with disdain'.[45] In addition, as Harriet Ritvo has shown, some species of Australian fauna, such as the platypus, disrupted classificatory systems and for this reason appeared 'in non-specialist contexts much more frequently than did those of any other exotic animal of similarly insignificant size and aspect'.[46] Furthermore, its combination of bird, reptile and mammal—according to late eighteenth-century naturalists' descriptions of the animal—took forward into the nineteenth century 'a Renaissance habit of interpreting American novelties as monstrous recombinations of familiar parts, analogous to the chimaeras and yales of medieval bestiaries'.[47] This

---

[43] Susan Stewart, *On Longing: Narratives of the Miniature, the Gigantic, the Souvenir, the Collection* (Durham and London: Duke University Press, 1993), p. 151.

[44] Stewart, *On Longing*, p. 151.

[45] Jim Endersby, 'Classifying Sciences: Systematics and Status in mid-Victorian Natural History', in Martin Daunton (ed.), *The Organisation of Knowledge in Victorian Britain* (Oxford: Oxford University Press, 2005), pp. 61–85, p. 61.

[46] Ritvo, *The Platypus and the Mermaid*, p. 4.

[47] Ritvo, *The Platypus and the Mermaid*, p. 132.

16   L. TALAIRACH

was a tradition on which Victorian children's writers were able to draw. As a matter of fact, scientists often tried to explain the unknown by referring to the known, using many comparisons and increasingly adding pictures to their descriptions. Revealingly, such tensions between the known and the unknown pervaded children's literature, in particular through the nonsense literature of the second half of the nineteenth century, as Chap. 5 underlines, by exploring the Victorian cabinets of curiosities devised by Edward Lear and Lewis Carroll and the monsters created by their nonsense. As Laura White argues, both Lear's and Carroll's nonsense was 'zoological' in that both more or less tackled 'predator-prey relationships',[48] and both hinted as well at 'reversals in hierarchy'.[49] White argues that Carroll 'obsessively joke[d] about the uneasy status of human beings at the apex of creation and the top of the food chain', mocking thereby 'what he saw as his age's growing softheadedness about the divide between the human and the animal'.[50] It will be argued here, *pace* White, that Lear's and Carroll's numerous references to the 'zoological enterprise'[51] of their time did not in fact reinforce the traditional human-animal divide, but, on the contrary, could serve to question playfully contemporary taxonomic systems.

As the last chapters of *Curious Beasties* will highlight, the Victorian period was marked not only by the explosion of children's periodicals, as studied in Chaps. 3 and 4, but also by that of fantasies, many of them aimed at children. Works such as Charles Kingsley's *The Water-Babies, A Fairy Tale for a Land Baby* (1863) and Lewis Carroll's *Alice's Adventures in Wonderland* (1865), both presented as 'fairy tales' by their authors, radically revamped the genre of the literary fairy tale, proposing new directions for children's literature and paving the way for numerous rewritings,

---

[48] White, *The* Alice *Books and the Contested Ground of the Natural World*, p. 120.

[49] White, *The* Alice *Books and the Contested Ground of the Natural World*, p. 122.

[50] White, *The* Alice *Books and the Contested Ground of the Natural World*, pp. 1, 3. White sees Lewis Carroll [Charles Dodgson] as exclusively Christian and Tory, supportive of 'many of the older views on the relationship between people and nature, including the absolute sovereignty of humankind over animals' (p. 2). According to White, Carroll thus often simply mocks or jokes about Darwinian accounts of the survival of the fittest or contemporary views about the reality of species extinction in the *Alice* books. Our point is not to look for evidence of Carroll's 'conservative and Christian attitudes toward the natural world' and how these 'inform his treatment of animals and anthropomorphism' (p. 4), but rather to look at Carroll's nonsense as a means for audiences to reflect upon the classification of curious 'beasties' as displayed in various Victorian venues.

[51] Ritvo, *The Platypus and the Mermaid*, p. 19.

1 INTRODUCTION   17

pastiches and imitations. Carroll's *Alice* books, moreover, were both reminiscent of earlier children's literature and animal fables, and often mocked them through satire, subverted them and turned them upside down—or inside out. Chapter 5 contends that Victorian nonsense offered a new language to deal with contemporary constructions of the world, proposing an original prism through which to examine the natural world and the place of humans in it, thus undermining the visions of empire and the totalising enterprise of the museum collection explored in the previous chapters. Drawing upon Jean-Jacques Lecercle's analysis of the discourse of natural history in nonsense and his definition of nonsense texts as 'reflexive texts',[52] Chap. 5 highlights how nonsense texts subvert the classification system and therefore challenge the imperial appropriation of animals. It also shows that if, as Eilean Hooper-Greenhill explains, the 'archetypal "cabinet of curiosity" is the German *Wunderkammer*, which is understood as a disordered jumble of unconnected objects, many of them fraudulent in character',[53] then nonsense provided the Victorians with a similar disordered jumble of fraudulent objects, as illustrated by many of the hybrids found in the *Alice* books.

As soon as she tumbles down the rabbit hole, Carroll's Alice is, interestingly, turned into an object which Wonderland animals examine, trying as they do to classify her. The little girl's shift from subject to object needs to be seen, as Jed Mayer argues, as part of 'a shift in her understanding of the place of the animals in Victorian society, and the complicity of spectatorship in rendering animals as objects of contemplation, whether for amusement, study, or experimentation'.[54] Meyer's conclusions regarding the relationship 'between peeping and marketing' in Christina Rossetti's 'Goblin Market' (1862), published around the same time, is significant in this context. The poem was certainly influenced by Dante Gabriel Rossetti's own collection of 'curious beasties', which included armadillos, kangaroos, wallabies and a wombat, among many other creatures, as well as the numerous creatures he enjoyed viewing at the London Zoological Gardens. For Mayer, because the reader is engaged in the act of peeping

[52] Jean-Jacques Lecercle, *Philosophy of Nonsense: The Intuitions of Victorian Nonsense Literature* (London and New York: Routledge, [1994] 2002), p. 2.

[53] Hooper-Greenhill, *Museums and the Shaping of Knowledge*, p. 79.

[54] Jed Mayer, '"Come Buy, Come Buy": Christina Rossetti and the Victorian Animal Market', in Laurence W. Mazzeno and Ronald D. Morrison (eds), *Animals in Victorian Literature and Culture. Contexts for Criticism* (London: Palgrave Macmillan, 2017), pp. 213–31, p. 221.

at the sisters, their privacy 'is made into a spectacle, and they resemble the animal figures with whom they share a frame'.[55] The humans are thus turned into a zoo exhibit whilst 'enjoy[ing] a symbiotic relationship with nature'. In this way, they are 'bestialized', Laura's 'behaviour and tastes [becoming] more animalistic'.[56] By becoming the objects of the reader's gaze, Rossetti's female characters are the entrapped spectators of beast-like goblins, recalling Lewis Carroll's avid consumer in *Alice's Adventures in Wonderland* in which Alice's untameable desire is also linked to her journey into a fantastic world peopled with strange and unknown creatures. As will been seen, the play on 'species hierarchies in which animals are reduced to objects of exchange and humans are bestialized'[57] similarly informs Carroll's fantasy, peopled by babies which metamorphose into pigs and little girls taken for serpents, as in Rossetti's poem, which features a heroine made 'of the same animal substance'.[58] In Carroll's and Rossetti's texts, furthermore, the 'curious beasties' (as exotic or fantastic creatures) engender the inversion of gaze: the fantastic zoos look back, deconstructing their own objectification in (sometimes) humorous and satirical ways—not only denouncing the commodification of animals but blurring as well the divide between humans and animals.

Thus, Chap. 5 shows how Edward Lear's writings for children, as much as Carroll's *Alice's Adventures in Wonderland* and *Through the Looking-Glass, and What Alice Found There* (1871), used nonsense to reflect the scientific context of the time and examine their era's 'zoological enterprise'. Their nonsense radically revisited the constructions of the natural world for young audiences that had informed the realistic children's narratives studied in Chaps. 3 and 4. In both prose and poetry for children, their nonsense becomes a means to accommodate resemblances and, as a result, to humorously supersede contemporary classificatory difficulties. Whilst Kingsley had popularised evolutionary theory in his parable, *The Water-Babies*, Carroll reversed Kingsley's enchanting vision of evolution to propose a much darker world ruled by blind chance and instability,

---

[55] Mayer, '"Come Buy, Come Buy": Christina Rossetti and the Victorian Animal Market', p. 224.

[56] Mayer, '"Come Buy, Come Buy": Christina Rossetti and the Victorian Animal Market', pp. 224–5.

[57] Mayer, '"Come Buy, Come Buy": Christina Rossetti and the Victorian Animal Market', p. 227.

[58] Mayer, '"Come Buy, Come Buy": Christina Rossetti and the Victorian Animal Market', p. 221.

following on Lear's representation of humans subjected to the forces of evolution. Carroll's and Lear's works, however, were also informed by contemporary knowledge about the natural world; both narratives presented to children the world of natural history as a realm devoted to the exploration, collection and exhibition of strange and unfamiliar creatures. Indeed, while Lewis Carroll's Alice believes her fall into the rabbit hole has sent her straight to the Antipodes, the mid-Victorian experimental fairy tale illustrates the impact of unsuspected creatures from the other end of the world on the popular imagination: Carroll's emphasis on his little girl's 'curiosity' and the 'curious' creatures she encounters in Wonderland or in Looking-Glass land takes readers into a museum world of unfamiliar objects, new and mythical monsters, and even composites. It will be seen, therefore, that Carroll's 'curious beasties'—from 'the white rabbit with pink eyes' to the Gryphon, half eagle, half lion and the Mock Turtle, half calf, half turtle—probe the era's taxonomic practices.

Lewis Carroll's imaginary tracking of a curious beast in *The Hunting of the Snark* (1876), published over a decade after his first *Alice* book, further demonstrated his interest in enigmatic creatures. Interestingly, Carroll refused one of Henry Holiday's illustrations for his 'Boojum', dismissing it as 'unimaginable', according to Holiday. But the latter retorted in 1898 that while 'unimaginable', Carroll's Boojum might not, in fact, be impossible. As the artist surmised, the creature might one day be discovered, or at least its remains found:

> Mr. Dodgson wrote that is was a delightful monster, but that it was inadmissible. All the descriptions of the Boojum were quite unimaginable, and he wanted the creature to remain so. I assented, of course, though reluctant to dismiss that I am still confident is an accurate representation. I hope that some future Darwin, in a new *Beagle*, will find the beast, or its remains; if he does, I know he will confirm my drawing.[59]

The idea that ever more improbable creatures might one day be found, derived from Darwinian theory, permeated both scientific and literary discourses, as Chap. 6 underlines. Indeed, some of the mythical or monstrous beasts found in nineteenth-century children's literature were informed by recent developments in popular science. Chapter 6 examines

[59] Henry Holiday, 'The Snark's Significance', *Academy* (29 January 1898), qtd. in Morton N. Cohen and Edward Wakeling (eds), *Lewis Carroll and his Illustrators. Collaborations & Correspondence, 1865–1898* (Ithaca and New York: Cornell University Press, 2003), p. 33.

how the exhibition of newly found extinct creatures—some of them fantastic composites made from the bones of different species and specimens—inspired children's writers. In the first half of the nineteenth century, the development of popular science owed much to the new disciplines of geology and palaeontology. Scientists were interested in disseminating their own scholarly specialities and competed with popularisers, eager to tap into the commercial potential of the mass market. Consequently, popular science publications and events became more and more sensational and spectacular. This sensationalisation of science increased in the second half of the nineteenth century and took a variety of forms, from public lectures and mass-market books to museums and travelling exhibitions.[60] As Chap. 6 emphasises, focusing on representations of the Crystal Palace dinosaur park, an open-air museum which epitomised contemporary pedagogical theories, the three-dimensional prehistoric 'beasties' devised by Benjamin Waterhouse Hawkins found their way into the children's literature of the second half of the nineteenth century and of the Edwardian period. This chapter therefore traces extinct 'beasties' through some examples of Victorian children's fiction and popular science books, and it closes with a study of Edith Nesbit's children's fiction. Nesbit's fantasies provide a fitting conclusion to our survey of children's literature in the long nineteenth century: often looking back to the Victorians and their museum culture, and deeply inspired by Lear's and Carroll's nonsense, Nesbit goes beyond their subversive construction of modern (scientific) knowledge. It will be seen that Nesbit uses her curious creatures to convey her stance on imperialism and consumerism and challenge the master narrative of British control over the natural world, proposing in so doing a much more political view of the world than that offered by comparable earlier authors.

From reflecting and processing contemporary knowledge about the animal kingdom and using it to maintain social order to denouncing the commodification of 'curious beasties' brought back from remote countries or past times, children's literature appropriated then new ways of thinking about the natural world the better to destabilise them or even

[60] Bernard Lightman, *Victorian Popularizers of Science: Designing Nature for New Audiences* (Chicago and London: The University of Chicago Press, 2007); Aileen Fyfe and Bernard Lightman (eds), *Science in the Marketplace: Nineteenth-Century Sites and Experiences* (Chicago and London: The University of Chicago Press, 2007); Ralph O'Connor, *The Earth on Show: Fossils and the Politics of Popular Science, 1802–1856* (Chicago: Chicago University Press, 2007).

1 INTRODUCTION 21

question them. Children's literature evolved dramatically, indeed, in a period marked by sensational scientific publications, from Buffon's *Histoire naturelle* to Darwin's *On the Origins of Species*, works which, in Diana Donald's words, 'brought animals to the forefront of consciousness, and caused them to be seen in a new light'.[61] By registering the changes in the understanding of the natural world and often bringing to light anthropocentric views of animals, children's literature played a key part in making both children and adults aware of contemporary ideologies central to the Western thought of the eighteenth and nineteenth centuries. Spanning over 150 years of children's literature, *Animals, Museum Culture and Children's Literature in Nineteenth-Century Britain: Curious Beasties* thus reveals how children's writers, educationalists and pedagogues used creatures that were hunted, collected and exhibited as emblems of Britain's imperial and capitalist systems. As we will see in the following chapters, by inviting children to engage with the material culture, breeding miniature imperialists or showing children the futility of attempting to master the world, the children's literature of the long nineteenth century often lay bare the ideologies informing the practice of natural history—ultimately reflecting, albeit through a glass darkly, the 'museumification' of the age.[62]

[61] Diana Donald, *Picturing Animals in Britain, 1750–1850* (New Haven and New York: Yale University Press, 2007), p. vii. The period from the publication of Buffon's *Histoire naturelle* to that of Darwin's *On the Origins of Species* corresponds to the timeframe of Donald's study.

[62] I am here adapting Barbara Black's 'museumified'; Black, *On Exhibit*, p. 170.

CHAPTER 2

Wild and Exotic 'Beasties' in Early Children's Literature

At leisure let us view, from day to day,
As they present themselves, the Spectacles
Within doors: troops of wild Beasts, birds and beasts
Of every nature, from all climes convened ...[1]

In Jane Austen's *Sense and Sensibility* (1811), one of the characters, M. John Dashwood, apologises to his sister for not having called upon her earlier, having been 'obliged to take Harry to see the wild beasts at Exeter Exchange',[2] a popular menagerie at the end of the eighteenth century. The sites where animals, in particular exotic animals, were collected and exhibited evolved dramatically in the course of the Georgian era (1714–1830). The 1820s represented a watershed in this respect, with the opening of the London Zoological Gardens in 1828 and the closures of the Exeter Exchange (or Exeter Change) in 1829 and the Tower menagerie in 1835. Animals, whether dead or living, were increasingly sold, exchanged and exhibited throughout the eighteenth and nineteenth centuries. They were central to anatomists' work, displayed

---

[1] William Wordsworth, 'Residence in London', *The Prelude* [1805], Book 7, in Stephen Gill (ed.), *The Oxford Authors: William Wordsworth* (Oxford and New York: Oxford University Press, 1990), pp. 468–86, p. 474, l. 244–7.

[2] Jane Austen, *Sense and Sensibility* [1811] (Oxford: Oxford University Press, [1980] 2008), p. 166.

© The Author(s), under exclusive license to Springer Nature Switzerland AG 2021
L. Talairach, *Animals, Museum Culture and Children's Literature in Nineteenth-Century Britain*, Palgrave Studies in Animals and Literature, https://doi.org/10.1007/978-3-030-72527-3_2

23

for entertainment and instruction and even consumed as luxury goods. It is thus not coincidental that their presence in society became reflected in the literature of the age, from William Wordsworth's reference to exhibited wild animals in *The Prelude* to Jane Austen's allusion to Exeter Change menagerie in *Sense and Sensibility*, a novel written in the 1790s. As this last example underlines, moreover, the places where animals were exhibited also catered for children, with contemporary advertisements urging parents to take their offspring to see these living attractions 'from all climes convened'.

Wild and exotic animals were also increasingly found in books for children, be they guidebooks, natural histories, or fables and tales aimed at young audiences. In fact, as this chapter will show, the children's literature of the second half of the eighteenth century was thoroughly suffused with such 'curious beasties'—those same animals which were being daily imported into Britain and which could be seen in the numerous travelling shows and menageries of the day. However, as knowledge about them increased during the course of the century, a change can be observed in how these exotic and sometimes wild creatures were represented in popular advertisements and children's literature. In eighteenth-century children's literature—both juvenile natural history and fiction—exotic animals 'were quickly recognized as promising didactic instruments'.[3] The children's literature of the age also reflected the intense circulation of animals in Georgian Britain, menageries and museums, coffee shops and even homes, participating therefore in the dissemination of knowledge about many of the species that were being imported into England during that period. Yet, whilst the children's literature of the late eighteenth century participated in rendering these animals visible at a time when more and more species were being discovered and brought to England, children's books increasingly reflected the creatures' various meanings: they could be used as purveyors of scientific knowledge, as conventional images of wildness or as symbols of polite elite leisure.

---

[3] Harriet Ritvo, *The Animal Estate: The English and Other Creatures in the Victorian Age* (Cambridge, Mass.: Harvard University Press, 1987), p. 131.

Displaying 'Curious Beasties'

From the sixteenth century onwards, more and more animals were imported into Britain. Brown bears could be seen in London at the beginning of the seventeenth century,[4] a rhinoceros was exhibited at the Belle Sauvage Inn for a few days in 1683,[5] and by the last decades of the century, even children were being taken to see the wild beasts of the Royal Menagerie.[6] In the eighteenth century, the expansion of the British empire and the growth of consumer society meant that shows and exhibitions of animals became popular venues for Britons keen to acquire knowledge and education. It was especially in the 1760s that menageries and animal dealers became visible in the British capital. They were mainly located along the Strand, in Piccadilly and in St James's Park,[7] and the competition between them was intense. In the last decades of the eighteenth century, two main menageries were competing in London: Exeter Change menagerie (also known as Pidcock menagerie, Polito's menagerie or even the Royal Menagerie) in the Strand, and the Royal Menagerie of the Tower of London. Exeter Change had opened in 1773, when Thomas Clark, an animal dealer, started advertising animal displays on the first floor of the building. The collection was bought by Gilbert Pidcock in 1793 and later owned by Stephan Polito, the two men managing travelling circuses and using the London premises as winter quarters for the animals. It was later purchased by Edward Cross, one of Polito's former employees, after the latter's death in 1814. The menagerie attracted the attention of many contemporary writers and artists, from Romantic poets

[4] Daniel Hahn, *The Tower Menagerie: The Amazing True Story of the Royal Collection of Wild Beasts* (London: Simon & Schuster, 2003), p. 86.

[5] Hahn, *The Tower Menagerie*, p. 142. See also Glynis Ridley, *Clara's Grand Tour: Travels with a Rhinoceros in Eighteenth-Century Europe* (New York: Grove Press, 2004) for references to Clara by the writers of the age, such as James Boswell, Oliver Goldsmith or Samuel Richardson (in the 4th edition of *Clarissa* of 1759).

[6] Hahn explains that Samuel Pepys's visit in 1662 is the earliest record of children being taken to see the lions; he relates this record to the new conception of childhood—a point confirmed by the development of children's literature at that time too; Hahn, *The Tower Menagerie*, p. 118.

[7] Christopher Plumb, 'Exotic Animals in Eighteenth-Century Britain', PhD. diss. (University of Manchester, 2010), p. 55, qtd. in Ingrid H. Tague, *Animal Companions: Pets and Social Change in Eighteenth-Century Britain* (University Park, Pennsylvania: The Pennsylvania State University Press, 2015), p. 32. See also Jessie Dobson, 'John Hunter's Animals', *Journal of the History of Medicine and Allied Sciences*, 17.4 (Oct. 1962): 479–86.

26   L. TALAIRACH

William Wordsworth and Lord Byron to artists, such as Edwin Landseer and Jacques-Laurent Agasse,[8] who painted animals in the collection. When Exeter Change was demolished in 1829, Cross sold some of his animals to the London Zoological Gardens (which had opened the previous year) and moved the rest to the Surrey Zoological Gardens in 1831, of which he became superintendent.[9] The Tower of London, on the other hand, had started housing the Royal Menagerie as early as 1110 and closed in 1835.

Both menageries reflected the growing power of the British empire, as naturalists explored and colonised the world, and as more and more goods, including both animals and plants, circulated globally, especially between the Old and New Worlds. Many of the contemporary advertisements for menageries foregrounded the rarity of their collections and appealed 'to the curious', emphasising the educational role of rare, wild or exotic menagerie animals. The advertisements also emphasised frequently the global movement of their collections (terms such as 'just arriv'd' or 'brought to Europe' were common), whilst the description of the animals played upon fears of the monstrous, as the following example illustrates:

To THE CURIOUS

Whatever deserves the Epithet of RARE, must certainly be worthy the Attention of the Curious.

JUST Arriv'd from the ISLAND of JAVA, in the East-Indies, and ALIVE, one of the greatest Rarities ever brought to Europe in the Age or Memory of Man, The GRAND CASSOWAR.

It is described by the late Dr. Goldsmith as follows, viz. The Head inspires some Degree of Terror like a Warrior; it has the Eye of a Lion, the Defence of a Porcupine, and the Swiftness of a Courser; but has neither Tongue, Wing nor Tail. Its Legs are stout like the Elephant, Heel as the Human Species, and three Toes before; it is upwards of six Feet high, and weighs above 200 lb. Its Head and Neck is adorned with a Variety of beauti-

---

[8] Landseer was also renowned for his paintings of animals from the Tower menagerie. Writers regularly mentioned the Tower menagerie: Hahn mentions a passage from Tobias Smollett's *The Expedition of Humphry Clinker* (1771), where the maid-servant Winifred Jenkins reports his visit to the Tower with his mistress and from William Makepeace Thackeray's *The History of Henry Esmond* (1852), set in the reign of Queen Anne (although written and published after the closure of the menagerie) where the hero is introduced to the Tower menagerie by his tutor; Hahn, *The Tower Menagerie*, pp. 156–7.

[9] After Cross's retirement in 1844, the business declined. Cross died in 1854 and all the animals were sold by auction in 1856.

ful Colours, the Top a Sky Blue, the Back Part Orange, the Front Purple, adorned on each side with Crimson, curiously beaded, and its Feathers resemble the Mane of a Horse – and what is more extraordinary, each Quill produces two Feathers.

The Dutch assert that it can devour Glass, Iron, Stones, and even burning Coals, without Fear or Injury.

This Bird laid a large Egg at Warwick, on the 14th of January last, which is of a green Colour, spotted with white.

Ladies and Gentlemen One shilling each.

PIDCOCK, the Proprietor of this BIRD, will be at Sheffield Fair the 28th Instant; and will visit all the other principal Towns in Yorkshire.[10]

Advertisements appealed to the sensational by describing 'rare' and 'extraordinary' creatures which seemed to defy all attempts at classification. The reliance upon the methods of comparative anatomy participated in the dramatic language chosen: the descriptive comparisons with more familiar animals engendered a profusion of material objects in the text which helped to map out the dramatic rise of unknown and exotic creatures brought to Britain. Moreover, the sensational depiction of the bird—comparing it to a warrior and able to fearlessly eat glass—is reminiscent of the world of the fair, where freaks and natural curiosities were commonly exhibited.[11] Another description of Pidcock's menagerie confirms the use of dramatic rhetoric and highlights the crowd of animals on display, underlining, in addition, the kind of interactions that visitors could expect to have with the animals ('will converse with any Person', and 'may approach them') and thus their physical proximity:

G. PIDCOCK'S

GRAND MENAGERIE of WILD BEASTS and BIRDS, all alive, is just arrived, and now exhibiting at the White Lion, Corn-Market, DERBY. This invaluable Collection consists of two Mountain Lion Tygers, Male and Female – two Satyrs, or Ætheopian Savages, ditto – a He Bengal Tyger – a Porcupine – an Ape – a Coata Munda – a Jackall – four Macaws – two Cockatoos, one of which will converse with any Person in Company; with a Number of other Curiosities not inserted.

N.B. The large Beasts are well secured, so that the most timorous may approach them with the greatest Safety.

[10] *Leeds Intelligencer* (16 Nov. 1779).

[11] Animals were often displayed in fairs alongside teratological specimens. See Pidcock's advertisement in the 1791 *Chester Chronicle* quoted below.

28   L. TALAIRACH

> Admittance 1s. each – a Price by no means adequate to the Variety of Curiosities exhibited.[12]

As this example shows, the many menageries that opened in the last decades of the eighteenth century offered new ways of engaging with animals, especially exotic ones. Audiences were invited to interact with the animals by touching or riding them.[13] In addition, advertisements often promoted their suitability for women and children. In the preceding example, for instance, the allusion to the fact that the large animals were confined safely behind bars might be a way of reassuring parents visiting with their children. This, of course, is symptomatic of the changes in the conception of childhood and the development of children's education in the second half of the eighteenth century. As Alan Richardson observes, 'games, toys, books, and apparel designed specifically for children [were] becoming increasingly available, at least among the middle and upper classes'.[14] But it also emphasises the thrilling experience promised by a visit to the menagerie.[15]

Likewise, the following advertisement from the *Chester Chronicle* insists upon the size, rarity and even monstrosity of some of the specimens, from curious (teratological) animals to human freaks:

> Just arrived from the Lyceum, and Exeter Exchange, Strand, London, and to be seen during the fair, in the market-place, two of the grandest assemblages of living rarities in all Europe: consisting of two stupendous and royal OSTRICHES, male and female. These birds exceed in magnitude and texture of plumage all the feathered TRIBE in the CREATION. They already measure upwards of NINE FEET high, although very young! – Also a BENGAL TYGER, a

[12] *Derby Mercury* (31 Dec. 1789).

[13] See Christopher Plumb, '"Strange and Wonderful": Encountering the Elephant in Britain, 1675–1830', *Journal of Eighteenth-Century Studies*, 33.4 (2010): 525–43, p. 534.

[14] Alan Richardson, *Literature, Education and Romanticism. Reading as Social Practice, 1780–1832* (Cambridge: Cambridge University Press, 2010), p. 9. J. H. Plumb calls the eighteenth century the 'new world of children' in England; J. H. Plumb, 'The New World of Children in Eighteenth-Century England', *Past and Present*, 67.1 (May 1975): 64–95.

[15] In *The Menageries. Quadrupeds, Described and Drawn from Living Subjects*, Vol. 1, The Library of Entertaining Knowledge (London: Charles Knight, 1829), p. 24, the writer complains about the uses of the Tower menagerie for children's entertainment: 'It had gradually declined in value for half a century; in some degree, perhaps, from the force of popular prejudice, which was accustomed to consider it only an occupation and amusement for children to make a visit to the "Lions in the Tower"'.

young LIONESS, a real spotted HYÆNA, a ravenous WOLF, two ring-tailed POR-CUPINES; an AFRICAN RAM, with four circular horns; and twenty other animals and birds, too numerous to insert. – Admittance, 1s. – Servants, half-price. – Likewise in the other exhibition is the ROYAL HEIFER with TWO HEADS, a beautiful COLT, of the race kind, foaled with only THREE LEGS, got by Sir Charles Bunbury's Diomed, out of Barcelli, which was the dam of Marcia, now the property of Lord Derby; also a RAM with SIX LEGS. – In addition to the animal curiosities one of the most extraordinary productions of the human species will be shewn, namely the double-jointed IRISH DWARF, who will engage to carry two of the largest men now existing, both at the same time. – Admittance, as above. – Birds and beasts bought, sold, or exchanged, by G. Pidcock. – The above collection will proceed to Warrington, Liverpool, Manchester, &c.[16]

As Ingrid H. Tague explains, although menageries 'explicitly appealed to elite buyers and those with educated tastes',[17] the prices charged by the menageries made it possible for the lower social classes also to see the animals on display. This made such exhibitions a popular form of entertainment, as illustrated by the above example which lists a collection of travelling 'curious beasties' for the fair.

The descriptions of the animals in private menageries differed from those of the animals in the Tower of London, however. The wild animals kept at the Tower had for centuries been exhibited above all as emblems of the monarch's prestige and power. Guidebooks therefore often underlined how the animals—the 'curiosities' of the Tower—were well cared for,[18] as well as the way in which captivity denoted that wild beasts had been successfully tamed. In *An Historical Account of the Curiosities of London and Westminster* (1768), for example, the exotic gifts kept at the Tower are displayed as symbols of British power and of the country's global hegemony; the guidebook invited visitors to enjoy 'the thrill of proximity to wild animals and the happy sense of secure superiority produced by their incarceration':[19]

---

[16] *Chester Chronicle* (14 Oct. 1791): 14–15.

[17] Tague, *Animal Companions*, p. 31.

[18] 'All the creatures that are here shewn are regularly fed with food proper for them, and as carefully attended, as if they were indeed of royal dignity. This takes off much of their savage nature, and makes them tame and submissive, and perhaps contributed not a little to disappoint the expectations of king James I when he made trial of the fierce nature of a lion'; [Anon.], *The Menageries. Quadrupeds, Described and Drawn from Living Subjects*, I, p. 27.

[19] Ritvo, *The Animal Estate*, p. 219.

Tho' these creatures are full grown, they live together in the most perfect amity and friendship: and the king of beasts permits his fair partner to share with him his empire without controul. ...

This lion is so perfectly tame and good-natured, that he will suffer his keepers to do any thing with him; and there is one circumstance, respecting his behaviour, which I must relate, as in tenderness he seems to have come near to human nature, and in friendship to have surpassed it.

...Dunco was too well bred to suffer his friend to go without some little ceremony, or marks of esteem: he first rubbed his great nose against the keeper's knees, then held him by the coat, as if he would have said, Do stay a little longer; and when he found no entreaties could prevail on William to take t'other nap, he courteously waited on him to the door.[20]

Relating the story of the keeper who slept with the lion—with the animal's paw resting on his breast—the writer stresses time and again the close relationship between animal and keeper, suppressing the wildness of the beasts from the text: 'Fanny, a young lioness: and tame she is, indeed more so than many of the human race'.[21] The fact that menageries publicised the management of wild animals much more than their fierceness thus reflected the Georgian discourse on exotic fauna in guidebooks (unlike advertisements for travelling shows, as shown above), the books being aimed at more genteel readers. This point is made explicitly in *The Menageries. Quadrupeds, Described and Drawn from Living Subjects* (1829), a natural history book presenting quadrupeds to juvenile audiences:

Ménagerie is derived from the French word ménager, from which we derive our English verb, to *manage*. The name *Ménagerie* was originally applied to a place for domestic animals, with reference to their nurture and training: it now means any collection of animals. It may be implied, therefore, that the animals in a Menagerie are not placed there merely for safe confinement, but that by care and kindness their noxious or ferocious propensities may there be restrained or subdued, and by constant discipline their habits may there be rendered useful, or at least inoffensive, to man.[22]

---

[20] [Anon.], *An Historical Account of the Curiosities of London and Westminster, in Three Parts* (London: Newbery & Carnan, 1769), pp. 14–15.

[21] [Anon.], *An Historical Account of the Curiosities of London and Westminster, in Three Parts*, p. 17.

[22] [Anon.], *The Menageries. Quadrupeds, Described and Drawn from Living Subjects*, I, p. 16.

The exotic creatures, showcasing the empire, were thus a means of promoting British power and dominion over the world. In such popular science books, the scientific value of the animals was often underlined, while in the following passage concerning quadrupeds,[23] the writer also points out their practical and economic utility:

> [W]e view these establishments, not as mere exhibitions for the gratification of a passing curiosity, but as the most effectual means of bringing to our very doors, those uncommon specimens of the animals of other climes, which, while they extend, as far as possible, our actual acquaintance with animal nature, may ultimately lead to the domestication of some of those races which, possessing many valuable qualities, have not yet been made available to the purposes of man even in his present state of civilization.[24]

Interestingly, the writer later stresses the role played by menageries in children's education, noting the importance of 'beholding' the creatures as instructive 'objects'. The emphasis on sight and on the direct experience offered by menageries was typical of contemporary pedagogical ideals.[25] Yet, the writer also compares the wild beasts of children's stories with those exhibited in the menageries, thus drawing links between fiction and reality:

> A better system of education has instructed us that there is nothing in nature beneath the attention of a reasonable being … and that if children feel the deepest interest in safely beholding those ferocious animals which form such attractive objects in many of the stories dedicated to their use, that interest may be readily carried far beyond the gratification of a passing curiosity, and may become the excitement to the acquisition of a great deal of real knowledge, capable of being presented in the most captivating form.[26]

[23] Other significant examples might be Edward Turner Bennett's *The Tower Menagerie, comprising the natural history of the animals contained in that establishment, with anecdotes of their characters and history* (London: Robert Jennings, 1829) and *The Gardens and Menagerie of the Zoological Society Delineated* (London: Thomas Tegg, 1830).

[24] [Anon.], *The Menageries. Quadrupeds, Described and Drawn from Living Subjects*, vol. 1, p. 9.

[25] Contemporary pedagogical theories, such as those of the Swiss educationalist Johann Heinrich Pestalozzi (1746–1827), stressed the importance of visual education. Although the link between animals and visual education in children's literature will appear throughout the book, the point will be particularly developed in Chap. 6.

[26] [Anon.], *The Menageries. Quadrupeds, Described and Drawn from Living Subjects*, I, p. 24.

The link between the animals on display in British menageries and the children's literature of the time is no coincidence. The advent of children's literature coincided with a period in which the 'curious' informed scientific investigation—a period in which, as Krzysztof Pomian has shown, 'curiosity was not only related to the desire to know, but also to see and to possess'.[27] As will be seen in this chapter, the children's literature of the late eighteenth century similarly used familiar and less familiar 'beasties' to disseminate knowledge about the natural world, throwing light on how scientific investigation and the 'culture of curiosity' were linked, inextricably linked, in fact, to museum culture. At the same time, however, Georgian fictional animals frequently played a significant role in raising public awareness of the living conditions and treatment of caged animals. In so doing, they showed how living creatures, both domestic and exotic, could be used in fiction to mould responsible and humane British citizens.

Eighteenth-Century Literary 'Beasties'

As more and more animals were imported into Britain, bestiaries became increasingly outmoded,[28] whilst guidebooks aimed at children, such as the first Tower guidebook, *Curiosities in the Tower of London* (1741), started combining the goals of education and entertainment. This two-volume guidebook was published by Thomas Boreman, known for his early children's books, and it included illustrations representing some of the animals kept at the Tower. Furthermore, at the same time as the menageries of the late eighteenth century were proposing new ways of engaging with animals, especially the exotic ones, new ways of representing animals were also appearing on the literary scene. As we will see, many of the 'beasties' depicted were related to fable literature—a genre known for playing a key role in the development of children's literature.

[27] Krzysztof Pomian, *Collectionneurs, amateurs et curieux. Paris, Venise: XVIᵉ–XVIIIᵉ siècles* (Paris: Gallimard, 1987), p. 74. My translation.
[28] Hahn, *The Tower Menagerie*, p. 84.

During the Middle Ages, fables, as much as bestiaries,[29] had played an important part in monastic education.[30] As Peter Travis explains, fables served to 'educat[e] both children and adults while interrogating human nature and dramatizing the politics and ethics of human society'.[31] In the eighteenth century, fables still offered rich materials for Georgian pedagogues, eager to mould children into responsible citizens. Beast fables were used to tackle the relationship between humans and animals or, rather, to capture how humans viewed themselves in relation to animals, sometimes satirically. Well before children's literature emerged as a fully distinct genre in the last decades of the eighteenth century, Aesop's *Fables*, first translated into English in 1484,[32] had been considered as 'tailored for children's reading'[33] and promoted by John Locke, whose educational theories and anti-cruelty messages were put forward in 'Some Thoughts Concerning Education' in 1693.[34] Alongside texts in the vein of Aesop's *Fables*, bestiaries were also commonly used for children until mid-century, providing young audiences with a little knowledge in natural history. According to Mary V. Jackson, printed natural histories became accessible to a wide audience between 1600 and 1770. Edward Topsell's *History of*

[29] Both developed contemporaneously, as Carolynn Van Dyke argues; Carolynn Van Dyke, 'Entities in the World: Intertextuality in Medieval Bestiaries and Fables', in Jane Spencer, Derek Ryan and Karen L. Edwards (eds), *Reading Literary Animals: Medieval to Modern Perspectives on the Non-Human in Literature and Culture* (New York: Routledge, 2019), pp. 13–28, p. 22.

[30] Elizabeth Carson Pastan, 'Fables, Bestiaries, and the Bayeux Embroidery: Man's Best Friend Meets the "Animal Turn"', in Laura D. Gelfand (ed.), *Our Dogs, Our Selves. Dogs in Medieval and Early Modern Art, Literature, and Society* (Leiden and Boston: Brill, 2016), pp. 97–126, p. 118.

[31] Peter W. Travis, 'Aesop's Symposium of Animal Tongues', *Postmedieval*, 2 (2011): 33–49, pp. 40–41, qtd. in Pastan, 'Fables, Bestiaries, and the Bayeux Embroidery', p. 118.

[32] See Peter Hunt (ed.) et al., *Children's Literature: An Illustrated History* (Oxford: Oxford University Press, 1995), p. 13.

[33] Richardson, *Literature, Education and Romanticism*, p. 113.

[34] 'One thing I have frequently observed in Children, that when they have got Possession of any poor Creature, they are apt to use it ill: They often *torment*, and treat very roughly young Birds, Butterflies and such other poor Animals … with a seeming kind of Pleasure. This I think should be watched in them … they who delight in the Suffering and Destruction of inferiour Creatures, will not be apt to be very compassionate or benign to those of their own kind … Children should from the beginning be bred up in an Abhorrence of *killing*, or tormenting any living Creature'; John Locke, 'Some Thoughts Concerning Education', in *The Educational Writings of John Locke*, ed. James L. Axtell (Cambridge: Cambridge University Press, 1968), pp. 13, 225–6.

Four-Footed Beasts (vol. 1, 1607; vol. 2, 1608) was an early example especially written for children and is 'thought to have influenced the first "modern" children's natural histories',[35] such as Thomas Boreman's *Description of Three Hundred Animals; viz. Beasts, Birds, Fishes, Serpents and Insects* (1730) and *Description of a Great Variety of Animals and Vegetables ... Extracted from the most considerable writers of natural history* (1736). These books were often heavily moralising in tone, in line with the religious slant of earlier bestiaries and animal fables,[36] or placed animals side by side with mythical creatures, such as griffins, manticores, mermaids or unicorns. Nonetheless, such works paved the way for later natural histories which would serve as manuals to instruct children.

A major work of natural history published around the same time, *A Short and Easie Method to give Children an Idea or True Notion of Celestial and Terrestrial Beings* (1710), seems to have been an early example of a science book for children,[37] whilst the adaptation of Buffon's *Histoire Naturelle* for children and the publication of Oliver Goldsmith's (1728–74) eight-volume *A History of the Earth and Animated Nature* (just like the *Curiosities in the Tower of London*, mentioned above) both offer a further illustration of the rise of natural histories for young audiences. Aesop's *Fables*, still in vogue at the end of the eighteenth century, increasingly bound natural histories with children's literature when Thomas Bewick illustrated *Select Fables, In Three Parts* (1784), published by Thomas Saint. His use of white-line engraving, a technique 'that revolutionized children's book illustrations',[38] gave a new lease of life to the classical fables and paved the way for Bewick's own *General History of Quadrupeds* (1790) and *A History of British Birds* (1797–1804), as well as other illustrated popular science books for children which were central to early children's literature and, significantly, represented animals which were true to life, rather than just moral or religious ciphers.

[35] Mary V. Jackson, *Engines of Instruction, Mischief and Magic: Children's Literature in England from its Beginnings to 1839* (Lincoln: University of Nebraska Press, 1989), p. 35.

[36] As Jackson points out, the earliest forms of bestiaries and animal fables were controlled by medieval monks, 'jealous guardians of learning, who added their levy of religious and moral freight'; Jackson, *Engines of Instruction, Mischief and Magic*, pp. 34–5.

[37] Jackson, *Engines of Instruction, Mischief and Magic*, p. 36.

[38] Jackson, *Engines of Instruction, Mischief and Magic*, p. 50.

John Gay's *Fables*, first published in two volumes in 1727 and 1738 and continually reprinted during the nineteenth century,[39] interestingly revisited many of the animals found in Aesop's *Fables*, offering new insights into the way in which animals were imported into Britain and used for people to observe, study or keep as pets. The fables were originally written for the six-year-old Prince William, later Duke of Cumberland, but they were not, however, merely moralistic, hence, perhaps, their enduring popularity. Gay's writing for adults generally conveyed his satire of early Georgian society and power relationships, and his *Fables* are no exception, underlining the dual audience to whom they were addressed. Several of them overtly mocked the work of contemporary naturalists and their (condescending) views of the natural world. Among the sixty-six fables published, wild and exotic creatures feature regularly, as clearly indicated by the titles of some of the fables,[40] alongside more familiar creatures, such as eagles, boars, rams, wasps, bulls, mastiffs, turkeys, stags, wolfs, hens and cats. Several of the fables deride human exploration of the world, like 'The Elephant and the Bookseller', where the speaking pachyderm casts doubts on the veracity of naturalists' reports and ultimately condemns humans' appetite for exotic species:

The man who with undaunted toils
Sails unknown seas to unknown soils.
With various wonders seast his fight:
What stranger wonders does he write!
We read, and in description view
Creatures which ADAM never knew:
For, when we risk no contradiction
It prompts the tongue to deal in fiction.
...
Man with strong reason is endow'd;
A beast fearce instinct is allow'd
...

[39] See, for instance, editions by Rivington and Longman or Darton and Harvey. The fables were illustrated by William Kent, John Wootton, Henri Gravelot and Thomas Bewick. The latter stressed the natural world more than his predecessors; Jenny Uglow, *Nature's Engraver: A Life of Thomas Bewick* (New York: Farrar, Straus & Giroux, 2006), pp. 74–5.

[40] Such as 'The Lion, the Tyger, and the Traveller', 'The Spaniel and the Chameleon', 'The Lion, the Fox, and the Geese', 'The Elephant and the Bookseller', 'The Monkey Who had Seen the World', 'The Lion and the Cub', 'The Scold and the Parrot', 'The Two Monkeys' and 'The Baboon and the Poultry'.

But is not man to man a prey?
Beasts kill for hunger, men for pay.[41]

Humans, here aligned with animals situated much lower in the chain of being, are obviously the target of Gay's satire, which frequently hinges upon themes such as that of exploration to undermine power relationships. The opposition between humans and wild or exotic animals often collapses, as in 'The Monkey who had seen the World' where a monkey is 'Resolv'd to visit foreign climes'.[42] The fable offers us access to the animal's point of view, as the monkey is captured by a man and imprisoned in a lady's room where it learns to mimic human manners. Gay's critique of British imperialism and animal collecting, as well as his stance on the commodification of exotic creatures, turned into consumer goods for wealthy ladies, permeates the fable. The story recalls satirical representations of women and pets, with monkeys often serving to map out fears related to female sexuality.[43] But the fable plays, above all, upon the proximity between apes and humans ('support your place,/ The next in rank to human race') and the monkey's capacity to *ape* human manners and grow into a 'polite' man:

Observe me right;
So shall you grow like man polite.[44]

Gay thus subverts the moralistic use of animals in fables the better to turn the mirror back on humans and mock both education and civilisation. Another significant example, 'The Baboon and the Poultry', features a woman collector of birds and other animals (dogs, parrots, apes, etc.) who takes a fancy to a baboon. The female character—'an ancient maid'—epitomises repressed sexuality whilst the verb 'cure' humorously illuminates unruly female physiology:

Once on a time, an ancient maid,
By wishes and by time decay'd,
To cure the pangs of restless thought,

---

[41] *Fables by the Late John Gay, in one volume complete* (London: J. F. Rivington & B. White, 1743), pp. 28–30.

[42] *Fables by the Late John Gay, in one volume complete*, p. 38.

[43] Tague, *Animal Companions*, p. 129.

[44] *Fables by the Late John Gay*, pp. 38–40.

In birds and beasts amusement sought:
Dogs, parrots, apes, her hours employ'd;
With these alone she talk'd and toy'd.
A huge Baboon her fancy took ...[45]

Monkeys appear time and again in other fables, such as 'The Scold and the Parrot' and 'The Cur and the Mastiff', showing how primates embodied, perhaps more than any other wild animals exhibited in the capital, the threat to human identity posed by imported fauna. 'The Two Monkeys', situated at the end of the first volume of fables, throws light on contemporary anxieties related to 'human uniqueness',[46] as the monkeys visit Southwark Fair and are contrasted with humans who behave less thoughtfully:

Men laugh at Apes, they men contemn;
For what are we, but Apes to them?

Two Monkeys went to Southwark fair,
No critics had a sourer air:
They forc'd their way through draggled folks,
Who gap'd to catch Jack-pudding's kikes;
Then took their tickets for the show,
And got by chance, the foremost row.
To see their grave, observing face,
Provok'd a laugh through all the place.[47]

Once again, the opposition between humans and animals becomes blurred, more especially so because Gay evokes the popular world of the fair and its uncouth visitors.[48] The relationship between humans and (exotic) animals throughout the fables, as well as Gay's satirical construction of natural science—particularly zoology—is representative of the eighteenth-century culture of sensibility. Diana Donald's analysis of the role of fables in contemporary representations of animals is significant in this context. She argues that fables, which were popular with eighteenth-century readers of

[45] *Fables by the Late John Gay*, p. 144.
[46] Tague, *Animal Companions*, p. 55.
[47] *Fables by the Late John Gay*, p. 101.
[48] For more on Gay's use of traditional lore to convey his social satire, see Dianne Dugaw, 'Folklore and John Gay's Satire', *Studies in English Literature 1500–1900*, 31.3 (Summer 1991): 515–33.

all ages, were the literary offspring of Aesop's *Fables* and permeated all kinds of texts, from natural history books and folk tales to proverbs and emblems. Yet, by 'depicting beasts and birds with human traits, they offered a quite different way of thinking about the animal mind from [that] encountered in the natural science and philosophy of the eighteenth century'.[49] In particular, the creatures 'constituted a locus where restrictive views of animal intelligence were frequently questioned, and where the very subjectivity that gave rise to sweeping theories about other species was itself subjected to destructive satire'.[50] In this way, the form of the fable enabled writers not only to use animals to represent humans but also, 'contrarily, ... to represent *themselves*'.[51] Thus, Gay's animals magnified both 'his subjects' animality' and 'their symbolising function'.[52]

Gay's satirical fables, published in the first half of the eighteenth century, exemplify, indeed, how the 'curious beasties' then being imported to Britain could become, as Jeffrey Jerome Cohen puts it, 'the vehicles through which desires for a differently configured (if obdurately anthropocentric) world are expressed'.[53] By bringing to light the 'injustices that anthropocentrism engenders',[54] Gay's fables marked therefore a key stage in the development of Georgian literature for children. Little by little, Georgian educationalists following in his footsteps changed 'the tenor of fable literature': the 'bloodthirsty predators, or at least their crueller successes, were gradually banished from it'.[55] Instead, children's writers cultivated benevolence and sympathy. We will see that the children's literature of the eighteenth century increasingly used 'beasties', be they 'curious' and exotic or much more familiar, to inculcate a sense of moral responsibility and respect for the natural world. At the same time, however, the development of children's literature reflected a growing consciousness of animal thoughts and feelings, as if sensing the complexities of the anthro-

---

[49] Diana Donald, *Picturing Animals in Britain, 1750–1850* (New Haven and London: Yale University Press, 2009), p. 114.

[50] Donald, *Picturing Animals in Britain*, p. 114.

[51] Donald, *Picturing Animals in Britain*, p. 114.

[52] Donald, *Picturing Animals in Britain*, p. 119.

[53] Jeffrey Jerome Cohen, 'Inventing with Animals in the Middle Ages', in Barbara A. Hanawalt and Lisa J. Kiser (eds), *Engaging with Nature: Essays on the Natural World in Medieval and Early Modern Europe* (Notre Dame, IN: University of Notre Dame Press, 2008), pp. 39–62, p. 40.

[54] Donald, *Picturing Animals in Britain*, p. 119.

[55] Donald, *Picturing Animals in Britain,* p. 118.

pocentrism mocked by Gay. In this way, argues Amy Ratelle, children's texts 'reflect[ed] and contribut[ed] to the cultural tensions created by the oscillation between upholding and undermining the divisions between the human and the animal'.[56] The relationship between the symbolising power of animals in early children's literature and the increased scientific knowledge of the real animals forms the heart of the following discussion, which argues that the growing visibility of curious 'beasties' in Georgian Britain enabled eighteenth-century educationalists to revisit animal fables and propose pedagogically innovative texts which recurrently addressed the human-animal relationship.

The market for children's literature was launched when the publisher John Newbery (1713–67) started to offer both amusement and instruction in his publications for children.[57] Though his *Lilliputian Magazine* (1751–52) was a commercial failure, many of his books sold well, illustrating the expansion of the children's book industry.[58] Newbery's first *Little Pretty Pocket Book* (1744) undoubtedly marked the invention of children's literature. However, the book was perhaps, as Mary Hilton argues, less influential than his *The Philosophy of Tops and Balls* (1761)—an adaptation of Locke's treatise, *Elements of Natural Philosophy*,[59] which drew on the traditional form of a conversation between teacher and child still to be found in mid-nineteenth-century works such as Charles Kingsley's *Madam How and Lady Why* (1869) and John Ruskin's *Ethics of the Dust* (1865).[60]

---

[56] Amy Ratelle, *Animality and Children's Literature and Film* (New York: Palgrave Macmillan, 2015), p. 4.

[57] Hunt (ed.) et al., *Children's Literature*, pp. 35–7. See also Sidney Roscoe, *John Newbery and his Successors, 1740–1814: A Bibliography* (Wormsley: Five Owls Press, 1973). As Mary V. Jackson argues, Thomas Boreman and Mary Cooper also contributed to establishing children's books as a separate genre. Boreman published seven books for children, including natural histories, such as *A Description of Three Hundred Animals* (1730) and *A Description of a Great Variety of Animals and Vegetables* (1736), mentioned above; Jackson, *Engines of Instruction, Mischief and Magic*, pp. 71–5.

[58] Hunt (ed.) et al., *Children's Literature*, pp. 35–7.

[59] See, in particular, James A. Secord, 'Newton in the Nursery: Tom Telescope and the Philosophy of Tops and Balls, 1761–1838', *Journal of History of Science*, 23 (1985): 127–51, reprinted in Laurence Talairach-Vielmas (ed.), *Science in the Nursery: The Popularisation of Science in Children's Literature in Britain and France, 1761–1901* (Newcastle: Cambridge Scholars Publishing, 2011), pp. 34–68.

[60] The fictional dialogue followed the conventions set by Galileo's *Discourses Concerning the Two New Sciences* (1638) or Robert Boyle's *Sceptical Chemist* (1661); Bernard Lightman, '"The Voices of Nature": Popularizing Victorian Science', in Bernard Lightman (ed.),

40   L. TALAIRACH

In 1765, Newbery published the anonymous *History of Little Goody Two-Shoes*, a moral tale which advocated the acquisition of practical knowledge by children, in particular that relating to animals, both domestic and exotic. *The History of Little Goody Two-Shoes* relates the story of Margery Meanwell, an orphan girl who walks with 'the Raven on one Shoulder, the Pigeon on the Other, the Lark on her Hand, and the Lamb and the Dog by her side',[61] and whose kindness to animals is rewarded at the end of the tale: she becomes a teacher and marries a rich man, reinforcing the moral lesson that children should be kind to animals.[62] Echoing Margery's story, yet providing an exotic counterpart to the main narrative at the end of the volume, is the story of Margery's brother, Tom Two-Shoes, a young adventurer who travels through Africa and hunts with a lion he has tamed. Ironically, however, although Tom's relationship with the lion may be compared with Margery's more familiar pets, the boy's activities in Africa, as a young naturalist, stand in sharp contrast to the discourse on animal protection developed in the main narrative, reflecting to some extent the eighteenth-century gendering of sensibility. The two narratives thus pro-

Victorian Science in Context (Chicago and London: The University of Chicago Press, 1997), pp. 187–211, p. 192.

[61] [Anon.], *The History of Little Goody Two-Shoes; otherwise called, Mrs. Margery Two-Shoes: with the Means by which she acquired her Learning and Wisdom, and in Consequence thereof her Estate* (London: T. Carnan & F. Newbery, 1772), p. 123.

[62] 'Mrs. *Margery*, you must know, was very humane and compassionate; and her Tenderness extended not only to all Mankind, but even to all Animals that were not noxious; as your's [*sic*] ought to do if you would be happy here, and go to Heaven hereafter. These are God Almighty's Creatures as well as we. He made both them and us; and for wise Purposes, best known to himself, placed them in this World to live among us; so that they are our fellow Tenants of the Globe. How then can People dare to torture and wantonly destroy GOD Almighty's Creatures? They as well as you are capable of feeling Pain, and of receiving Pleasure, and how can you, who want to be made happy yourself, delight in making your fellow Creatures miserable? Do you think the poor Birds, whose Nest and young Ones that wicked Boy *Dick Wilson* ran away with Yesterday, do not feel as much Pain as your Father and Mother would have felt, had any one pulled down their House, and ran away with you? To be sure they do. Mrs. *Two-Shoes* used to speak of those Things, and of naughty Boys throwing at Cocks, torturing Flies, and whipping Horses and Dogs, with Tears in her Eyes, and would never suffer any one to come to her School who did so'; *The History of Little Goody Two-Shoes*, pp. 68–9. Julia Briggs argues that *Goody Two-Shoes* is however 'an important feminist fable … ahead of its time in celebrating the pleasures of the independent woman', since the female character's marriage is 'treated as a mere interlude'; Julia Briggs, '"Delightful Task!" Women, Children, and Reading in the Mid-Eighteenth Century', in Donelle Ruwe (ed.), *Culturing the Child, 1690–1914. Essays in Memory of Mitzi Myers* (Lanham, Toronto, Oxford: Scarecrow Press, 2005), pp. 67–82, p. 78.

pose different discourses on animals, while showing little, if any, concern for the more 'curious beasties' Tom kills in Africa. Although the tension at the heart of the book is seemingly resolved by the gendered narratives (the girl looks after familiar animals; the boy tames, hunts and kills wild beasts), the stories of Little Goody Two-Shoes and Tom Two-Shoes offer a striking instance of the ways in which children's literature of this period used the prism of gender to offer two contrasting forms of moral teaching when it came to animals, in a way that was both like and unlike earlier fable literature. The one was predicated on teaching (female) children the quality of sensibility when it came to familiar creatures; the other, steeped in the tradition of masculine colonial natural history, considered the wild and the exotic as something that was there for Man (and particularly *men*) to study and label—and perhaps shoot.

The presence of 'curious beasties' in Newbery's early publications for children not only reveals the significance of animals in children's literature but also helps map out the trajectory of the genre itself. Many later children's books were written by women who turned towards the natural sciences either through popular science books and articles or through realistic stories informed by natural history and contemporary knowledge of the natural world. As we will see in the following chapters, children's interest in natural history—especially natural history collecting—resulted largely from books written by (middle-class) women, which invited children to observe the natural world, and to collect and classify specimens. Among the most famous children's writers of the period, many penned works of popular science. These include Anna Laetitia Barbauld (1743–1825), Sarah Trimmer (1741–1810), Priscilla Bell Wakefield (1751–1832), Margaret Bryan (1780–1818) and Jane Marcet (1769–1858). Wakefield's *Introduction to Botany, in a Series of Familiar Letters* (1796), which popularised botany for girls, offering an overview of Linnaean taxonomy (albeit a desexualised one), and Marcet's *Conversations on Chemistry* (1805), which inspired Michael Faraday, were, just like Barbauld, Trimmer and Bryan, key texts and figures in children's education and the development of children's literature. Although many of their publications remained highly didactic and moralising, often using science more to bend children's minds to bourgeois standards of behaviour than for its own sake, they nonetheless reflected the development of scientific knowledge of animal species and even at times integrated in their work the latest findings in natural history.

Unlike Bewick, who never anthropomorphised animals, as seen earlier, these women educators frequently used talking animals to teach children about social mores, thus exemplifying the way in which animals became significant educational tools in attempts to inculcate moral values and respect for others in young minds.[63] Following Locke, many women pedagogues favoured empirical fact over fiction, even when the animals were anthropomorphised. As noted above, their works were also informed by works of natural history and by new descriptions, definitions and understandings of animals. Interestingly, pedagogical natural history writing in this period aimed above all at developing sympathy and empathy in children—ideas that matched the eighteenth-century cult of sensibility referred to earlier. Domestic pets and exotic wild animals (bearing in mind that wild animals sometimes *were* pets) were all put to work by eighteenth-century educators in their efforts to address the human-animal relationship. However, representations of the traffic and circulation of exotic animals in Britain sometimes became intertwined with the issue of slavery, since, as we will see, humans as well as exotic fauna were imported, collected and exhibited.

'Tyger Tyger, Burning Bright': Petted, Performing and Caged 'Beasties'

Sarah Trimmer's children's books are certainly representative of the period and of the use made of animals by eighteenth-century educators, whether familiar or of the more 'curious' variety. As an influential children's book reformer, Trimmer remains best known today for her strict condemnation of the use of imagination in children's literature. She was the editor of the *Family Magazine; or a repository of religious instruction and rational amusement* (1788–89), which contained moral tales and sermons. She also wrote on education: her periodical *Guardian of Education* (1802–06) directly addressed parents and governesses and sought to instruct them about the qualities—or dangers—of recently published children's literature. As Peter Hunt underlines, many of the female writers 'who dominated the children's market at this time were themselves heavily involved in education'; women like Hannah Moore, Anna Laetitia Barbauld, Mary

[63] This point is developed in the first chapter of Tess Cosslett's *Talking Animals in British Children's Fiction, 1786–1914*; Tess Cosslett, *Talking Animals in British Children's Fiction, 1786–1914* (Aldershot: Ashgate, 2006), pp. 9–36.

Sherwood, Sarah Trimmer and Mary Pilkington either managed schools or were governesses.[64] Trimmer founded several Sunday schools and charity schools, and wrote textbooks and manuals for women intending to set up their own schools. Her methods and tastes position her poles apart from those educationalists who advocated Rousseauian approaches to education. Her works, characterised by a prevailing moralising tone, belonged, rather, to the older didactic tradition. They favoured 'useful knowledge' and tied science closely to religious teaching, rather than seeking to develop children's rational faculties.[65] Yet, whatever Trimmer's intentions, her works were often as entertaining as they were moralising and were paradigmatic instances of the mid-eighteenth-century enthusiasm for natural history.

As noted above, most textbooks and guides dealing with natural philosophy were written by women and aimed at a female audience.[66] As Mary Hilton points out, throughout the eighteenth century, women writers were leaders in the educational field and authored many popular texts. Defining themselves as the 'nation's teachers', they 'reached out from the literary world to construct a variety of intellectual and pedagogical practices, propagating culture through a variety of educational formats, from treatises and popular guides, to conduct books, stories and handbooks'.[67] Interestingly, these books also taught children to behave with Christian benevolence towards the natural world: Trimmer and her contemporaries repeatedly argued that humankind's position at the head of Creation should not give them the right to kill or torment inferior creatures.

Trimmer's *Fabulous Histories: Designed for the Instruction of Children Respecting Their Treatment of Animals* (1786) (later published as *The History of the Robins*) placed side by side a family of humans and a family of robins—the robins standing as representatives of proper human behaviour. Though mainly metaphorical, her narrative is at times informed by

[64] Hunt (ed.) et al., *Children's Literature*, pp. 54–5. See also Aileen Fyfe, 'Reading Children's Books in Late Eighteenth-Century Dissenting Families', *The Historical Journal*, 43.2 (2000): 453–73.

[65] Aileen Fyfe, 'Tracts, Classics and Brands: Science for Children in the Nineteenth Century', in Julia Briggs, Dennis Butts and M. O. Grenby (eds), *Popular Children's Literature in Britain* (Aldershot: Ashgate, 2008), pp. 209–28, p. 220.

[66] Mary Hilton, *Women and the Shaping of the Nation's Young: Education and Public Doctrine in Britain 1750–1850* (Aldershot: Ashgate, 2007), p. 110.

[67] Hilton, *Women and the Shaping of the Nation's Young*, p. 2.

44   L. TALAIRACH

information drawn from natural history, such as when a mockingbird is mentioned in the text and a footnote indicates that '[t]he mock-bird is a native of America but is introduced here for the sake of the moral'.[68] More significantly still, the scientific world and contemporary modes of study of the natural world are recurrently alluded to throughout the narrative, particularly through references to the collection and exhibition of animals.

The book opens with a preface which encourages 'Christian benevolence' and 'compassion' towards 'the Animal Creation' (n.p.). However, while Trimmer condemns children 'tormenting inferior creatures', she simultaneously warns the young not to display 'immoderate tenderness to them' (pp. ix–x). This introductory passage, reminiscent of Lockean anti-cruelty messages, illustrates the part played by animals in Georgian Britain.[69] As Ingrid Tague has shown, moreover, pets were part and parcel of eighteenth-century consumer society.[70] But pet-keeping in eighteenth-century Britain included also many wild animals,[71] which simultaneously symbolised 'the flowering of a consumer society, the spread of the British Empire, and a fascination with the natural world'.[72] Many aristocrats bought and collected newly available exotic animals and displayed them as

[68] Sarah Trimmer, *Fabulous Histories: Designed for the Instruction of Children Respecting Their Treatment of Animals, in two volumes comprised in one* [1786], 10th edn (London: John Sharpe, 1815), II, p. 54. All further references are to this edition and will be given parenthetically in the text.

[69] Trimmer's *Fabulous Histories* reflects Lockean pedagogy through its moral stance on the treatment of animals. The book was not only reprinted throughout the eighteenth and nineteenth centuries but also cited in other educational books advocating kindness to animals, such as in Mary Wollstonecraft's *Original Stories from Real Life; with Conversations Calculated to Regulate the Affections, and Form the Mind to Truth and Goodness* (1788, second edn. illustrated in 1791).

[70] Tague, *Animal Companions*.

[71] In 1759, a shop advertised for '[a]n uncommon good Talking Parrot, a Pair of Pintail Parroquets, some Mule Birds, Canary Birds, with a Variety of Song Birds, a beautiful Pair of red-legg'd Partridges, all Sorts of Foreign and English Pheasants, Guinea Fowls, Pea Fowls, the large wild Turkies, the black Fowl with white Toppings, the Silk Fowl from India, small Bantams, a Pair of breeding Swans, white Muscovy Ducks, the large Roan Ducks, &c. a Pair of white Doves, Turtle-Doves, Foreign Pigeons, a Monkey from Guinea full of Tricks, likewise a very small Marmoset Monkey, a Virginia Squirrel, India Sheep and Goats, a Boar and Sow from ditto, a remarkable large Irish Wold Dog, small Italian Greyhound Puppies, some very small Puppies of the true King Charles's Breed, some Cocking Spaniels, the Pomeranai or the Fox Dogs, &c.'; *Public Advertiser*, 7542 (9 January 1759), qtd. in Tague, *Animal Companions*, pp. 14–15.

[72] Tague, *Animal Companions*, p. 15.

living consumer goods in their menageries. The latter became significant places where (sometimes amateur) naturalists described animal behaviour and attempted to classify the birds and beasts. As objects of education and knowledge *and* emblems of luxury, exotic pets nevertheless blurred to some extent the distinction between humans and nonhumans, and the boundary was perhaps made even more unstable by humanoid apes. Thus, Trimmer and her contemporaries used familiar and less familiar creatures to convey their discourse on the natural world—the 'curious beasties' helping children's educators to place human-animal relations at the heart of Georgian pedagogy.

Fabulous Histories is a good illustration of how exhibited animals could be used to explore the relationship between humans and animals and at the same time uphold human superiority. This is the case, for instance, for some performing animals mentioned in the narrative. As Antonia Losano notes, performing animals were intended to entertain by 'bring[ing] welcome assurance of the inherent superiority of humanness as … nonhuman creatures [were compelled] to execute human behaviour'.[73] In Trimmer's book, the learned pig exhibited in London is able to spell words, a feat probably managed after having suffered 'great barbarities [*sic*]' (I, p. 97), as the mother, Mrs Benson, surmises. But the anti-cruelty message works in tandem with Mrs Benson's assurance that animals are never 'capable of attaining human sciences' (I, p. 96) and that it is therefore likely that the pig, while able to identify letters, cannot actually spell words.

The (potential) blurring of boundaries between humans and animals is pursued later on in the work. This time, however, Trimmer uses the practice of keeping exotic animals as pets as a way of questioning human superiority. At the beginning of the second volume, Mrs Benson's daughter, Harriet, is brought to Mrs Addis's. As they enter the house, Harriet's mother notices 'a very disagreeable smell, and was surprised with the appearance of a parrot, a paroquet [*sic*], and a macaw, all in most superb cages. In the next room she came to were a squirrel and a monkey, and had each a little house neatly ornamented' (II, p. 3). The variety of exotic species referred to in the second part of the story is telling. The little girl is instantly attracted by 'the animals which Mrs Addis had collected together' (II, p. 4) and is described as being 'curious'—the term revealing

---

[73] Antonia Losano, 'Performing Animals/Performing Humanity', in Laurence W. Mazzeno and Ronald D. Morrison (eds), *Animals in Victorian Literature and Culture: Contexts for Criticism* (London: Palgrave Macmillan, 2017), pp. 129–46, p. 131.

the young girl to be an enlightened child, used to training her senses to acquire knowledge. Her sharp sense of observation is emphasised: '[S]he observed in one corner a lap-dog lying on a splendid cushion; and in a beautiful little cradle, which she supposed to contain a large wax doll, lay, in a great state, a cat with a litter of kittens' (II, p. 3). But the scene also aims at deriding the lady's fondness for her pets, both domestic and exotic, as Mrs Addis spends more time (and money) on her animals than on her own daughter and is unable to speak of anything but 'the perfection of her birds and beasts' (II, pp. 8–9).[74] The monkey makes a mess of the room whilst the lapdog is commodified and denounced as a symptom of luxury: 'a miserable object, full of diseases, the consequences of luxurious living' (II, pp. 11–12). The monkey serves here to undermine Mrs Addis's humanity. As Tague argues, '[i]mporting exotic animals might reinforce notions of British superiority, but contact with apes and monkeys, particularly the humanoid apes, threatened to destabilize comforting notions of humanity's unique place at the top of the great chain of being'.[75] Asserting 'human uniqueness'[76] was thus made all the more difficult by the numerous discoveries and infinite variety of the natural world—especially so when faced with primates. It is significant, however, that when the monkey eventually dies, its status as one of Mrs Addis's 'dear darlings', merely consumed as goods, means that it is quickly replaced:

> Mrs Addis lost her Parrot by the disorder with which it was attacked while Mrs Benson was visiting at the house; and before she had recovered the shock of this misfortune, as she called it, her grief was renewed by the death of an old Lap-Dog. About a year afterwards her Monkey escaped to the top of the house from whence he fell and broke his neck. The favourite Cat went mad, and was obliged to be killed. In short, by a series of calamities, all her dear darlings were successively destroyed. She supplied their places with new favourites, which gave her a great deal of fatigue and trouble. (II, pp. 158–9)

---

[74] Exotic animals were purchased by the aristocracy and gentry as well as by the middle classes; see Christopher Plumb's thesis, *Exotic Animals in Eighteenth-Century Britain* (University of Manchester, 2010), p. 61. For more on women's excessive fondness for animals at the expense of humans, see Diana Donald, *Women against Cruelty: Protection of Animals in Nineteenth-Century Britain* (Manchester: Manchester University Press, 2020), pp. 22–4.

[75] Tague, *Animal Companions*, p. 10.

[76] Tague, *Animal Companions*, p. 55.

Mrs Addis is made to pay the price of her consumption of pets. She constitutes, therefore, an adult female equivalent to Master Jenkins who, in the first part of the narrative, tortures (familiar and common) animals and as a grown-up is eventually killed by the horse to which he had been cruel (II, p. 158). Moreover, Trimmer's use of the animal's point of view participates in her denunciation of the fashion for collecting exotic 'beasties', such as when the robins—the main 'characters' of the story—fly away and discover the house of 'a gentleman, who, having a plentiful fortune, endeavoured to collect all that was curious in art and nature, for the amusement of his own mind and the gratification of others' (II, p. 119). The aviary—a 'dismal prison' (II, p. 114)—and the 'menagery' are labelled as 'articles of taste' (II, p. 119), but Trimmer ironically intermingles the gentleman's high aesthetic standards and his compulsion to acquire (and literally consume) goods when the robins witness one of the partridges being plucked, roasted and eaten.

Another passage dealing with caged animals furthers Trimmer's discourse on wild animals and more generally hints at the rising concern of Georgian educators with the treatment of animals (later manifested through animal welfare legislation to improve the conditions of working animals),[77] when one of the male characters is condemned for wishing to catch birds and keep them in a cage:

And would you really confine these little creatures in a cage, Frederick, merely to have the pleasure of feeding them? Should you like to be always shut up in a little room, and think it sufficient if you were supplied with victuals and drink? ... Though these little animals are inferior to you, there is no doubt but they are capable to enjoyments similar to these; and it must be a dreadful life for a poor bird to be shut up in a cage, where he cannot so much as make use of his wings; where he is separated from his natural com-

---

[77] The legal protection of animals was launched by the passing of a bill to 'Prevent the Cruel Treatment of Cattle' in 1822 as well as the foundation of the Society for the Prevention of Cruelty to Animals two years later; it did not cover cruelties to domestic or wild animals, however. See David Perkins, *Romanticism and Animal Rights* (Cambridge: Cambridge University Press, 2003), pp. 16–19. On animal welfare legislation, see also J. Turner, *Reckoning with the Beast: Animals, Pain and Humanity in the Victorian Mind* (Baltimore and London: Johns Hopkins University Press, 1980), R. S. White, *Natural Rights and the Birth of Romanticism in the 1790s* (Basingstoke: Palgrave Macmillan, 2005), Samuel F. Pickering, Jr., *John Locke and Children's Books in the Eighteenth Century* (Knoxville: The University of Tennessee Press, 1981) and Donald, *Women against Cruelty*, pp. 55–93.

panions; and where he cannot possibly receive that refreshment which the air must afford to him when at liberty to fly to such height. (I, pp. 42–3)

Even though the narrative concedes that caging animals may in some cases be a way of protecting the weakest,[78] it nonetheless criticises overtly the culture of collecting. Seen from the point of view of the robins ('Oh! why will the human race be so wantonly cruel?' [I, p. 63]), *Fabulous Histories* thus denounces human cruelty towards the animal world, as in the following example which uses the stock egg-thief motif:

> I have known some of them whose appearance was as engaging as that of our young benefactors, who were, notwithstanding, barbarous enough to take eggs out of a nest and spoil them; nay, even to carry away nest and all before the young ones were fledged, without knowing how to feed them, or having regard to the sorrows of the tender parents. (I, p. 74)

Master Jenkins, who collects eggs, without hatching or cooking them, explains that he 'blow[s] out the inside, and then run[s] a thread through them, and give[s] them to Lucy to hang up amongst her curiosities' (I, pp. 79–80). The children's cruelty and their objectification of the animal world (the eggs become mere 'curiosities' on display) reveal their lack of taste, intimating as a result how Trimmer's discourse on animals hinges upon structures of power and class: 'And so, said Miss Harriet, you had rather see a string of empty egg-shells, than hear a sweet concert of birds singing in the trees? I admire your taste truly!' (p. 80).

Thus, by comparing familiar and more 'curious' or exotic creatures which fall prey to the cruelty of both children and adults and to their society's culture of collecting, Trimmer anchors her children's book in Enlightenment Britain. With her discussion of collecting, exhibiting and studying exotic animals,[79] Trimmer instils Lockean pedagogical principles

---

[78] 'The case is very different in respect to Canary-birds, my dear, said Mrs Benson; by keeping them in a cage, I did them a kindness. I considered them as little foreigners who claimed my hospitality. This kind of bird came originally from a warm climate; they are in their nature very susceptible of cold, and would perish in the open air in our winters; neither does the food which they feed on grow plentifully in this country; and as here they are always bred in cages, they do not know how to procure the materials for their nests abroad. And there is another particular which would greatly distress them were they to be turned loose, which is, the persecution they would be exposed to from other birds' (I, pp. 44–5).

[79] The reference to Benson's father's microscope is a case in point: the instrument is mentioned but instantly related to the issue of pain (II, p. 73).

in a typically bourgeois discourse where exotic animals are constructed as symbols of aristocratic taste. The case resembles that of one of her contemporaries, Anna Laetitia Barbauld, whose 'Petition for a Mouse', written in 1771 (and published in 1773), uses a speaking mouse kept prisoner in a cage to denounce scientific experiments on live animals.[80] As another key Georgian educator, who had founded with her husband the Palgrave School and taught natural sciences,[81] Barbauld followed in the footsteps of those of her contemporaries who used the animal's viewpoint[82] in their children's fiction. This is notably the case in several of her tales included in *Evenings at Home; or, the juvenile budget opened* (1792–96), a collection of narratives combining natural history and more 'scientific' topics, such as chemistry or astronomy, with poetry and moral stories, written with her brother John Aikin.[83]

[80] The poem was written as a reaction against Dr Priestley's use of live mice in his experiments on air. See Mary Ellen Bellanca, 'Science, Animal Sympathy and Anna Barbauld's "The Mouse's Petition"', *Eighteenth-Century Studies*, 37.1 (Fall 2003): 47–67 and Julia Saunders, '"The Mouse's Petition": Anna Laetitia Barbauld and the Scientific Revolution', *The Review of English Studies*, 53.212 (Nov. 2002): 500–16. Bellanca ultimately argues, however, that '[i]n the end, "The Mouse's Petition" is neither a definitive pronouncement on animal experiments nor a broadside attack on science' (p. 61).

[81] Bellanca, 'Science, Animal Sympathy and Anna Barbauld's "The Mouse's Petition"', p. 50.

[82] The use of the animal's point of view in late Georgian children's fiction could be found in many other examples, such as Elizabeth Sandham's *The Adventures of a Poor Puss* (1809). As Jane Spencer argues, Georgian children's writers' attempt at representing 'fictional minds' was also in line with contemporary fiction, such as Henry Fielding and Samuel Richardson; Jane Spencer, 'Creating Animal Experience in Late Eighteenth-Century Narrative', *Journal for Eighteenth-Century Studies*, 33.4 (2010): 469–86, p. 470.

[83] This is the case in 'The Young Mouse, A Fable' (I, 18–20), 'The Discontented Squirrel' (I, pp. 43–8), 'The History and Adventures of a Cat', (I, pp. 105–19), 'The Little Dog, A Fable' (I, pp. 119–24) and 'The Travelled Ant' (V, pp. 101–16). Aikin and Barbauld, *Evenings at Home; or, the Juvenile Budget Opened. Consisting of a Variety of Miscellaneous Pieces, for the Instruction and Amusement of Young Persons* (London: J. Johnson, 1792). All subsequent references are to this edition and will be given parenthetically in the text. Barbauld's actual contribution to *Evenings at Home*—fourteen pieces out of the ninety-nine pieces that appeared in the six volumes—was in fact much more limited than readers tended to believe at the time of its publication. Her pieces appeared, moreover, only in the first two volumes of the series. See William McCarthy, *Anna Letitia Barbauld: Voice of the Enlightenment* (Baltimore: The Johns Hopkins University Press, 2008), p. 324. McCarthy refers to Barbauld's niece, Lucy Aikin, who named the fourteen pieces written by her aunt in 1825. See also Daniel E. White, 'The "Joinerina": Anna Barbauld, the Aikin Family Circle, and the Dissenting Public Sphere', *Eighteenth-Century Studies*, 32.4 (Summer 1999): 511–33, p. 516. Fyfe contends, however, that Barbauld wrote two-thirds of the first two volumes, but this would make a total of twenty-three to twenty-four pieces contributed by

In both *Lessons for Children* (1778–79) and *Evenings at Home*, Barbauld frequently mixes familiar and less familiar creatures. *Lessons for Children* features a vast array of different animals, from squirrels, cats, rats, owls, frogs and lambs to monkeys, tigers, camelopards, ostriches, rhinoceros, chamois, reindeer and lions. In *Evenings at Home*, however, a work aimed at older children, Barbauld and Aikin play rather on the contrast between domestic and wild animals, the better to denounce the contemporary fashion for exotic pets. In 'A Mouse, Lapdog and Monkey – A Fable', for example, the story serves to convey Barbauld and Aikin's moral lesson: the mouse is led to compete with more desirable animals—creatures which typify the development of pet-keeping in eighteenth-century Britain. The monkey and lapdog, both used for the amusement of their masters, are fed according to their capacity to entertain their owners. The mouse ultimately realises this, since its lack of exoticism condemns it to starvation:

> When the family was again seated, a lapdog and a monkey came into the room. The former jumped into the lap of his mistress, fawned upon everyone of the children, and made his court so effectually that he was rewarded with some of the best morsels of the entertainment. The monkey, on the other hand, forced himself into notice by his grimaces. He played a thousand little mischievous tricks, and was regaled, at the appearance of the dessert, with plenty of nuts and apples. ...
>
> Alas! how ignorant was I, to imagine that poverty and distress were sufficient recommendations to the charity of the opulent. I now find, that whoever is not master of fawning and buffoonery, is but ill qualified for a dependant, and will not be suffered even to pick up the crumbs that fall from the table. (I, pp. 99–100)

To prevent any sympathy for, or identification with, the monkey,[84] Barbauld and Aikin prefer therefore to use the much more commonplace

Barbauld; Fyfe, 'Reading Children's Books in Late Eighteenth-Century Dissenting Families', p. 455.

[84] In 'The History and Adventures of a Cat' (*Evenings at Home*, I, pp. 105–19, p. 117), the chained monkey is revealingly also described as a dangerous animal, which bits the cat's tail. Barbauld and Aikin's choice of more common animals to elicit children's sympathy is similar to Trimmer's, as seen above, or later Edward Augustus Kendall's *Keeper's Travels in Search of his Master* (1798). Kendall's *Keeper's Travels in Search of His Master* relates the travels of a dog which has lost his master and goes on a quest to find him again, experiencing accidents along the way. The narrative frequently uses the dog's point of view to express his feelings, as well as his 'sympathetic sensibility' (p. 23), climaxing with a poem written by

mouse to elicit feelings in their child readers, as Dorothy Kilner had done a few years before, as will be seen below. However, Barbauld and Aikin also choose more 'noble' wild creatures like the elephant to denounce the fate of exotic animals brought, used as pets or exhibited, such as in 'Tit for Tat, A Tale' (III, 29–31). This poem offers a factual description of the elephant, while also evoking the animal's feelings (III, p. 30). As the volumes progress, however, the fables tend to disappear, especially those involving animals, making room for more specialised articles on science (e.g. botany, mineralogy, astronomy, chemistry and physics) and moral stories. As in *Lessons for Children*, exotic animals (the cast in the six volumes includes tigers, monkeys, crocodiles, wild boar, elephants, lynxes, rhinoceroses, hippopotami, lions, antelopes and wild cats) feature alongside more common ones (mice, goldfinches and linnets, crows, swallows, tortoises, geese, horses, hogs and other farm animals). There are also encounters with peoples unfamiliar to their readers, like the Hottentots, such as in 'Travellers' Wonders' (I, pp. 22–31) and 'The Ship' (I, pp. 134–50). In fact, the issue of humanity percolates into the narratives as a whole when they take the animal's viewpoint or tackle the treatment of animals,[85] as well as when readers are confronted with different 'races' of Men and are invited to distinguish 'man' from 'brutes' (III, 'On Man', pp. 1–12) or question human slavery.[86] The parallel between human and

Keeper's master using 'I' for the dog's voice. Interestingly, dogs are contrasted with more exotic animals, such as elephants or apes:

'The understanding of dog', [the magistrate explained], 'surpasses that of all other animals, except man and the elephant'.

'Are not apes and monkeys very sensible?'

'They are reckoned among the most stupid of quadrupeds'; answered the magistrate, 'the appearance of understanding in them, is entirely in consequence of the resemblance which their form bears to that of man: but this fimiliarity [*sic*] is, in fact, a convincing proof of their total want of capacity. Because, if they possessed this, in addition to the advantages of exterior conformation, they would never be surpassed by the dog, and the elephant, and even the horse; whose shape and organisation differs so widely from ours'; Edward Augustus Kendall, *Keeper's Travels in Search of His Master*, 2nd edn (London: Newbery, [1798] 1799), pp. 65–7.

[85] See, in particular, in 'The Transmigrations of Indur' (II, pp. 1–34), where the main character's 'benevolence towards animals in distress' (p. 2) leads him to a series of transformations into different animals which are often killed by humans, and 'What Animals are Made For' (IV, pp. 147–56).

[86] See 'Master and Slave' (VI, pp. 81–8): 'Humane! Does it deserve that appellation to keep your fellow-men in forced subjection, deprived of all exercise of their free-will' (p. 85); or 'A Globe Lecture' (VI, pp. 123–44): 'If a black dog is as much a dog as a white one, why

52 L. TALAIRACH

animal slavery, as Tague demonstrates, reveals how '[a]dvocates of the nascent humane movement ... linked abolition to their goal of improving the treatment of animals'. In this way, animal slavery became 'an ideal testing ground for ideas about the morality of all forms of enslavement'.[87] Indeed, the parallels made in children's literature between humans and animals reflect the 'new ideals of humanity' which emerged in the last decades of the eighteenth century, what David Perkins refers to as the 'plea on behalf of animals [being] part of a much broader civilizing, educative, and disciplinary effort'.[88] Developing sympathy in children by encouraging them to identify with animals, whether domestic or wild, partook therefore of a wider educational programme in line with Enlightenment values.

This project is particularly visible in a number of other works from the same period. They include Thomas Percival's *A Father's Instruction; Consisting of Moral Tales, Fables, and Reflections Designed to Promote the Love of Virtue, a Taste for Knowledge, and an Early Acquaintance with the Works of Nature* (1776), whose articles, in their variety and themes, are sometimes very close to Barbauld's.[89] *A Father's Instruction* merges (natu-

---

should not a black man be as much a man. I know nothing that colour has to do with mind' (pp. 131–2). Slaves appear here and there in other narratives, such as 'The Birthday Gift' (VI, pp. 118–22).

[87] Tague, *Animal Companions*, p. 54.

[88] Perkins, *Romanticism and Animal Rights*, p. 31.

[89] One article especially tackles the issue of animal experimentation, echoing Anna Barbauld's 'Petition for a Mouse': 'Cruelty in Experiments'. The main character, Euphronius, carries out a scientific experiment with his friend, Dr Priestley, and puts small fishes into mephitic water. The fish drop dead and his father warns him: 'Beware, my son! said Euphronius, of observing spectacles of pain and misery with delight. Cruelty, by insensible degrees, will steal into your heart; and every generous principles of your nature will then be subverted. The Philosopher, who has in contemplation the establishment of some important truth, or the discovery of what will tend to the advancement of *real science*, and to the good and happiness of mankind, may perhaps be justified, if he sacrifice to his pursuit the life or enjoyment of an inferior animal. But the emotions of humanity should never be stifled in his breast; his trials should be made with tenderness, repeated with reluctance, and carried no farther than the object in view unavoidably requires. Wanton experiments on living creatures, and even those, which are merely subservient to the gratification of curiosity, merit the severest censure. They degrade the man of letters into a brute; and are fit amusements only for the Cannibals of New-Zealand'; Thomas Percival, *A Father's Instruction; Consisting of Moral Tales, Fables, and Reflections Designed to Promote the Love of Virtue, a Taste for Knowledge, and an Early Acquaintance with the Works of Nature*, 8th edn, 2 vol. (London: Warrington, 1793), I, pp. 65–7, pp. 66–7. All subsequent references are to this edition and will be given parenthetically in the text. The more general proximity between *A Father's Instruction* and

2 WILD AND EXOTIC 'BEASTIES' IN EARLY CHILDREN'S LITERATURE   53

ral) science and sensibility, combining articles in natural history with others warning against cruelty towards animals. The focus on sensibility permeates many of the articles, several of them denouncing the imprisonment of animals. Like his contemporaries, Percival's articles deal as much with domestic or familiar species as with wilder or more exotic ones (such as sloths, chameleons, lions, rhinoceroses, tigers, hyenas, crocodiles, camels, hippopotami, orang-utans, wolverines, elephants and ostriches). The animals featured are placed in different contexts, such as in travel journals[90] and animal exhibitions.[91] Whilst some of the creatures are used allegorically, often to represent a particular character trait (sloths are not industrious, tigers represent ferocity more than courage, etc.), Percival also dwells on their feelings in order to arouse our sympathy for them. 'The Canary Bird and Red Linnet' (II, pp. 301–9), for example, contrasts a caged bird with a free linnet, and several articles denounce cruelty towards insects, horses or birds, condemning, for instance, bird-nesting. Hence, the importation of wild animals (through references to the traffic in birds, e.g. as in 'A Winter Evening's Conversation', II, pp. 137–50) or their exhibition ('Beauty and Deformity', II, pp. 269–71), both characteristic features of Enlightenment science are presented as inimical to the development of sympathy and sensibility in the British people.

Dorothy Kilner's *Life and Perambulation of a Mouse* (1784), in which a mouse addresses the narrator and asks her to write her story, differs in the choice of animal, choosing not the exotic, but a creature that is commonplace and generally vilified. As in Barbauld and Aikin's 'A Mouse, Lapdog and Monkey – A Fable', Kilner aims at developing the sensibility of the young by encouraging them to identify with a small and familiar creature rather than with exotic beasts. Yet, as we shall see, her construction of an insignificant animal, unworthy of being collected or exhibited, unlike more extraordinary and unusual 'beasties', functions in ways very similar to what could be found in guidebooks and advertisements for animal shows. Indeed, Kilner's 'history'[92] of the mouse is intended to 'instruct

---

Evenings at Home was also due, as Fyfe notes, to the fact that both Percival and the Aikins were Unitarians; Fyfe, 'Reading Children's Books in Late Eighteenth-Century Dissenting Families', p. 457.

[90] See 'Parental Affection', I, pp. 14–17 and 'A Winter Evening's Conversation', II, pp. 137–50.

[91] See 'Beauty and Deformity', II, pp. 269–71.

[92] Kilner's mouse probably influenced Lewis Carroll whose *Alice's Adventures in Wonderland* (1865) revisits earlier children's literature and is informed by the culture of

54   L. TALAIRACH

and improve' the reader, as much as 'to amuse and divert' (I, p. vi). The presentation of the mouse as an 'object' worthy of interest and a purveyor of knowledge is telling: in the narrative, the mouse is constructed in increasingly sensational terms, as if its life were as *extra*ordinary as those of the unfamiliar animals presented in menageries and natural histories. But as the narrative follows the mouse's adventures, relating how the animal and its family try to find food, it provides very little zoological information about mice, focusing instead on the animal's sufferings at 'the barbarous hands of man' (II, p. 14). The point of view of the mouse is stressed not only by the pronoun 'I' and the use of some of the conventions of sentimental or Gothic literature throughout the narrative but also by the way the narrative analyses and interprets the mice's feelings, such as when the main character, the mouse Nimble, imagines what his brother may feel once made prisoner:

> I crept softly after her, to see what would be the fate of my beloved brother. But what words can express my horror, when I saw her holding it in one hand close to the candle, while in the other she held the child, singing to her with the utmost composure, and bidding her to look at *mousy*! *mousy*!
>
> What were the actions or sensations of poor Softdown at the dreadful moment I know not; but my own anguish, which it is impossible to describe, was still augmented every moment by seeing her shake the trap almost topsy-turvy, then blow through the trap at one end, at which times I saw the dead creature's tail come out between the wires on the contrary side, as he was striving, I suppose, to retreat from her. At length, after she had thus tortured him for some time, she set the trap on the table so close to a large fire, that I am sure he must have been much incommoded by the heat, and began to undress her child. (I, pp. 32–4)

The compassion expressed by Nimble and his identification with his brother's suffering mirrors the ways in which the text invites its readers to adopt the animal subject's position and sympathise with this 'inferior'

---

natural history, as will be seen in Chap. 5. Kilner's narrator first assures the readers that 'in earnest, [she] has never heard a Mouse speak in all [her] life', just like Alice who wonders whether the mouse she encounters in the pool of tears can speak. A scene in Kilner's book also depicts the mouse, thoroughly wet, drying its hair. Moreover, whilst Kilner's book warns children against cruelty towards animals, Alice's innocent frightening of the mouse with mentions of her cat may be read as an ironic reversal of the story. [Dorothy Kilner], *The Life and Perambulation of a Mouse*, 2 vol. (London: John Marshall, 1784), I, p. xii. All subsequent references are to this edition and will be given parenthetically in the text.

creature. The mouse is later crushed and thrown into the fire, and the narrative focuses at length on Nimble's feelings at that moment:

> My very blood runs cold within me at the recollection of seeing *Softdown's* as it spirted from beneath the monster's foot; whilst the craunch of its bones almost petrified me with horror. At length, however, recollecting the impossibility of restoring my beloved brother to life, and the danger of my own situation, I, with trembling feet, and palpitating heart, crept softly back to my remaining two brothers, who were impatiently expecting me behind the closet. (I, 37–8)

The descriptions of the mouse's blood running cold, its being 'petrified … with horror', just like the 'trembling feet' and 'palpitating heart', do not simply sensationalise the scene. The detailed physiological reactions of the mouse collapse the distinction between human and animal feelings, paving the way for the father's moral lesson on the subject of animal cruelty:

> I am not condemning people for killing vermin and animals, provided they do it expeditiously, and put them to death with as little pain as possible; but it is the putting them to needless torment and misery that I say is wicked. Had you destroyed the mouse with one blow, or rather given it to somebody else to destroy it, (for I should not think a tender-hearted boy would delight in such operations himself) I would not have condemned you; but to keep it hanging the whole weight of its body upon its tail, to swing it about, and by that, to hold it terrified over the cat's jaws, and to take *pleasure* in hearing it squeak, and seeing it struggle for liberty, is such *unmanly*, such *detestable* cruelty, as calls for my utmost indignation and abhorrence. (I, p. 47)

The way in which Kilner magnifies feelings, emotions and sensations the better to blur the boundaries between humans and animals, and her emphasis on the consciousness of animals,[93] inviting readers to identify with them, shows how the children's writer aims above all at developing

---

[93] As Julie A. Smith explains, nineteenth-century Romantic interest in consciousness in animals may have been a 'central influence on animal autobiography'; Julie A. Smith, 'Representing Animal Minds in Early Animal Autobiography: Charlotte Tucker's *The Rambles of a Rat* and Nineteenth-Century Natural History', *Victorian Literature and Culture*, 43 (2015): 725–44, p. 727. See also Cosslet, *Talking Animals in British Children's Fiction, 1786–1914*, Jacqueline Colombat, 'Mission Impossible: Animal Autobiography', *Cahiers Victoriens et Edouardiens*, 39 (April 1994): 37–49, Christine Kenyon-Jones, *Kindred*

children's sensibilities. In the eighteenth century, as Anne C. Vila observes, sensibility became 'the essential link between the human body and the psychological, intellectual, and ethical faculties of humankind'.[94] If, as Tague puts it, '[i]n the arts, the concept of sensibility, with its emphasis on heightened emotion and compassion for the suffering of others, could easily be relevant to animals as well as to humans',[95] then children's stories, such as Kilner's *Life and Perambulation of a Mouse*, illustrate how the choice of a mouse as a literary subject could serve the author's purpose. Here the size of the animal, its apparent irrelevance or even status as vermin functions as so many ways of surprising even more the young audience and intensifying compassion for the animal. In other words, the sensationalism of the narrative supersedes the sensationalism of the creature (or lack thereof), thereby reinforcing the lesson in sensibility.

Kilner's presentation of a small and common creature, in terms reminiscent of the dramatic discourse generally used for more impressive animals, would be echoed half a century later by A. L. O. E. (Charlotte Maria Tucker) in *The Rambles of a Rat* (1857), a tale which also adopts the viewpoint of a much disliked animal. *The Rambles of a Rat* relates a series of adventures experienced by a British rat who travels the world. The parallel between British pests and wilder species is foregrounded when the animal visits his rodent 'relatives' kept at the London Zoological Gardens in order to discover more about his family: 'the race of Mus'.[96] At the zoo, the rat encounters German hamsters and Canadian muskrats, all of which complain about their living conditions, from lack of food to captivity, whilst the creature also learns that gentlemen's hats and gloves are made with rats' fur. Interestingly, the rat encounters as well zoo visitors, similarly looking at the different species exhibited. But when a father and his son, who are also visiting the Zoological Gardens, later get to look at (and pity) a starving family in London, the narrative draws a parallel between the animal and human gaze on poverty, starvation, destitution and death. By collapsing boundaries between rats and humans, on the one hand, and the exotic and the familiar, on the other hand, Tucker's story builds on Kilner's

---

Brutes: Animals in Romantic Period Writing (Aldershot: Ashgate, 2001) and Perkins, *Romanticism and Animal Rights*.

[94] Anne C. Vila, *Enlightenment and Pathology* (Baltimore and London: Johns Hopkins University Press, 1998), p. 2.

[95] Tague, *Animal Companions*, p. 68.

[96] A. L. O. E. [Charlotte Maria Tucker], *The Rambles of a Rat* [1857] (London, Edinburgh and New York: T. Nelson and Sons, 1864), p. 50.

original sensationalisation of the mouse. This idea is confirmed at the end of the story, when a rat's head is displayed in a museum as a 'great curiosity', just like any exotic 'beastie' in a menagerie:

> The tooth which had been Furry's torment in life, was destined to make him famous after death. Learned men – I know not how many – examined the head of the rat, looked, wondered, consulted together; and the end of the matter was, that is was placed as a great curiosity in some building which is called a museum. There, amidst fine vases and ancient weapons, old manuscripts and precious stones, and noble busts of the wise and great, is the head of poor old Furry preserved, with the mouth wide open, to display the extraordinary tooth! (pp. 81–2)

In this Victorian example, just like in Kilner's earlier *Life and Perambulation of a Mouse*, therefore, the animal's (in)significance is magnified the better to elicit compassion. Kilner's and Tucker's method nevertheless differs slightly from Trimmer's: in *Fabulous Histories*, kindness to animals has direct social implications; it aims to shape ideal class relations and offer 'an orderly world in which both human and animal inferiors were kept in their proper place'.[97] If the mother in Trimmer's *Fabulous Histories* praises the children's 'humane behaviour towards animals' (p. 11), they must nonetheless learn to control their tender feelings so as not to 'give to inferior creatures what is designed for mankind' (p. 12). Trimmer's education in sensibility is thus less radical than in the work of Kilner, Barbauld or Tucker, who all use the mouse and its 'relatives', and whose plea for these small common pests (or common pets, like dogs) could potentially have political implications. As David Perkins argues, for advocates of the animal cause, the claim for animal rights could easily be extended 'to subordinate classes of humans'.[98] Still, in all these examples, references to the traffic and commodification of animals, in particular to exotic 'beasties', highlighted the fact that 'curiosity'—and scientific investigation more generally—was synonymous with the desire to know, to see and to possess. This point has already been made, but it can be added that in doing so, these writers nonetheless inextricably bound the culture of science to the notion of sensibility. William Darton's *Trifles for Children*, published in three volumes in 1796–98, is another relevant example in this context. The book offers a mix of familiar and less familiar creatures whilst

[97] Tague, *Animal Companions*, p. 146.
[98] Perkins, *Romanticism and Animal Rights*, p. 43.

condemning cruelty to animals, from children killing insects to adults teaching tricks to wild animals, such as dancing bears. Very much anchored in museum culture, Darton's *Trifles for Children*, moreover, adds recurrent references to the circulation of wild animals brought to England. He even gives readers 'News from the Tower!', which include, in the third volume, the escape of a leopard from its den.

As these examples show, the type of discourse that lay behind the use of animals in Georgian children's literature could vary, from Gay's burlesque and satirical fables to the children's writers of the second half of the eighteenth century, who—especially during the 1780s et 1790s—condemned the 'subversive energies'[99] that informed more popular literary forms and aimed, rather, at developing the sensibilities of children. All, however, whether featuring enormous pachyderms or the humble mouse, challenge the view of eighteenth-century children's literature as designed merely to purvey religious lessons and conform to prevailing moral standards. This conception is in keeping with recent scholarship on eighteenth-century children's literature which argues that imaginative reading and the religious instruction of children were not necessarily incompatible.[100] In the earlier part of the century, education and amusement had been seen as occupying quite distinct roles in the life of a child. Pedagogical texts therefore tended to be characterised by dry didacticism and moralising whilst imaginative reading was banned, most particularly fairy tales.[101] However, the chapbook form showed that literature for children did not have to be merely didactic like Isaac Watts's *Divine Songs, Attempted in Easie Language for the Use of Children* (1715).[102] This is not to suggest that early Georgian educationalists did not believe strongly that children should be educated by didactic stories in the vein of Maria Edgeworth's narratives and sermons[103] and protected from the dangers of the imagination in order to become industrious and responsible citizens. However, in

[99] Richardson, *Literature, Education and Romanticism*, p. 152.

[100] See Richardson, *Literature, Education and Romanticism*, p. 59.

[101] As Jeanie Watson shows, the fairy tale was 'never truly banished' although it remained 'outside the mainstream of acceptable children's literature'; Jeanie Watson, '"The Raven: A Christmas Poem". Coleridge and the Fairy Tale Controversy', in James Holt McGavran (ed.), *Romanticism and Children's Literature in Nineteenth-Century England* (Athens and London: The University of Georgia Press, [1991] 2009), pp. 14–33, p. 17.

[102] Hunt (ed.) et al., *Children's Literature*, p. 26.

[103] Maria Edgeworth (1767–1849) was regarded as a classic writer for children. She was very much involved in educational issues.

recent decades, the historiography of English children's literature has proposed a much more nuanced view of their ideas.[104] As Alan Richardson argues, the dualistic model ('reason' vs 'imagination'), often emphasised in accounts of English children's literature, fails to take into consideration the way in which late Georgian texts for juvenile audiences developed new educational strategies which challenged such dichotomy between imagination and reason. The emphasis on rationality in children's literature inspired, in fact, 'a revaluation of ... traditional "imaginative" forms for children'.[105] A study of the many curious creatures brought back from the Antipodes which feature in the literature of the period confirms this new account of English children's literature. It shows, moreover, it will be argued here, that the 'beasties' frequently stood at the crossroads between reason and imagination, fact and fantasy, instruction and amusement, thereby invalidating any simple dichotomy.

The light verse published by John Harris at the turn of the nineteenth century, when he took over Newbery's firm, is a case in point. *Harris's Cabinet of Amusement and Instruction* (1807–09) seemed directly descended from older satirical material in the vein of Gay's *Fables* and popular chapbooks, showing how the latter continued to inform mainstream literature for children.[106] Indeed, references to animals, birds or insects, sometimes humorously anthropomorphised, suffuse the collection, such as in Sarah Catherine Martin's *The Comic Adventures of Old Mother Hubbard and Her Dog* (1805), William Roscoe's *The Butterfly's Ball, and the Grasshopper's Feast* (1807) and Catherine Dorset's *Peacock 'At Home'* (1807). Roscoe's *The Butterfly's Ball, and the Grasshopper's Feast* describes a party given for insects and a few other small animals including a dormouse and a mole. The poem mentions a variety of insects

[104] For recent scholarship on eighteenth-century children's literature showing how the rationalist tradition, as represented by Richard and Maria Edgeworth, John Aikin, Mary Wollstonecraft or Anna Laetitia Barbauld, was often misread, see Mitzi Meyers, 'Impeccable Governesses, Rational Dames, and Moral Mothers: Mary Wollstonecraft and the Female Tradition in Georgian Children's Books', *Children's Literature*, 14 (1986): 31–59; Richardson, *Literature, Education and Romanticism*; Donelle Ruwe (ed.), *Culturing the Child, 1690–1914. Essays in Memory of Mitzi Myers* (Lanham, Toronto, Oxford: Scarecrow Press, 2005); McGavran (ed.), *Romanticism and Children's Literature in Nineteenth-Century England*.

[105] Richardson, *Literature, Education and Romanticism*, p. 31.

[106] As Alan Richardson argues, chapbooks 'would survive ... in underground form, and eventually resurface in "official" children's literature over the next two centuries'; Richardson, *Literature, Education and Romanticism*, p. 152.

60   L. TALAIRACH

and gives a little information on some of them, mainly through the adjectives used to describe them (hence the blind mole, for instance).[107] Likewise, *The Comic Adventures of Old Mother Hubbard and Her Dog* is essentially a comic work, providing no zoological information. Yet by playfully comparing humans and animals through the figure of the dog, the narrative confuses boundaries between species (the dog smokes a pipe, sits in a chair and is given beer to drink), thus revealing the evolution of the relation between humans and animals at a time of global expansion and the discovery of new creatures.

This is even more clearly illustrated in the other children's books published by Harris, which refer to less familiar animals which nevertheless behave (and stand) like humans. Thus, a monkey shaves a bear in Roscoe's *The Butterfly's Ball*; Dorset's *The Lion's Masquerade*[108] (1807) features wild and exotic 'beasties', such as lions, a jackal, orang-utans, a baboon, a monkey, a camel, an elephant and a tiger; and W. B.'s *The Elephant's Ball and Grand Fete Champetre* (1808) contains over twenty wild animals, including elephant, buffalo, bison, elk, antelope, lion, jackal, leopard, tiger, rhinoceros, hyaena, wolf, bear, panther, lynx, wolverine, camel, sloth, monkey and orang-utan. These in turn are mixed up with more familiar species such as dogs, cats, rats, mice, asses and squirrels, along with imaginary creatures, including a pard and a unicorn.[109]

Dorset's *Peacock 'At Home'* is also interesting in the way in which later editions of the book built upon the subtle balance between instruction and entertainment which informed the first edition. *Peacock 'At Home'* concerns a peacock which decides to organise a ball and sends a Valentine's card to all the feathered guests. Though primarily intended to amuse, the tale does mention nonetheless over seventy species of birds, along with their geographical origin or migration patterns, and provides information

---

[107] William Roscoe, *The Butterfly's Ball and the Grasshopper's Feast* (London: Newbery, [1807] 1808). F. J. Harvey Darton underlines that despite being 'an accomplished botanist', Roscoe did not aim at popularising science; F. J. Harvey Darton, *Children's Books in England: Five Centuries of Social Life*, 2nd edn (Cambridge: Cambridge University Press, [1932] 1970), pp. 205–6.

[108] The lion, jealous of the Butterfly's Ball and the Grasshopper's Feast, gives a 'grand masquerade' (p. 5) where all types of quadrupeds are invited; [Catherine Dorset], *The Lion's Masquerade* (London: J. Harris & B. Tabart, 1807). *The Lioness's Ball; being a Companion to the Lion's Masquerade* (London: C. Chapple, B. Tabart, J. Harris, Darton & Harvey, c. 1808) is another similar example.

[109] [W. B.], *The Elephant's Ball and Grand Fete Champetre* (London: J. Harris, 1808). In the illustrations, all the creatures are bipeds, however.

in footnotes for two of them in the 1807 edition: the Greater Honeyguide, from Africa, and the Numidian Crane.[110] John Harris's following editions of *Peacock 'At Home'*, which by 1841 also included *The Fancy Fair; or Grand Gala at the Zoological Gardens*, contained seventeen extra footnotes on some of the lesser-known species of birds, an addition which effectively transformed Dorset's text into a popular science book. More interestingly, perhaps, the now coloured plates featured a bird which was absent from the text in half of the illustrations. The bird may have represented the guillemot or the penguin (both mentioned only once in the story), yet the position and size of the bird, as well as the shape of its wings, make it look rather like a great auk, which became extinct in 1844 and was much sought after for museum collections in the years preceding its disappearance. The bird would later tell the story of its own extinction in Charles Kingsley's *The Water-Babies, A Fairy Tale for a Land Baby* (1863) (Fig. 2.1).[111]

The recurring presence of a bird frequently displayed in natural history museums and private collections, and valued for its rarity, confirms that the children's publications of John Harris (and his successors) were a series of entertaining books which drew strongly upon the scientific reality of the day and the exhibition of unknown species from far-away lands. Between 1807 and 1810, as Mary V. Jackson has shown, numerous 'Harris Papillonnades' appeared; many of them, she observes, 'dwindl[ing] into mere mechanical applications of the form, used chiefly to teach botany, ornithology, or the like'.[112] Some of Harris's titles, however, such as *The*

[110] [Catherine Dorset], *The Peacock 'At Home'* (London: Newbery, 1807).

[111] See Chap. 6 for more on the story of the last Gairfowl.

[112] Jackson mentions, for instance, the following titles: W. B.'s *The Elephant's Ball and the Grand Fete Champetre* (1807), *The Lioness's Ball* (1808), *The Lobster's Voyage to the Brazils* (1808); A. D. M.'s *The Butterfly's Birthday, St. Valentine's Day, and Madame Whale's Ball* (1808); Mary Cockle's *The Fishes Grand Gala* (1808); Dorset (?)'s *The Lioness's Rout* (1808); Theresa Tyro's *The Feast of the Fishes; or, the Whale's Invitation to His Brethren of the Deep* (1808), *Flora's Gala* (1808), *The Horse's Levee; or, the Court of Pegasus* (1808); Roscoe's *Butterfly's Birthday* (1809), *The Mermaid 'At Home'* (1809); Dorset's *The Peacock and Parrot on Their Search for the Author of the Peacock at Home* (1838). Other examples from other presses were also to be found, such as *The Cat's Concert* (London: C. Chapple, 1808), *The Fishes' Grand Gala* (London, 1808), *Madame Grimalkin's Party*, *The Jackdaw's 'At Home'* (London: Didier and Tebbet, 1808; A. K. Newman, c. 1810), *The Eahen 'At Home'; or, the Swan's Bridal Day* (J. L. Marks, c. 1840), *The Lion's Parliament; or, The Beasts in Debate* (London: J. B. Batchelor, 1808), *The Water King's Levée; or, the Gala of the Lake* (London: W. Lindall, 1808), *The Eagle's Mask. By 'Tom Tit'* (J. Mawman, c. 1807), R. C. Barton's

Fig. 2.1 'The Dowager Lady Toucan first cut it', *The Peacock 'At Home'*, *by a Lady and The Butterfly's Ball: An Original Poem, by Mr. Roscoe* (London: John Harris, 1834)

Fancy Fair; or Grand Gala at the Zoological Gardens (1832), overtly referred to places where such animals could be seen, from the London Zoological Gardens to the Surrey Zoological Gardens, which had opened the previous year. In the latter example, furthermore, most of the animals then being exhibited at the zoological gardens are mentioned,[113] even if placed outside their cages and generally anthropomorphised. In addition, as Tess Cosslett has pointed out, in some of the titles, such as Mary Cockle's *The Fishes Grand Gala* (1808), 'the notes often refer back to the

---

Chrysallina; or, The Butterfly's Gala, Addressed to Two Little Girls (T. Boy, 1820); Jackson, *Engines of Instruction, Mischief and Magic*, p. 213.

[113] Among the animals are monkeys, kangaroos, emus, pumas, lamas, antelopes, springboks, civet cats, opossums, moose deer, a zebu, a camel, sloths, hyenas and zebras.

2 WILD AND EXOTIC 'BEASTIES' IN EARLY CHILDREN'S LITERATURE   63

illustrations, implying they are accurate representations of the creatures, even though some of the pictures show animals dressed up in human clothes'.[114] Even more significantly, perhaps, in *The Lobster's Voyage to the Brazils* (1808), 'the notes refer the reader to other sources of information: "see Shaw's Naturalist Miscellany Plate 124", "Dictionary of Natural History published by Harris, St Paul's churchyard", "specimen in the British museum", "Chaetodon Enceledus Linnaei Syst Nat p 462"' and so on.[115] Harris's publications highlight how later reprintings increasingly stressed popular science in their representations of less familiar animals and invited children to draw links between the publications they were reading and the animals (both living and stuffed) that could be seen in museums. This illustrates how children's literature increasingly featured wild and exotic creatures which resonated with colonial ideology and its collections of riches, as animals moved into new spaces.

Mary Lamb's 'The Beasts in the Tower', published around the same period as Harris's first publications for children, represents a significant landmark, since the animals are used less to educate children's sensibilities than to subvert forms of hegemonic power and domination. The poem, included in Charles Lamb's *Poetry for Children* (1809), deals with the captives kept in the Tower of London. The exotic 'beasties', incarcerated in cages, map out imperial Britain, but, more importantly, they stress the way in which the animals have been tamed and enslaved:

> Within the precincts of this yard,
> Each in his narrow confines barr'd,
> Dwells every beast that can be found
> On Afric or on Indian ground.
> How different was the life they led
> In those wild haunts where they were bred,
> To this tame servitude and fear,
> Enslav'd by man, they suffer here! ...[116]

---

[114] Cosslett, *Talking Animals in British Children's Fiction, 1786–1914*, p. 58.

[115] Mary Cockle, *The Fishes Grand Gala* (London: J. Harris, 1808), pp. 12–16; [Anon.], *The Lobster's Voyage to the Brazils* (London: J. Harris, 1808), p. 17; qtd. in Cosslett, *Talking Animals in British Children's Fiction*, p. 58. Incidentally, Cosslett later argues that these books certainly inspired Charles Kingsley's *The Water-Babies*; Cosslett, *Talking Animals in British Children's Fiction*, p. 118.

[116] Mary Lamb, 'The Beasts in the Tower', in Charles and Mary Lamb, *Poetry for Children* [1809], ed. by William MacDonald (London: J. M. Dent & Co., 1903), pp. 132–4, l. 1–8, p. 132. All further references are to this edition and will be given parenthetically in the text.

64   L. TALAIRACH

However, Lamb also underlines the threat posed by these wild animals, a lioness, a bear, a wolf, a hyena, a panther, a tiger, a lion, all of which are typically found at the Tower: 'Perils and snares on every ground/Like these wild beasts beset us round' (l. 57–8, p. 134). Demonstrating 'the process of dominion in microcosm',[117] the poem therefore uses contemporary wild 'beasties' displayed in London to reflect on the problematic relationship between captives and captors: 'This place methinks resembleth well/The world itself in which we dwell' (l. 56–7, p. 134), the narrator says, as if suddenly thrusting the viewer/reader into the cage and dissolving its bars. Chase Pielak's study of how the poem undermines the 'fantasy of control'[118] is interesting. He highlights how the poem revolves around a grammatical ambiguity ('cabin'd.../The smallest infant'), which interchanges the captive and the viewer, thus placing the child inside the cage:

> How strong his muscles! he with ease
> Upon the tallest man could seize,
> In his large mouth away could bear him,
> And in a thousand pieces tear him:
> Yet cabin'd so securely here,
> The smallest infant need not fear. ... (l.31–6, p. 134)

This natural history lesson, as the narrator invites children to observe the wild animals and names them successively, thus turns into a discourse which goes beyond sympathising with the caged creatures or denouncing the exhibition of animals. Through its play on captivity, the poem potentially undermines, in Mary Ellen Bellanca's words, 'the fantasy of global domination at stake in the construction of the early nineteenth-century menagerie',[119] and reflects fears related to the discovery, capture and exhibition of exotic species which 'challenged anthropocentric complacency'.[120]

As a last example of early children's literature featuring 'curious beasties', Lamb's poem shows how discourses on animals changed as collections (of animals) became more and more institutionalised. Although, as

---

[117] Chase Pielak, *Memorializing Animals during the Romantic Period* (Farnham: Ashgate, 2015), p. 24.

[118] Pielak, *Memorializing Animals during the Romantic Period*, p. 26.

[119] Pielak, *Memorializing Animals during the Romantic Period*, p. 25.

[120] Mary Ellen Bellanca, 'Science, Animal Sympathy, and Anna Barbauld's "The Mouse's Petition"', *Eighteenth-Century Studies*, 37.1 (Fall 2003): 47–67, p. 49.

argued in the introduction to this chapter, the Tower of London started housing the Royal Menagerie as early as in 1110, the space epitomises here British hegemony and control over the world, much more than Mrs Addis's collection of exotic pets in Sarah Trimmer's *Fabulous Histories*, for instance. Thus, as the importation and exhibition of wild animals came to reflect patterns of upper-class urban leisure, served various educational purposes and facilitated the construction of gentility, their didactic function gradually shifted to a more political one. As exemplified by the children's writings of Trimmer, Barbauld and Aikin or Percival, encounters with wild and exotic creatures helped eighteenth-century educators to familiarise children with the realities of Enlightenment science. The curious and less curious 'beasties' were also significant tools in the project to mould the sensitive and compassionate citizens of tomorrow. Both familiar animals and more curious creatures were used to convey Lockean anticruelty messages and advocate benevolence towards the natural world. The familiarity of some animals ensured children's identification with the creatures, whilst reference to wilder beasts enabled Georgian pedagogues to address the relationship between humans and animals, thereby revealing humans' (changing) position in the chain of being. Some of the works did provide some natural historical information, yet the discourse on sympathy and empathy prevailed, marking a first stage in the representation of wild and exotic 'beasties' in children's literature. At the turn of the nineteenth century, however, the animals became more and more located in places where these 'curious beasties' were actually kept and invested with scientific knowledge. As street shows, travelling exhibitions and private menageries competed with public venues (from the London Zoological Gardens to numerous public museums), the institutionalisation of animal collections changed the meanings and potential uses of the 'curious beasties' on display. As will be seen in the following chapters, Victorian children's literature drew upon museum culture even more strikingly to propose stories in which the management of wild creatures conveyed pleas for animal welfare legislation or much more brutal narratives informed by imperialistic ideologies.

CHAPTER 3

Victorian Menageries

As he pays his money at the gate of the London Zoological Gardens, the visitor who has retained that freshness which is one of the greatest of earthly blessings, is irresistibly taken back to his old childish days. The click of the turnstile that admits him, seems to have snipped a score of years off his life, and, always sniffing from afar that faint musty odour of exaggerated mousiness which pervades the place, he feels that he is returning to the days of lessons and holidays, of a coercion whose strongest restrictions were liberty itself to the restraints of later life, and that he has entered a region of wonder and delight, of lions, tigers, bears – and buns. Have we not, all, some cherished memory of L for lion in the spelling-book, illustrated by a small woodcut of an animal with a human profile like that on a George the Third shilling? ... Have we forgotten, either, T for tiger, or W for the wolf that killed Red Riding-Hood?[1]

'Give me your definition of a horse'.
(Sissy Jupe thrown into the greatest alarm by this demand.)
'Girl number twenty unable to define a horse! ... Girl number twenty possessed of no facts, in reference to one of the commonest of animals! Some boy's definition of a horse'. ... 'Bitzer,' said Thomas Gradgrind. 'Your definition of a horse'.

---

[1] [Anon.], 'Our Eye-Witness and a Salamander', *All the Year Round* (19 May 1860): 140–44, p. 140.

© The Author(s), under exclusive license to Springer Nature
Switzerland AG 2021
L. Talairach, *Animals, Museum Culture and Children's Literature in Nineteenth-Century Britain*, Palgrave Studies in Animals and Literature, https://doi.org/10.1007/978-3-030-72527-3_3

'Quadruped. Graminivorous. Forty teeth, namely twenty-four grinders, four eye-teeth, and twelve incisive. Sheds coat in the spring; in marshy countries, sheds hoofs, too. Hoofs hard, but requiring to be shod with iron. Age known by marks in mouth'.[2]

Thomas Gradgrind's pedagogical method in Charles Dickens's *Hard Times* (1854), filling his students' heads with facts and training them to 'weigh and measure any parcel of human nature' (p. 48), lies at the heart of the novel's critique of utilitarianism. In Dickens's fictional industrial city, the boundaries between humans, animals and machines have collapsed, such as when Bitzer's eyelashes look 'like the antennae of busy insects' (p. 50). Unlike Sissy Jupe, Gradgrind's own 'model' children have 'at five years dissected the Great Bear like a Professor Owen' (p. 54),[3] and keep 'a little conchological cabinet, and a little metallurgical cabinet, and a little mineralogical cabinet; all the specimens ... arranged and labelled' (p. 55). But when Gradgrind's children ask to have a peep at the local circus instead of looking at their shells and minerals, their father suspects that they may have been given a story-book or read something that has developed their imagination. The association of the circus and its travelling menagerie with the realm of imagination is telling. Indeed, in Coketown, ironically, steam-engines work 'monotonously up and down, like the head of an elephant in a state of melancholy madness' (p. 65). The town, of 'unnatural red and black', looks like 'the painted face of a savage' and bells are kept in bird-cages (p. 65). Through its comparisons to wild and undomesticated fauna, the text thus functions like a disembodied Thomas Gradgrind, repressing, and ultimately suppressing, the wildness of 'curious beasties' so as to turn them into mechanical contrivances.

Charles Dickens's description of factories in *Hard Times* owes a lot to the live animals that were exhibited in London in the nineteenth century. The 'elephant in a state of melancholy madness' is undoubtedly a reference to Chunee (or Chuny), the Indian elephant exhibited at Edward Cross's Exeter Change in the 1810s, put down in March 1826. Dickens had

[2] Charles Dickens, *Hard Times* [1854], ed. by David Craig (Harmondsworth: Penguin, 1982), pp. 49–50.

[3] In the novel, anatomy embodies the power of science/knowledge and the laying bare of the body/imagination. Later on, the anatomist is described as being able to 'strike his knife into the secrets of [one's] soul' (p. 241), and the characters' heads are 'crammed with all sorts of dry bones and sawdust' (p. 168).

written articles on Chunee in the *Morning Chronicle*, portraying the animal as an unpredictable beast driven by a 'sanguinary desire'—its madness illustrating the creature's untamed instincts and thus wildness:

> The death of the elephant was a great shock to us; we knew him well; and having enjoyed the honour of his intimate acquaintance for some years, felt grieved – that in a paroxysm of insanity he should have so far forgotten all his estimable and companionable qualities as to exhibit a sanguinary desire to scrunch his faithful valet, and pulverize even Mrs Cross herself, who for a long period had evinced towards him that pure and touching attachment which woman alone can feel.[4]

Dickens's construction of one of the capital's most iconic captives as an insane and sanguinary beast reveals the fears that underlay 'the fantasy of global domination'[5] represented by nineteenth-century menageries, as seen in Chap. 2. This idea was emphasised even more strongly in his fiction, since the metaphor of the 'head of an elephant in a state of melancholy madness' occurs time and again throughout *Hard Times*, as if omnipresent and ubiquitous—the textual repetitions hammering out the industrial city's monotony and reducing the animal to mere productivity and economic value. Although, as Ronald D. Morrison observes, Dickens's attitudes towards animals 'remain difficult to characterize',[6] Dickens's

---

[4] Charles Dickens, 'Scotland Yard – Sketches by Boz, n° II (New Series)', *Morning Chronicle* (4 Oct. 1836): n.p. As Tamara Ketabgian explains, this article, later published in the series of *Sketches by Boz*, did not however include the passage. Ketabgian contrasts Dickens's two articles which refer to elephants: [Charles Dickens]'s 'Sketches on London', *Morning Chronicle* (7 Feb. 1835): n.p. and 'Scotland Yard – Sketches by Boz, n° II (New Series)', above mentioned. The latter describes the sudden madness of the beast, while the former 'treats the elephant's regularity not as a self-evident trait, but as a representation of irrational instincts and feelings, produced by a body that acts most mechanically when it is bereft of reason'; Tamara Ketabgian, *The Lives of Machines: The Industrial Imaginary in Victorian Literature and Culture* (Ann Arbor: The University of Michigan Press, 2011), p. 62. See also Richard Altick, *The Shows of London* (London and Cambridge, Mass.: Belknap Press of Harvard University Press, 1978), p. 316, and John Butt and Kathleen Tillotson, *Dickens at Work* (London: Methuen, 1957), p. 55.

[5] Chase Pielak, *Memorializing Animals during the Romantic Period* (Farnham: Ashgate, 2015), p. 25.

[6] Ronald D. Morrison, 'Dickens, *Household Words*, and the Smithfield Controversy at the Time of the Great Exhibition', in Laurence W. Mazzeno and Ronald D. Morrison (eds), *Animals in Victorian Literature and Culture: Contexts for Criticism* (London: Palgrave Macmillan, 2017), pp. 41–63, p. 42.

irony and lack of sympathy for the dead elephant contrasts sharply with his use of animals 'as an index of human suffering and, more abstractly, ... an index for the condition of England'[7] in both his fiction and journalism. It also stands in stark contrast to the discourse on animal protection which reflected Victorian middle-class sensibilities and which characterised late eighteenth- and early nineteenth-century children's literature, as seen in Chap. 2. Indeed, Dickens's portrait of Chunee appears to run counter to the emotion actually spurred by the elephant's death in London at the time. The animal's fits of violence, probably due to musth and aggravated by toothache, hardly matched the cruelty of Chunee's execution: the animal was shot with 152 musket balls and finished off with a sword as it lay dying in its cage.

Unlike Dickens, Thomas Hood's 'Address to Mr. Cross, of Exeter 'Change, on the Death of the Elephant' (1826)[8] portrayed the beast as a 'gentle giant' or 'kindly brute' with a child-like character which 'tenderly fondled' visitors:

> Oh, Mr Cross!
> Permit a sorry stranger to draw near,
> And shed a tear
> (I've shed my shilling) for thy recent loss!
> I've been a visitor,
> Of old, a sort of a Buffon inquisitor
> Of thy menagerie – and knew the beast
> That is deceased!
> I was the Damon of the gentle giant,
> And oft have been,
> Like Mr Kean,
> Tenderly fondled by his trunk compliant;
> Whenever I approach'd, the kindly brute
> Flapp'd his prodigious ears, and bent his knees, –
> It makes me freeze
> To think of it! – No chums could better suit,
> Exchanging grateful looks for grateful fruit, –
> For so our former dearness was begun.

[7] Morrison, 'Dickens, *Household Words*, and the Smithfield Controversy at the Time of the Great Exhibition', p. 41.

[8] Thomas Hood, 'Address to Mr. Cross, of Exeter 'Change, on the Death of the Elephant', *The New Monthly Magazine and Literary Journal*, 16 (London: Henry Colburn, 1826), pp. 343–4.

I bribed him with an apple, and beguiled
The beast of his affection like a child;
And well he loved me till his life was done
(Except when he was wild):
It makes me blush for human friends – but none
I have so truly kept or cheaply won!

In Hood's poem, the animal's wildness is marginalised and bracketed ('(Except when he was wild)'). Moreover, the poet compares his grief to that of Cross's other exotic and wild captives (bison, hyena, monkey, panther, lion, kangaroo, vulture and boa), further breaking down the distinction between the human visitor and the exhibited animals—an idea which reaches a climax at the end when the poet invites Cross to shoot him as well:

The very beasts lament the change, like me;
The shaggy Bison
Leaneth his head dejected on his knee!
Th' Hyaena's laugh is hush'd, and Monkey's pout,
The Wild Cat frets in a complaining whine,
The Panther paces restlessly about,
To walk her sorrow out;
The Lions in a deeper bass repine, –
 The Kangaroo wrings its sorry short fore paws,
Shrieks come from the Macaws;
The old bald Vulture shakes his naked head,
And pineth for the dead,
The Boa writhes into a double knot,
...
'Tis worse to think
How like the Beast's the sorry life I've led! –
A sort of show
Of my poor public self and my sagacity,
To profit the rapacity
Of certain folks in Paternoster Row,
A slavish toil to win an upper story –
And a hard glory
Of wooden beams about my weary brow!
Oh, Mr C.!
...
Shoot *me*!

In spite of the contrasting responses of Dickens and Hood to Chunee's death, the story of this iconic beast offers a striking illustration of the desperate conditions in which many captive animals were kept in this period and the cruelty of those charged with their care.[9] The elephant's fate, just like that of many other 'curious beasties' kept in menageries, also reflected the commercial possibilities offered by the exhibition of live animals, particularly exotic ones, with many entrepreneurs eagerly seeking to stake their share of the massive profits to be made in the industry. As a result, like Mary Lamb's caged tiger examined in Chap. 2, wild and exotic animals, many of them predators, appear increasingly throughout the Victorian period as metaphors for industrial labour and symbols of imperial power. The 'humane ideology' which permeated many Victorian discussions of animals expressed, as Morrison points out, 'the policing of the lives of the poor, the enforcement of middle-class values, and the exploitation of the symbolic value of animals to promote a sense of British superiority over its imperial rivals'.[10] Debates about the treatment of animals thus served to explore cultural hierarchies, and representations of animals became significant political tools for writers. As the British empire was both thriving and endangered, powerful and vulnerable, the numerous collections of 'curious beasties' brought back to England reflected what Nigel Rothfels terms the 'overpowering [British] urge to overpower'.[11]

It is hard, however, to separate the imperialist objectives that lay behind the making of such collections of live animals from the way in which these

---

[9] See [Anon.], 'Fifty Years Ago. Some of an Old Man's Recollections of London in His Childhood', *Aunt Judy's Christmas Volume for Young People* (London: Bell and Daldy, 1874), pp. 748–52, pp. 750–51. In this article, the narrator mentions the live animals he used to see in Fleet Street, such as a bison and a polar bear (p. 749), the latter suffering from poor living conditions: 'The other sight was very painful; for the large white bear, accustomed to icebergs, was panting from the heat of a low-roofed crowded chamber, and buckets of water were being thrown over the poor beast to refresh it'. The animal shows are part of other entertainments, such as Mrs Salmon's Waxworks (p. 749), the article reaching a climax with the sight of Exeter Change menagerie and Chunee.

[10] Morrison, 'Dickens, *Household Words*, and the Smithfield Controversy at the Time of the Great Exhibition', p. 41.

[11] Nigel Rothfels, *Savages and Beasts: The Birth of the Modern Zoo* (Baltimore: John Hopkins University Press, 2002), p. 22. Rothfels rephrases here Harriet Ritvo's views on British collections of exotic animals. Ritvo stresses the deliberate contrasts constructed by menageries between the wild beasts confined in small cages in artificial settings and the surrounding urban landscapes, calling the exhibitions of exotic animals 'shows of power'; Harriet Ritvo, *The Animal Estate: The English and Other Creatures in the Victorian Age* (Cambridge, Mass.: Harvard University Press, 1987), pp. 217–18.

creatures embodied the Victorians' thirst for knowledge, as 'symbols of the power of science and witnesses of how Progress could edify man while taming the natural order'.[12] This chapter will argue that live animals, when kept behind bars and exhibited in shows and zoological gardens, actively participated in the development of children's literature in the course of the Victorian period. On the one hand, they evoked far-away lands and enabled child readers to travel the world from the safety of their own nurseries and experience imaginary adventures of the kind feared by Thomas Gradgrind. On the other hand, they stood for imperialism, colonial experience and the ideology of white domination, while reflecting the development of scientific knowledge about unknown species. As pedagogical methods increasingly relied upon the visual, therefore, references to animal displays and shows became ideal educational tools, both instructive and entertaining, moulding young middle- and upper-class children into miniature imperialists.

LIVE ANIMALS AND VICTORIAN POPULAR SCIENCE

When I first undertook my labours in Natural History, my strongest motive was to lead the minds of youth to the study of that delightful pursuit, the surest foundation on which Religion and Morality was efficiently to be implanted in the heart, as being the unerring and unalterable book of the Deity. My writings were intended chiefly for youth; and the more readily to allure their pliable, though discursive, attention to the Great Truths of Creation, I illustrated them by figures delineated with all the fidelity and animation I was able to impart to mere woodcuts without colour; and as instruction is of little avail without constant cheerfulness and occasional amusement, I interspersed the more serious studies with *Tale*-pieces of gaity and humour; yet even in these seldom without an endeavour to illustrate some truth, or point some moral; so uniting with my ardent wish to improve the rising generation, the exercise of my art and profession, by which I lived.[13]

When Thomas Bewick presented the first volume of his *History of British Birds* (1797) to his young readership and their parents, he aimed to offer an educational book which would be as faithful to reality as possible in

---

[12] David Blackbourn and Geoff Eley, *The Peculiarities of German History: Bourgeois Society and Politics in Nineteenth-Century Germany* (Oxford: Oxford University Press, 1984), p. 200, qtd. in Rothfels, *Savages and Beasts*, p. 22.

[13] Thomas Bewick, *A History of British Birds, Vol. 1, containing the history and description of land birds* [1797] (Newcastle: Edw. Walker, 1826), p. iii.

74 L. TALAIRACH

order to improve young minds. The 'delightful pursuit' he proposed, however, would be interspersed with '*Tale*-pieces of gaity and humour', ensuring a simultaneously instructive and entertaining reading experience. Bewick's '*Tale*-pieces' were humorous vignettes, with a series of comic scenes illustrated in the margins, including adult characters on rocking-horses or riding boars, monkeys holding wooden spoons and cooking, mice riding carts and gentlemen bowing to geese.[14] These vignettes signalled the audience for whom the book was intended. Yet Bewick's distinctive use of the textual and the visual marked his clear-cut separation of instruction from entertainment: the text was meant to instruct; the images to entertain.

This dichotomy between the textual and the visual was nonetheless challenged by Bewick's systematic reference to the real bird specimens he used for his descriptions. These included stuffed birds lent by ornithologist Prideaux John Selby (1788–1867), specimens borrowed from the Wycliffe Museum, from Mr Leadbeater (Bird and Animal Preserver to the British Museum), from the Newcastle Museum, from the Museum of Ravensworth Castle and from the Edinburgh Museum.[15] All these specimens were thus real exhibits, in some cases on show to the general public. As argued in the previous chapter, Bewick's work and illustrations radically changed the type of natural histories which were published for young audiences. By both illustrating the classics of children's literature, such as Aesop's fables (*Select Fables, In Three Parts* [1784]), and authoring natural histories (*General History of Quadrupeds* [1790]; *A History of British Birds* [1797–1804]), Bewick promoted a much more realistic type of animal representation. In addition, his use of the visual, increasingly central in the pedagogical theories of his day, presented animals in an entirely new way to young readers. Although his illustrations of birds never represented the creatures in a museum but instead placed them in a natural environment, many of Bewick's successors chose to focus on live 'beasties', especially when confined behind bars and exhibited to the (young) public.

Indeed, early Victorian children's literature witnessed major changes in the ways in which wild and exotic fauna were portrayed. These changes

---

[14] The cruelty of many of the vignettes must also be noted: the illustrations added here and there to entertain young readers include hunters aiming at or ambushing their quarry, cats being hanged or strangled, dogs running away with pans attached to their tails and characters practising bird-nesting.

[15] Thomas Bewick, *A History of British Birds, Vol. 2, containing the History and Description of Water Birds* [1804] (Newcastle: Edw. Walker, 1826).

were certainly not unrelated to the opening of the London Zoological Gardens in 1828[16] and the closures of the Exeter Change menagerie in 1829 and the Tower menagerie in 1835, as previously seen. Compared to earlier natural histories aimed at children, such as those published by Milner and Sowerby (*Buffon's Natural History, Containing a Full and Accurate Description of the Animated Beings in Nature* [1800]), many books on animals now featured above all birds and beasts kept in menageries, especially following the publication in 1829 of the first guide to the London Zoological Gardens, James Bishop's *Henry and Emma's Visit to the Zoological Gardens in the Regent's Park* (a work which would be republished throughout the Victorian period). This trend would continue up until the end of the century, when the lives of the animals kept at the London Zoological Gardens were recorded in publications such as Charles John Cornish's *Life at the Zoo* (1895).[17]

Everything about Bishop's *Henry and Emma's Visit to the Zoological Gardens in the Regent's Park*, published by Thomas Dean, indicated that the book was designed for children. Dean's London publishing firm, established at the turn of the nineteenth century, was known for its children's books and toy books printed using the lithographic process. The London Zoological Gardens guidebook was priced at one shilling (making the book more expensive than Dean's sixpenny toy books which could include up to fifteen coloured engravings) and contained twelve coloured engravings. Throughout the multiple editions, the animals depicted varied according to the animals displayed at the Regent's Park site, many of them likely to be 'curious' and therefore of interest to children (the pigeons, in contrast, look 'too common to be curious'

---

[16] The London Zoological Gardens were opened to the general public in 1846; Sally Shuttleworth, *The Mind of the Child: Child Development in Literature, Science, and Medicine, 1840–1900* (Oxford: Oxford University Press, [2010] 2012), p. 245.

[17] C. J. Cornish's *Life at the Zoo* was not exclusively aimed at children, since some of the chapters were originally written for and published in *The Spectator*. Rev. Theodore Wood's *The Zoo (fourth series)* (London: Society for Promoting Christian Knowledge, 1895) is another example, although the book reads much more like a natural history. The Society for Promoting Christian Knowledge published several volumes of articles originally penned by the Rev. J. G. Wood (Theodore Wood's father), published in *Good Words*, *The Sunday Magazine* and *The Child's Pictorial*, and aimed at child audiences. J. G. Wood also published a series of articles entitled 'The Zoological Gardens' for *The Boy's Own Magazine* in the 1870s and 1880s.

76   L. TALAIRACH

(p. 7),[18] suggesting that the interest of the Zoological Gardens lies precisely in its exhibition of unfamiliar animals). The guidebook presented a comprehensive list of animals on display from various parts of the world.[19] Guided by an adult, the child characters learn about the origins and habits of several of the 'beasties' kept at the Zoological Gardens, as well as about the usefulness of some of the animals: tortoises are described as useful to make combs and ostriches serve for ladies' head-dresses, for example.

Moreover, the thrills of both entertainment and danger, to which the narrative and its illustrations draw attention, serve to promote the attractiveness of the site (and the text). The guidebook mentions repeatedly that the animals are kept behind bars and that it is therefore safe for children to approach and stroke or feed the animals, such as when they see the black and brown bears climbing on their poles:

> Their fears, however, ceased, when they observed that they could not come near them, and then they were much amused at their tricks. Henry's father, with a long stick kept for that purpose, handed the bears an apple and a bun, which they readily took and eat [*sic*], and looked about for more. (p. 6)

Similarly, when the children visit the reptile house, the sight of the venomous snakes makes the little girl fear for her safety ('in boxes, wired at the top, were one or two snakes; and notwithstanding that Emma was assured that they could not hurt her, she could hardly be persuaded to look at them' [p. 28]). The stress on the danger represented by the 'beasties' is also prominent in the illustrations. These ironically often foreground the cages in which the animals are kept much more than the creatures themselves. Thus, the polar bear is confined in a small domed

---

[18] James Bishop, *Henry and Emma's Visit to the Zoological Gardens in the Regent's Park. Interspersed with a Familiar Description of the Manners and Habits of the Animals Contained Therein. Intended as a Pleasing Companion to Juvenile Visitors of this Delightful Place of Recreation and Fashionable Resort* (London: Dean and Munday; A. K. Newman and Co., [1829] 1830), p. 7. All subsequent references are to this edition and will be given parenthetically in the text.

[19] The 1830 edition mentions an Indian rhinoceros, elephants, brown bears, grizzly bears, polar bears, tapirs, lemurs, opossums, pigeons, parrots, macaws, cockatoos, lamas, zebras, cormorants, Persian sheep, mouflons, a Corsican goat, nilghaus, antelopes, zebus, emus, ostriches, jackals, lions, tigers, leopards, a lynx, hyenas, a raccoon, arctic foxes, sloth bears, gnus, panthers, leopards, armadillos, kangaroos, cassowaries, monkeys, apes, baboons, tortoises, crocodiles, beavers, snakes, eagles, gulls, alongside British birds, rabbits, hares and so on.

cage, and the otter and squirrels are hardly visible behind their bars. In addition, the caged animals are out of scale, as if dwarfed, while bizarrely, the visitors often look taller than the cages. The presence of apparently over-sized visitors may have hinted at humankind's superiority over the creatures, as a means of reassuring readers and visitors. But it also emphasises the act of seeing itself, presenting the Zoological Gardens above all as a show and inviting the readers to look at the zoo's visitors looking at 'curious beasties'. The visual stress upon entertainment more than scientific education is confirmed by representations of less dangerous animals. When animals like the deer, the ostrich and the kangaroo are featured, the low gates indicate that the visitors can touch the creatures, while the Indian elephant, protected by a slightly higher gate, interacts with a visitor with his trunk. The bear pole and monkey pole are also oddly scaled, this time highlighting the entertainment offered to visitors by the creatures. The attractions available are mentioned several times in the text: the children have 'had the additional pleasure of witnessing several curious feats which these Elephants performed before them' (pp. 35–6). This illustrates both how Regent's Park successfully combined education and entertainment and how the guidebook managed to conflate them through its text and illustrations.

Henry and Emma's Visit to the Zoological and *The Menageries. Quadrupeds, Described and Drawn from Living*, both published in 1829 and both aimed at children, reflect the cultural significance of the London Zoological Gardens (or London 'Zoo', as it was later called) in the years following its opening. As seen in Chap. 2, *The Menageries. Quadrupeds, Described and Drawn from Living* stresses the importance of the menagerie, and the appearance and behaviour of the animals found there. Indeed, the work's very title reproduces textually the capture of the wild 'beasties', as if to shift the emphasis to the place where the animals are to be viewed. In this sense, the menagerie becomes the site where animals are no longer *represented* but *seen*, and where the world becomes miniaturised:

> The people see in these menageries a great number of rare animals, brought together from distant parts of the earth, whose habits are very curious and surprising: but they never see the Griffin, which is represented as half beast and half bird; nor the Centaur, which the poets have described as half horse and half man; nor the Phœnix, which is drawn as a bird, and is stated to perish by fire at the end of a hundred years, and then to rise against from its own ashes. The people thus gradually learn to disbelieve the existence of

78 L. TALAIRACH

these things, because the fables to which they have trusted never receive a confirmation from any living specimen; whilst, on the other hand, the statements of intelligent travellers and naturalists, which they may have also heard of, are abundantly proved by the evidence of their own senses.[20]

The references here to mythical creatures and the way in which the author contrasts them with the 'curious beasties' seen by travellers and naturalists, or in menageries, is significant because it presents unfamiliar animals as the new focus of the child's curiosity and the means to acquire knowledge. The focus on sight both reflects contemporary pedagogical methods based upon the visual and establishes a link with earlier constructions of scientific investigation driven by passion, wonder and curiosity which similarly relied heavily upon the senses. The 'curious beasties' (like the 'curious' child) are pivotal to a model of empirical investigation in which observation is construed as essential to the development of scientific knowledge. However, such observation remains controlled by the text (and, in some cases, the illustrations) which mediates the gaze: the viewer's 'curiosity' is, in fact, as caged as the beasties. The museum of live animals metonymically hints at the unknown places, countries and worlds where the creatures come from, but frames and artificially restrains the wildness of knowledge by imposing an organising structure upon it:

> The principal subjects of our descriptions and drawings are to be found in the Menageries of London and its neighbourhood. We shall especially direct our attention to the Collection in the Garden of the Zoological Society, which already contains many interesting specimens, and which may be justly regarded as the foundation of a National Vivarium, – for so such a collection was called by the Romans. It may be objected to this mode of writing a book on Quadrupeds, that the individual animals are seen under artificial restraints, and that, in circumstances so opposed to their ordinary states of existence, we can form no adequate notion of their natural habits. To this we reply, that the observation of Quadrupeds in Menageries is the only mode by which a sufficient number can be viewed alive at one time.[21]

Throughout the popular science book, moreover, the management of knowledge and of the animals march hand in hand. As in the guide to the

---

[20] [Anon.], *The Menageries. Quadrupeds, Described and Drawn from Living Subjects*, vol. 1, The Library of Entertaining Knowledge (London: Charles Knight, 1829), p. 21.

[21] [Anon.], *The Menageries*, p. 8.

London Zoological Gardens, the emphasis is laid upon the creatures' domestication, as part of the scientific project informing the Regent's Park collection of live animals. The menageries are also a means of testing the creatures' instincts and taming them so they can live harmoniously with one another. Thus, the education of animals and of children becomes uncannily similar, as in the following passage:

> There is a little Menagerie in London where such odd associations may be witnessed upon a more extensive scale, and more systematically conducted, than in any other collection of animals with which we are acquainted. Upon the Surrey side of Waterloo Bridge, or sometimes, though not so often, on the same side of Southwark Bridge, may be daily seen a cage of about five feet square, containing the quadrupeds and birds which are represented in the annexed print. The keeper of this collection, John Austin, states that he has employed seventeen years in this business of training creatures of opposite natures to live together in content and affection. ... It is impossible to imagine any prettier exhibition of kindness than is here shown: – the rabbit and the pigeon playfully contending for a lock of hay to make up their nests; the sparrow sometimes perched on the head of the cat, and sometimes on that of the owl, – each its natural enemy; and the mice playing about with perfect indifference to the presence either of cat, or hawk, or owl. ... This is an example, and a powerful one, of what may be accomplished by a proper education, which rightly estimates the force of habit, and confirms, by judicious management, that habit which is most desirable to be made a rule of conduct. The principle is the same, whether it be applied to children or to brutes.[22]

The merging of (animal) domestication and (children's) education permeated many texts aimed at the young, explaining perhaps the popularity of caged creatures during the Victorian period. Jane Loudon's *The Young Naturalist's Journey; or, the Travels of Agnes Merton and Her Mamma* (1840) is a case in point. The popular science book relates a little girl and her mother's expedition by train to visit a menagerie. Mrs Merton and her daughter see monkeys, mongooses, lemurs, Virginian partridges and various sorts of other birds, and they learn about flying squirrels, hawks, falcons and chameleons. The species they encounter have been brought to England from all over the world—Africa, America, the East Indies, Madagascar, Italy and Spain. The book, inspired by the *Magazine of Natural History* and aiming to adapt some of the papers published in

---

[22] [Anon.], *The Menageries*, pp. 19–20.

80   L. TALAIRACH

the magazine, provides descriptions of the 'very curious'[23] creatures encountered by the two women, accompanied by illustrations, and information on the creatures' origins, their eating habits and advice about how to keep them.

However, throughout the narrative, what seems to interest the two female characters most is whether the creatures may be tamed, since most of the captive animals they are shown illustrate how wild species may be domesticated. The issue of human power over nature lies thus at the heart of the narrative. Indeed, domestication is presented as a means of saving endangered species, as one of the characters tells the mother and her daughter at the end of the journey: '[T]he same circumstances which led to the extinction of the Dodo are now tending towards that of the Kangaroo. [A]s no attempt is made to tame them, or breed them in confinement, in time the race must become extinct' (p. 177). From the education of hawks for falconry, with its risks of failure, to that of rats, used as pets, the collection of animals presented to the reader illustrates the human capacity to control the natural world. This is epitomised by the book's final chapter which concludes the work with a visit to Mr Trelawney's museum. The museum proudly exhibits stuffed birds such as the Nuthatch which had once been part of the live menagerie. The character of Mr Trelawney, who 'keep[s] the papers that are sent with the different animals composing [his] menagerie and museum' (p. 140), symbolises the extent to which control over the British colonies was often little more than an attempt at 'collect[ing] and collat[ing] information', and the making of an 'archive',[24] as argued in the introduction. Caged 'beasties' displayed for children hence represented in ideal form what Thomas Richards refers to as the Victorian 'obsess[ion] with the control of knowledge'[25] as a reaction to the increased production of knowledge represented by the British empire.

The visit to Mr Trelawney's house, who also owns a 'collection of animals' (p. 80), kept in cages or as pets[26]—or both, since he sometimes

---

[23] Jane Loudon, *The Young Naturalist; or, the Travels of Agnes Merton and her mamma*, 3rd edn (London: Routledge, Warne, and Routledge, [1840] 1860), pp. 2, 86, 124, 166. All further quotations are from this edition and will be given parenthetically in the text.

[24] Thomas Richards, *The Imperial Archive: Knowledge and the Fantasy of Empire* (London and New York: Verso, 1993), p. 3.

[25] Richards, *The Imperial Archive*, pp. 4–5.

[26] Like, for example, the slow lemur, the Diana monkey, the ungka ape and the sparrow-hawk.

lets out some of the animals, like the chameleon—confirms the discourse on 'curious beasties' that informs Loudon's narrative. The aim here is not limited to providing the little girl and young readers with knowledge of natural history. As Barbara Gates has argued, the narrative shows a prototypical middle-class family who 'approved of the domestication of exotic species but then liked to imagine that their wildness might reemerge'.[27] But as rats and monkeys are being turned into pets, the little girl's education is interestingly paralleled with the animals' adaptation, even more so as the young protagonist is rewarded with a tame-looking and long-domesticated monkey, which she knows is likely to grow fierce in the presence of strangers. The discovery of the creatures' inner nature while they are pets links the little girl's education with domestication, suggesting thereby that she must tame her own nature—if only superficially—in order to adapt successfully to the adult world. Moreover, together with references to debates over classification, to the arguments between scientists and to the evolution of classification from Linnaeus to Cuvier and later naturalists, the wildness of each species is systematically gauged according to the creature's appetite. Although most of the species may appear wild and docile, they always become fierce and ravenous when they see food, acting stealthily and cunningly, and revealing the inner beast that slumbers in them. The wild creatures' hunger thus emblematises the power of their instincts. In this way, instinctual behaviour stands as the ultimate sign of bestiality, as the mother tells her daughter, hoping that this lesson in natural history will ensure that her little girl will not in her turn be 'rightly punished for [her] gluttony' (p. 70).

The example of Loudon's popular science book presenting collections of live 'beasties' to a young girl is significant because Loudon uses the menagerie to map out the process of a child's education. The fear of the wild, whether kept in a cage or repressed, informs the natural history lesson as much as it symbolises the raising of the child. Sally Shuttleworth's study of child development in Britain is significant in this context. For Shuttleworth, as a 'science of child development' emerged in the second half of the nineteenth century, 'exploring the mental and physiological processes of normal child development, often from the point of birth',[28] it was commonly linked to studies of animal psychology, especially through

---

[27] Barbara T. Gates, *Kindred Nature: Victorian and Edwardian Women Embrace the Living World* (Chicago and London: University of Chicago Press, 1998), p. 46.

[28] Shuttleworth, *The Mind of the Child*, p. 221.

explorations of simian development.[29] As products of the Victorian museum culture,[30] 'baby shows', where babies were exhibited by their mothers,[31] echoed contemporary animal shows, both exploring the relationships between children and animals. Such links between children and animals were reflected in the children's literature of the time—from museum guidebooks depicting stuffed creatures, as in E. W. Payne's *More Pleasant Mornings at the British Museum; or, the Handy-Work of Creation. Natural History Department* (c. 1858), where the tameness of the strange and exotic creatures exhibited is underlined,[32] to works of fiction. In each case, literary culture constantly stressed the 'curious beasties" docility, once imported into England. Whilst the domestication of new species lay at the root of the scientific project of the London Zoological Gardens (and of many other European zoological gardens eager to develop new sources of food), it was also strongly emphasised in educational books presenting live menagerie creatures to young audiences. Even abecedaria, which pre-date the development of children's literature, such as through hornbooks, featured more and more caged animals during the Victorian period, as will be seen in the next section.

F for Ferocious? Abecedaria in the Nineteenth Century

As Mary V. Jackson has argued, the earliest ABC books played a crucial part in the rise of literacy in England. Combining elementary reading lessons with selections from Scripture, they were under Puritan control

---

[29] As she explains, the interest in the relationship between simian and child development started before the publication of Charles Darwin's *On the Origin of Species by Means of Natural Selection, or the Preservation of Favoured Races in the Struggle for Life* (1859); popular culture had associated children with monkeys at least as early as in the Elizabethan era; Shuttleworth, *The Mind of the Child*, p. 245.

[30] According to Susan J. Pearson, 'baby shows' rose during 'the golden age of the dime museum … and the freak show'; Susan J. Pearson, '"Infantile Specimens": Showing Babies in Nineteenth-Century America', *Journal of Social History*, 42.2 (Winter 2008): 341–70.

[31] Shuttleworth, *The Mind of the Child*, p. 237.

[32] 'Case 77 contains the toucans, with their extraordinary beaks. They are peculiar to the New World, and resemble our hornbills in habit and food, but are easily tamed. They do not live long when brought to England'; E. W. Payne, *More Pleasant Mornings at the British Museum; or, the Handy-Work of Creation. Natural History Department* (London: The Religious Tract Society, c. 1858), p. 130.

and at first much more utilitarian than entertaining.[33] Battledores, which like hornbooks before them, were used as reading or spelling tools from the middle of the eighteenth century, often consisted of alphabets,[34] which became increasingly secular in the course of the nineteenth century and included more and more entertaining illustrations. The history of battledores, as Vicky Anderson has shown, was closely related to the rise of children's literature: battledores were invented in 1746 by a business associate of John Newbery,[35] the founder of children's literature, as seen in Chap. 2, and are thus very pertinent to this discussion. Battledores frequently presented ABCs of animals, which reflected current pedagogical methods based upon the visual, hence their popularity. At the same time, however, because abecedaria aimed to develop reading skills, they also presented the 'curious beasties' as letters and therefore *as* representation. As will be seen in the following examples, such tensions which informed the representations of 'curious beasties' became even more telling in Victorian ABC books featuring animals found in menageries.

Indeed, as the number of toy books published in the second half of the nineteenth century increased, 'curious beasties' appeared more and more in the richly coloured publications, many of which were inspired by the live animals then being exhibited in Britain. Numerous Victorian toy books were published by firms such as Dean and Son, and Routledge and Warne.[36] During the period, the books evolved to include more than the original six illustrations (sold for sixpence), always featuring curious animals, which enabled the publishers to combine a study of English and natural history whilst remaining entertaining. As the offspring of bestiaries and fables, but shorn of their scriptural and/or moralising content, the toy

---

[33] Mary V. Jackson, *Engines of Instruction, Mischief and Magic: Children's Literature in England from its Beginnings to 1839* (Lincoln: University of Nebraska Press, 1989), pp. 29–30.

[34] Battledores were generally made of cheap wood or cardboard, unlike hornbooks, which were used from the sixteenth to the eighteenth century, and were made with sheets of paper or parchment mounted on a wooden paddle; the name was due to the thin piece of cow's horn which was used to cover the paper so as to protect it. Both were primers for children. By the 1830s, hornbooks were no longer used in Britain.

[35] Vicky Anderson argues that the 'invention of the battledores was claimed by the Salisbury bookseller Benjamin Collins, business associate of John Newbery, who began to manufacture a royal battledore in association with Newbery in 1746'; Vicky Anderson, *The Dime Novel in Children's Literature* (Jefferson and London: McFarland and Company, 2005), p. 45.

[36] Frederick Warne (1825–1901) first worked with the engraver and printer Edmund Evans (1826–1905) before Evans started publishing toy books himself.

books and picture books (with illustrations gradually dominating the text) increasingly represented the types of animals children could see at the London Zoological Gardens.

Both exotic and more familiar 'beasties' competed on the pages of abecedaria published in the first decades of the nineteenth century. *Martin's New Battledore of Natural History* (c. 1810) shows E for Elephant, I for Ibex, J for Jerboa, K for Kangaroo, L for Lynx, M for Monkey and N for 'Nyl-ghaw',[37] whilst Darton's battledore, *Birds, Beasts, Fish and Insects to Teach Little Folks to Read* (1823) includes a Z for Zebra, many of the other animals being much more common.[38] In the 1870s and 1880s, Frederick Warne and Co. published several alphabet and picture books depicting the animals of the London Zoological Gardens. A significant example is Ernest Griset's *The Alphabet of Animals* (c. 1880), which shows jaguars and kangaroos alongside unicorns, yet places a chapter on 'The Zoological Gardens' in the middle of the alphabet book:

> When you go to the Zoo, you must look for all the animals in this Alphabet. You will find the Ape and the Elephant and the Fox there; and the Lynx, the Porcupine, the Rhinoceros, and the Tiger. You will also see the Lion, the King of Beasts, who can be tamed, and is then as good and kind as a dog; and the Leopards, that jump about so nimbly.
>
> There also you will see the Brown Bears climbing up their pole, and you will know then that they can easily climb trees also Give them a bun, for they are as fond of sweets as children are.
>
> Then go and look at the Seal in his tank of water. You will see how fond he is of his keeper. The Seal is a very sensible creature. His home is near the North Pole, and his thick soft fur makes him able to bear the cold seas, and to lie on the ice without being frozen.[39]

Extinct and extraordinary animals were also regularly included between their pages. Griset's ABC has an X for 'Xtinct Animal!' ('You won't regret, That creatures like this one no longer are met'[40]), whilst Carton Moore-Park's *An Alphabet of Animals* (1899), published at the very end of the century, associates X with 'Two Extraordinary animals: the anteater and

---

[37] [Anon.], *New Battledore of Natural History* (London: G. Martin, c. 1810).

[38] [Anon.], *Birds, Beasts, Fish and Insects to Teach Little Folks to Read* (London: W. Darton, 1823).

[39] Ernest Griset, *The Alphabet of Animals* (London: Frederick Warne and Co., c. 1880), n.p.

[40] Griset, *The Alphabet of Animals*, n.p.

the ornithorynchus'.[41] More curious animals appeared as well around the same period in picture books such as *The Animal Picture Book for Kind Little People* (?1873), which features duckbills,[42] tapirs, kangaroos,[43] armadillos and pangolins; a dodo, a capybara, a sloth, an ant-eater, an opossum, a chinchilla and a jerboa. Warne's picture books, *The Zoological Gardens* (c. 1875 and c. 1880), both foreground even more strikingly the Regent's Park 'beasties'. Although the caged animals children could see in the London Zoological Gardens were all depicted in the wild in the illustrations, Warne capitalised upon the selling potential of the images: later editions played upon the visual more and more while cutting back on the text and the natural historical information. Indeed, by the time of the firm's 1880 picture book, the text had disappeared altogether.[44] Likewise, the American abecedaria published by McLoughlin around the same period, *The ABC of Animals* (c. 1880), offers a 'show' of animals, with the brown bear opening the pageant and the zebra closing it.[45] Featuring bears, elephants, jackals and kangaroos, the picture book also visually represents animals behind bars, such as the brown bear and the jackal, with families with children depicted looking at the creatures. The significance of sight is stressed throughout the alphabet: 'N is a Nilghau from India, you see', 'Q's for a Quagga you here may see pass', 'V the Vicugna that here you behold' (n. p.). The animals are moreover construed as commodities to be ridden ('E is an Elephant which you may ride' [n. p.]), purchased ('J is a Jackal the Zoo has just bought' [n. p.]) or killed: 'I is for Ice, where the White Bears are caught' (n. p.), the illustration depicting a hunter killing a white bear with a spear.

[41] Carton Moore-Park, *An Alphabet of Animals* (London: Blackie and Son, 1899), n.p. Similar examples may be found in American editions of alphabet books. In *The ABC of Animals* (New York: McLoughlin Bros., c. 1899), K is for kiwi and kangaroo, O for opossum, Q for quagga and U for Urson, a Canadian porcupine.

[42] 'This remarkable animal is found in Australia. It is about the size of a large rabbit, and generally lives on the borders of streams. It can swim well, and has webbed feet. But the strange fact about it is that, with four feet, it has a bill like that of a bird. It thus partakes of the nature both of the birds and quadrupeds. Its Latin name is Ornithoryncus'; [Anon.], *The Animal Picture Book for Kind Little People* (London: Ward, Lock and Tyler, ?1873), p. 18.

[43] Kangaroos are above all presented as food: 'His flesh is eaten by the natives'; *The Animal Picture Book for Kind Little People*, p. 39.

[44] [Anon.], *The Zoological Gardens* (London: Frederick Warne and Co., c. 1875); [Anon.], *The Zoological Gardens* (London: Frederick Warne and Co., c. 1880).

[45] [Anon.], *The ABC of Animals* (New York: McLoughlin Bros., c. 1880). All subsequent references are to this edition and will be given parenthetically in the text.

86   L. TALAIRACH

Popular science books offering a more extensive discussion of exotic and wild creatures also regularly featured the Regent's Park beasts. In Laura Valentine's *Aunt Louisa's Zoological Gardens* (c. 1876), for example, the caging of the curious animals is stressed on the book cover, which shows a giraffe, a monkey and a goat behind a gate, while the following twenty-four illustrations show, in contrast, animals in the wild.[46] The reference to the London Zoological Gardens introduces the book's natural history lesson: Valentine states in the preface that '[m]ost English children have visited the Zoological Gardens or seen Wild Beasts in Menageries. We think they will like to read about these animals in their wild state ...' (n. p.). Interestingly, the animals are always defined through their amenability to taming: the lion, orang-utan, kangaroo and giraffe can be successfully tamed while the zebra, tiger and cheetah remain potentially dangerous, even when apparently domesticated.[47] The wildness of the creatures is also constructed as more interesting when caged: 'We ought to be very thankful that there are no wild beasts in our country, except in the Zoological Gardens, in London, where it is a pleasure to look at them, as we know they cannot get at us to hurt us' (p. 4). The economic value of the animals is also highlighted: 'The skin of the Polar Bear is very valuable' (p. 4); 'The Elephant when tamed becomes a very useful servant. ... The valuable ivory tusks of the Elephant cause him to be eagerly hunted' (p. 2); or again, 'The Kangaroo is hunted for its flesh and skin; the flesh is very good to eat' (p. 5).

Similar in this respect are Mrs R. Lee's *Little Neddie's Menagerie* (c. 1884), whose title refers to the site where 'beasties' are exhibited,[48] and *My Zoo Animals* (c. 1890), where most of the animals are depicted behind bars.[49] Mary Seymour's *Little Arthur at the Zoo and the Animals He Saw There* (1892)—whose frontispiece pictures a lion sleeping behind bars— and her *Little Arthur at the Zoo and the Birds He Saw There* (1892) once more use the animals kept at the London Zoological Gardens to introduce

---

[46] Laura Valentine, *Aunt Louisa's Zoological Gardens* (London: Frederick Warne and Co., c. 1876). All subsequent references are to this edition and will be given parenthetically in the text.

[47] 'The Lion can be tamed, and becomes very fond of his master', 'The Tiger is so treacherous that, even when apparently quite tamed, he can never be trusted', and so on. Valentine, *Aunt Louisa's Zoological Gardens*, pp. 1–2.

[48] Mrs R. Lee, *Little Neddie's Menagerie* (London: Griffith and Farran, c. 1884).

[49] [Anon.], *My Zoo Animals* (London, New York: Raphael Tuck and Sons, c. 1890).

a natural history lesson, providing here and there illustrations of the Regent's Park menagerie.[50]

The spectacular 'beasties' presented in these alphabet books thus combined a study of English with a study of natural history. Consequently, the framing of Britishness for young readers—and miniature citizens—hinged upon the mastery of both language and (natural) science, as taught by alphabet books. In this way, education became synonymous with mastering the word and mastering the world—and its wild fauna. This also implied controlling its representation, as the unknown beasties were turned into signs—letters in an alphabet—often challenging the taxonomic system, as will be seen further in Chap. 5. As so many metonymical representations of the British empire, the live 'curious beasties' on display participated therefore in the construction of the empire both as representation (the 'beasties' are bound to letters on a page) and as a form of control (the 'beasties' are caged and/or commodified), with the show or museum display embodying (both literally and figuratively) the taming of meaning and/or 'beasties'.

As a result, abecedaria which used the inhabitants of the London Zoological Gardens, like other popular science books presenting animals through alphabets, illustrate how the menagerie increasingly appeared 'as a cultural form', in Koenigsberger's terms—revealing, in ways similar to adult literature, 'a sense of the empire as the preeminent expression of the English spirit, but also as something that England's domestic cultures struggle to grasp in its total aspect'.[51] The idea that the menagerie became 'an institution generating and managing narratives of empire'[52] can be clearly seen in many children's stories and books published during the Victorian period. Koenigsberger's study of the role played by the menagerie in the novel and its mediation of 'the novel's relation to empire and to Englishness'[53] is particularly illuminating when one examines children's literature and its constant plays upon visualisation and miniaturisation.

---

[50] Mary Seymour, *Little Arthur at the Zoo and the Animals He Saw There* (London, Edinburgh and New York: Thomas Nelson and Sons, 1892); Mary Seymour, *Little Arthur at the Zoo and the Birds He Saw There* (London, Edinburgh and New York: Thomas Nelson and Sons, 1892). The natural history lessons using zoo animals are very much like Rev. Theodore Wood's *The Zoo (fourth series)* (London: Society for Promoting Christian Knowledge, 1895) mentioned above, published around the same period.

[51] Kurt Koenigsberger, *The Novel and the Menagerie: Totality, Englishness, and Empire* (Columbus: Ohio State University Press, 2007), p. x.

[52] Koenigsberger, *The Novel and the Menagerie*, p. x.

[53] Koenigsberger, *The Novel and the Menagerie*, p. x.

As Frederick Burr Opper's *Museum of Wonders, and What Young Folks Saw There Explained in Many Pictures* (1884) illustrates, the world and its wonders were displayed in cages, the better for children to master them.[54] The museum culture that informed all types of representations of 'curious beasties' was also, therefore, a means of playing with scale, a form particularly well adapted to young audiences. The show made the invisible visible, both metonymically representing the vast British empire and caging it.

As will be seen in the next section, Victorian children's literature boomed at mid-century, as lower costs of production and distribution opened the market even further. Together with the Kay-Shuttleworth reforms in the 1850s, which gave a spur to education, and with more government expenditure on schools,[55] the development of children's weeklies and monthlies offered a growing range of natural history publications for children. From Samuel Beeton's *Boy's Own Magazine* (founded in 1855), W. H. G. Kingston's *Magazine for Boys* (founded in 1859) and the *Boy's Own Paper* (founded in 1879) to Alexander Strahan's *Good Words for the Young* (founded in 1868) and Margaret Gatty's *Aunt Judy's Magazine* (founded in 1866), children's magazines placed less emphasis on religious topics and offered young readers instead a mix of fiction and secular instruction. Nature studies and articles on explorers and scientists were a regular feature of such publications, thus blending instruction and entertainment.[56] Richly illustrated, these publications testified to the increasing dependence on pictures in multiple forms of popular science. Good examples of such publications are *Aunt Judy's Magazine* and *Good Words for the Young*, both of which addressed well-educated middle-class children, disseminating contemporary views about the natural world in the decades that followed the advent of evolutionary theory. Through their varied character, moreover, they foregrounded the material culture of natural history, a culture which promoted collections and exhibitions of curious live specimens, and invited child readers to identify with the characters and engage, like them, with the natural world.

---

[54] Frederick Burr Opper, *Museum of Wonders, and what young folks saw there explained in many pictures* (London, New York: George Routledge and Sons, 1884).

[55] David Elliston Allen, *The Naturalist in Britain: A Social History* (Princeton: Princeton University Press, [1976] 1994), p. 123.

[56] Richard Noakes, 'The *Boy's Own Paper* and Late-Victorian Juvenile Magazines', in Geoffrey Cantor, Gowan Dawson, Graeme Gooday, et al. (eds), *Science in the Nineteenth-Century Periodical: Reading the Magazine of Nature* (Cambridge: Cambridge University Press, 2004), pp. 151–71, p. 155.

Showcaging the Empire? 'Curious Beasties' and Victorian Children's Periodicals

Aunt Judy's Magazine and *Good Words for the Young* were launched in May 1866 and November 1868, respectively. Founded around the same time, they also seemed to address similar audiences. Unlike Beeton's and Kingston's magazines (*Boy's Own Magazine, Magazine for Boys* and *Boy's Own Paper*), the titles of Gatty's and Strahan's periodicals were not overtly gendered, aiming to attract both young middle-class girls as well as boys. Gatty founded and edited *Aunt Judy's Magazine* from the first volume until her death in 1873. Her daughters, Juliana Horatia Ewing and Horatia Katherine Frances Gatty, subsequently took over the editorship—from 1873 to 1875 in the former case, and until 1885 in the latter. The Gattys all contributed pieces to the magazine: Margaret Gatty and Juliana Horatia Ewing were established children's writers in their own right and the latter remained a major contributor of serials until the end of the magazine. *Aunt Judy's Magazine* also published contributions by celebrated writers, from Lewis Carroll to Hans Christian Andersen. *Good Words for the Young*, founded by Alexander Strahan and first edited by Norman Macleod, was edited by George MacDonald from 1869 up until the final instalments in 1873. Among its regular contributors were Dinah Maria Mulock Craik, Charles Kingsley, Hans Christian Andersen and George MacDonald himself—*At the Back of the North Wind* being serialised from the first issue of the magazine, together with Kingsley's popular science work, *Madam How and Lady Why; or, First Lessons in Earth Lore for Children*. Both magazines targeted middle-class audiences; both were richly illustrated. George Cruikshank and Randolph Caldecott regularly illustrated pieces for *Aunt Judy's Magazine*, while Arthur Hughes and W. S. Gilbert were frequently commissioned to provide work for *Good Words for the Young*. Serialised stories, musical pieces, poems and articles on natural history were found in the monthly issues. The contents of both periodicals remained firmly moralistic in tone.

In the first issue of *Aunt Judy's Magazine*, Margaret Gatty assured parents that they 'need[ed] not fear an overflowing of mere amusement' and that '[o]f natural history, too, [they] hope[d] to find something to say in most numbers'.[57] In fact, natural history informed most of the stories,

---

[57] Margaret Gatty, 'Introduction', *Aunt Judy's Magazine for Young People* (London: Bell and Daldy, 1866), pp. 1–2, p. 2.

poems, articles and music pieces published in the magazine. In her very first attempt at children's literature, *Parables from Nature* (1855–71), Gatty had proposed short stories combining scientific information, morality and religion. As she explained in her preface to *Parables from Nature*, her 'effort to gather moral lessons from some of the wonderful facts in God's creation' led her to an interest in both the physical and the spiritual. Many of the titles of her short stories indicate the moral or religious stance of the narrative, such as 'A Lesson of Faith', 'The Law of Authority and Obedience', 'A Lesson of Hope' and 'The Circle of Blessing'. Gatty's tales were nonetheless rooted in the facts of natural history: 'A Lesson of Faith' relates the metamorphosis of a caterpillar into a butterfly; 'The Law of Authority and Obedience' presents the lives of bees, the poisonous flowers they must not visit and the dangers that a honey-moth may represent for the hive; 'The Unknown Land' features a sedge warbler whose characteristics are drawn from Gilbert White's *The Natural History of Selborne* (1789); 'Whereunto' has a crab, a starfish and a limpet whose 'advantages' are compared, while 'Cobwebs' explains how spiders build their webs. Such a mix of morality and science could also be found in *Aunt Judy's Magazine*, although, as we will see in the next chapter, the magazine was also concerned with the *practice* of natural history, rather than simply listing facts about animals and plants.

Both *Aunt Judy's Magazine* and *Good Words for the Young* typified the evolution of children's literature in the second half of the nineteenth century. Mid-Victorian periodicals for children moved away from didactic stories, sermons and the 'maternal' tradition of women popularisers of science of the beginning of the nineteenth century, who had often used letters, dialogues and conversations in a domestic setting.[58] Instead, the children's magazines invited young readers to travel to far-away places and discover new species and peoples, opening onto the vast British empire—and its curious peoples and creatures.[59] In *Aunt Judy's Magazine*, many of

[58] Bernard Lightman, *Victorian Popularizers of Science: Designing Nature for New Audiences* (Chicago and London: The University of Chicago Press, 2007), pp. 95–166.

[59] As shown with the example of Margaret Gatty, science popularisers (especially women) nevertheless continued to mediate science through a religious lens. In the first four decades of the nineteenth century, as Bernard Lightman argues, drawing on Jonathan Topman's and Aileen Fyfe's work (Jonathan Topman, 'Science, Natural Theology, and the Practices of Christian Piety in Early Nineteenth-Century Religious Magazines', in Geoffrey Canto and Sally Shuttleworth (eds), *Science Serialized: Representation of the Sciences in Nineteenth-Century Periodicals* (Cambridge, M.A.: MIT Press, 2003), pp. 33–66, p. 38; Aileen Fyfe, *Science and Salvation: Evangelical Popular Science Publishing in Victorian Britain* (Chicago and London: The University of Chicago Press, 2004), p. 7) women popularisers presented a 'theology of nature'

the natural history articles were penned by Margaret Gatty herself, but recurring names included E. Horton, Ruth Mervyn, Catherine C. Hopley[60] and E. S. H. Bagnold.[61]

more than they adopted natural theology as their model. The aim of their publications was to look at nature 'as a world full of divine purpose', not to attempt 'to present philosophical arguments to prove God's existence' (Lightman, *Victorian Popularizers of Science*, p. 24). While prototypical during the first half of the century, the 'narrative of natural theology', in which women popularisers foregrounded a view of the natural world fraught with religious significance, was furthered by mid-Victorian pedagogues and shared by many Anglican parsons involved in popularisation, as illustrated by the contributors to *Aunt Judy's Magazine* and *Good Words for the Young*. The most famous were Francis Orpen Morris (1810–93), Charles Alexander Johns (1811–74), Rev. Thomas W. Webb (1806–85), Rev. Ebenezer Cobham Brewer (1810–97), Charles Kingsley (1819–75), William Houghton (1828–95) and Rev. George Henslow (1835–1925); Lightman, *Victorian Popularizers of Science*, pp. 39–94. Lightman adds that many of the women popularisers were Anglicans (e.g. Jane Loudon, Lydia Becker and Elizabeth Twining) and some were evangelical Anglicans (e.g. Margaret Gatty and R. M. Zornlin) (Lightman, *Victorian Popularizers of Science*, p. 147). Their presentation of nature as 'charged with religious significance' was a response to science's increasing materialism (Lightman, *Victorian Popularizers of Science*, p. ix). The secularisation of science in the nineteenth century meant that the women popularisers who followed in the footsteps of their foremothers' religious and moral teaching were increasingly at odds with contemporary scientific authority. While some of them often attempted to reconcile the Biblical religious truths with current scientific theories (such as Anne Wright or Jane Loudon) (Lightman, *Victorian Popularizers of Science*, p. 155), others, such as Margaret Gatty, went so far as to denounce contemporary scientific discourse, in particular evolutionary theory. *Aunt Judy's Magazine* included contributions by renowned popularisers of science, such as Rev. J. G. Wood (1827–89), a natural historian and microscopist at the peak of his popularity in the 1860s, who was well known for his contributions to popular science for children, as in the *Boy's Own Paper* or the *Boy's Own Magazine*, as mentioned above, and for the moral and religious tone of his scientific writing. The acclaimed populariser's numerous articles on natural historical topics stressed Wood's desire to reveal the presence of divine power in the natural world, hence his prominent role in children's weeklies and monthly magazines in the second half of the nineteenth century. Wood's Christian conception of the natural world is well illustrated in his late 1870s pedagogical series 'On Killing, Setting, and Preserving Insects', published in the *Boy's Own Paper* (Rev. J. G. Wood, 'On Killing, Setting, and Preserving Insects. I–Killing', *Boy's Own Paper*, 1 (1879): 431–2). In these articles, scientific knowledge is combined with theological preoccupations, inviting children to care about the creatures he taught them how to dissect. See Richard Noakes, 'The *Boy's Own Paper* and Late-Victorian Juvenile Magazines', in Geoffrey Cantor, Gowan Dawson, Graeme Gooday, et al., *Science in the Nineteenth-Century Periodical: Reading the Magazine of Nature* (Cambridge: Cambridge University Press, 2004), pp. 151–71, p. 167.

[60] Catherine Hopley (1817–1911) was a naturalist who travelled to the United States and worked at the London Zoological Gardens in London where she developed a passion for reptiles.

[61] Probably Eliza Sophia Helen Bagnold (1822–c. 1890), born in India and daughter of John Bagnold, lieutenant, 13th Bengal Native Infantry.

The magazine also published pieces by Victorian travellers. George Carrington, who spent a decade in Australia and authored *Colonial Adventures and Experiences* (1871) and *Behind the Scenes in Russia* (1873), was a regular contributor.[62] Lord Wharncliffe (Edward Stuart Wortley), a British soldier, also contributed pieces to the magazine on his experiences in India.[63] Interestingly, in spite of these contributors' direct experience with the exotic and wild fauna of the places they lived in or had visited, the 'curious beasties' which pepper their articles and stories, such as kangaroos,[64] iguanas, parrots, opossums[65] and elephants,[66] are hardly ever described. More often than not, in fact, the wild and exotic creatures are presented merely as savage animals eager to destroy plantations or harm humans. This type of representation of exotic animals reflected colonial discourse; the wildness of the beasties, just like that of the natives, served to widen the gulf between the coloniser and the colonised. In addition, although exotic birds, like cockatoos, bell-birds[67] and parrots, as well as elephants, could also be found in the poetry and in some fictional pieces

---

[62] George Carrington, 'A Day in the Australian Bush', *Aunt Judy's Yearly Volume for Young People* (London: Bell and Daldy, 1869), pp. 224–7; George Carrington, 'Adventures among the Blacks in Australia', *Aunt Judy's Yearly Volume for Young People* (London: Bell and Daldy, 1869), pp. 345–52; George Carrington, 'A Family-Man for Six Days', *Aunt Judy's Christmas Volume for Young People* (London: Bell and Daldy, 1868), pp. 114–21; George Carrington, 'Gold Digging', *Aunt Judy's Christmas Volume for Young People* (London: Bell and Daldy, 1871), pp. 147–53; George Carrington, 'With the Children in Australia', *Aunt Judy's Yearly Volume for Young People* (London: Bell and Daldy, 1869), pp. 173–8; George Carrington, 'Commonplace Journey Notes', *Aunt Judy's Christmas Volume for 1873* (London: Bell and Daldy, 1873), pp. 105–10; 172–7; 301–6; 359–64; 403–8; 487–93; George Carrington, 'The City of the Sultans', *Aunt Judy's Christmas Volume for Young People* (London: Bell and Daldy, 1870), pp. 275–80; George Carrington, 'An Abstract View of a Sea Voyage', *Aunt Judy's Yearly Volume for Young People* (London: Bell and Daldy, 1869), pp. 312–17. George Carrington's *Behind the Scenes in Russia* (1873) was also reviewed in *Aunt Judy's Magazine* in 1874 ('Book Notices', *Aunt Judy's Christmas Volume for Young People* (London: Bell and Daldy, 1874), p. 317).

[63] Lord Wharncliffe, 'A Day's Elephant Hunting in Ceylon', *Aunt Judy's May-Day Volume for Young People* (London: Bell and Daldy, 1867), pp. 210–15.

[64] Carrington, 'A Day in the Australian Bush', p. 224; [Anon.], 'Rather a Long Walk', *Aunt Judy's Yearly Volume for Young People* (London: Bell and Daldy, 1869), pp. 351–8, p. 354.

[65] Carrington, 'With the Children in Australia', p. 177.

[66] Lord Wharncliffe, 'A Day's Elephant Hunting in Ceylon'; [Anon.], 'Scene in an Elephant Kraal. Extract from a Letter of the Bishop of Colombo', *Aunt Judy's Yearly Volume for Young People* (London: Bell and Daldy, 1869), pp. 367–8.

[67] Pan [Miss Anna Clara Shute], 'The Bell-Bird. Poem', *Aunt Judy's Christmas Volume for Young People* (London: Bell and Daldy, 1870), pp. 337–8. Interestingly, a footnote added at

3 VICTORIAN MENAGERIES   93

which depicted travelling characters,[68] most of them appeared in non-fictional narratives.

Even though the 'curious beasties' in the wild were seldom presented in detail, descriptions of caged creatures were legion. The articles by E. S. H. Bagnold, which present beasts from India, are a case in point. From the title onwards, the proximity between animals and humans is stressed. Bagnold's 'Friends and Acquaintances' series praises 'a love of animals'[69] and encourages 'friendships with beasts and birds' (p. 559). Yet, the less familiar animals she mentions, such as monkeys, which 'seiz[e] and destroy[–] the infants of the natives, but ... rarely touch the better guarded white children' (p. 562), and elephants, are presented as commodities ('valuable animal[s]' [p. 564]) that are necessarily caged and tamed. The caging and/or taming of these creatures enables the author to pause and describe or define the animal, subjecting it to the Westerner's gaze.[70] As Bagnold explains, her parents not being 'rich enough to keep any [elephants] of their own' (p. 562), she chooses to relate the story of Chunee, the Indian elephant exhibited at Exeter Change. She underlines the poor conditions of the beast's captivity, yet interestingly does not compare Chunee to elephants in the wild but to those kept at the London Zoological Gardens:

> His position being certainly worse, Chuny probably suffered even more severely from the dull tenor of his prison life – with want of natural food, occupation, and sufficient exercise – than did the great elephant at the Zoological Gardens, whose touching history readers of 'Friends in Fur and

the end of the poem explains what a bell-bird is, anticipating therefore its mention in the story which follows.

[68] Alfred Gatty, 'The Prince of Sleona', *Aunt Judy's Magazine for Young People. The Christmas Volume for 1866* (London: Bell and Daldy, 1866), pp. 28–36; 94–99; 151–66; 216–25; 277–90; 344–54; Margaret Gatty, 'Nights at the Round Table', *Aunt Judy's Magazine for Young People. The Christmas Volume for 1866* (London: Bell and Daldy, 1866), pp. 42–55; 99–111; [Annie Keary], 'The Cousins and their Friends', *Aunt Judy's Magazine for Young People. The Christmas Volume for 1866* (London: Bell and Daldy, 1866), pp. 3–15; 65–79; 129–44; 204–13; 269–76; 357–66.

[69] E. S. H. Bagnold, 'Friends and Acquaintances', *Aunt Judy's Christmas Volume for Young People* (London: Bell and Daldy, 1874), (I) pp. 501–508; (II) 559–66; (III) 631–36; (IV) 654–63; (V) 728–38, p. 559.

[70] In very similar way, the article by the Viscountess Enfield, entitled 'A Little Natural History', deals with Australian cockatoos, yet the writer weaves her lesson around a cockatoo ('Cocky') bought at Exeter Change. The Viscountess Enfield, 'A Little Natural History', *Good Words for the Young* (1 Nov. 1870): 6–9.

Feathers' will remember. It was at last considered necessary to destroy the valuable animal, and the duty being assigned to a detachment of the Guards, poor Chuny was shot in his cage. (p. 564)

Bagnold's story of Chunee, furthermore, is mediated through a book she has read ('I was lately reminded of this half forgotten tale on reading in the very interesting memoir of Charles Mayne Young, written by his son', p. 562), just like the elephant of Regent's Park, similarly recalled through Gwynfryn's [Dorothea Jones] stories in *Friends in Fur and Feathers* (1869); and in stories of real pets published in Charlotte Yonge's *Monthly Packet* and Gatty's *Aunt Judy's Magazine*, republished in volume form by Bell and Daldy in 1869.

Bagnold's presentation of these 'curious beasties' through other texts—memories of representations—is very much characteristic of the Victorian period. As Takashi Ito argues, in the first part of the nineteenth century, the London Zoo was mostly known through the images and stories that circulated, especially as entry was at first limited to professionals. In the second half of the nineteenth century, however, the part played by newspapers and the popular press turned the zoo into 'a narrative space' through the dramatisation of the zoo's stories and the sensational depiction of new species, aimed at encouraging frequent visits and increasing attendance,[71] as demonstrated by the case of J. Bishop's *Henry and Emma's Visit to the Zoological Gardens in the Regent's Park*, examined above. But this transformation of the zoo into 'a narrative space' also reveals the extent to which animals, as objects typifying Victorian museum culture, functioned as signifiers crystallising the imperial myth and encapsulating colonial ideology. Bagnold's text presents animals that have, in fact, been captured twice: the creatures, already caged, are additionally framed by the intertextual references. The readers' mediated contact with the wild involves therefore distanciation in the same way as the animals' commodification suggests their management and control. Likewise, when Bagnold relates the experiences of her 'sailor uncle' (p. 564), the traveller's encounter with wild or exotic beasts ironically takes place in a menagerie where he recognises a monkey that had been kept as a pet by a sailor:

'Why, Jacko, what have they been doing to you?' asked my uncle, advancing in spite of the precaution, and reaching towards the little beast.

[71] Takashi Ito, *London Zoo and the Victorians, 1828–1859* (Woodbridge: The Boydell Press, 2014), p. 122.

> At the words, after a moment of startled hesitation, – the monkey, with a strange excited chatter, dropped down to meet the outstretched hands, threw its long hairy arms round the visitor's neck, and laying its head upon his shoulder, broke into a fit of alternate sobbing and laughter, or what seemed something like it, – its monkey equivalent. Poor Jacko had belonged to a sailor on board a ship in which the Captain had made a voyage as passenger, many months, nay, I believe some few years before. After being domesticated and fondled, the creature was sold to the proprietor of this menagerie, and – a prisoner among strangers, and strange, hateful conditions (a *thing* kept to be stared at) – Jacko, lately the happy, merry playfellow and plaything of a jolly crew, earned the character given him by his keeper. (p. 365)

This time, the menagerie—a symbol of the commercial exploitation of 'curious beasties' and their objectification ('a *thing*')—breeds wildness, it seems, because of the animals' poor living conditions. The contrast between the 'prisoner among strangers, and strange, hateful conditions' and 'the happy, merry playfellow and plaything of a jolly crew' is stressed by the use of polyptotons which highlight the connections between the animal's feelings and its living conditions. Bagnold's discourse on animal collections is never monolithic, however, since while encouraging the ownership and caging of 'curious beasties' and underlining their (commercial) value, she also condemns their living conditions in travelling shows and menageries, as we will see.

As in *Aunt Judy's Magazine*, many of the stories and articles published in *Good Words for the Young* reflected the visibility of animals in Victorian Britain. However, although the magazine's publications foregrounded a fairly similar anthropocentric view of animals, they stressed much more gendered constructions of education which were deeply rooted in imperialistic ideologies. Compared to *Aunt Judy's Magazine*, a higher proportion of stories in *Good Words for the Young* dealt with travels to, and adventures set in, far-away lands, or concerned the lives and interests of the most frequent contributors to the magazine. The mix of travel writing and adventure fiction reflects the magazine's stress on mimetic realism; what John Miller terms 'the collation and representation of ethnographic, geographical, and natural historical detail in the spirit of a strictly scientific commitment to factual veracity'.[72] Whilst imperial adventures, distinctly characteristic of the second half of the nineteenth century, highlight the

---

[72] John Miller, *Empire and the Animal Body: Violence, Identity and Ecology in Victorian Adventure Fiction* (London: Anthem Press, 2014), p. 28.

'political investments of adventure fiction', the genre was 'more than just fictional entertainment', since the stories' verisimilitude stressed their 'ideological importance; the sense that they [were] somehow "true to life" lending authority to their pro-imperial didacticism'.[73] Paving the way for the 'nature fakers' who proposed what John Burroughs refers to as 'mock natural history' in later wild animal stories,[74] many adventure story writers featuring 'curious beasties' intermingled fact and fiction. The disjunction that Miller sees at the heart of adventure fiction, a genre blending romance and realism, escapism and a 'didactic educational agenda', what he terms the 'paradox at the form's core',[75] reveals that the factual truth (here *zoological* truth, embodied by 'curious beasties') was not necessarily opposed to fiction—or even, in some cases, fancy.[76]

Encounters with wild nature were recounted through references to fictional hunters, as in *King George's Middy* by William Gilbert (1804–90). The story reads at times like an imitation of *Robinson Crusoe* or *Gulliver's Travels* and relates the story of a midshipman trapped off the coast of Africa. The idea that real knowledge about 'beasties' resulted from direct observation and the blurring of fact and fiction here strengthens the myth of colonial masculinity. The character's situation echoes Gilbert's personal life, since before becoming a Royal Navy surgeon, he was a midshipman with the East India Company. Likewise, the stories by Cupples Howe ('The Deserted Ship'[77]) and S. W. Sadler ('How We Took Our First Slaver', 'A Night in an African Cruiser' and 'Another Cruise in the "Planet"'[78]) all involve sea travel and frequent encounters with wild animals. In 'Another Cruise in the "Planet"', the characters shoot alligators and water snakes and encounter a hippopotamus. George Cupples's 'Going Up the Hooghly'[79] also depicts men at sea (in Bengal), while in 'A Paddle in a Canoe', 'an Australian Missionary' describes black people

---

[73] Miller, *Empire and the Animal Body*, pp. 23–4.

[74] John Burroughs, 'Real and Sham Natural History', in Ralph Lutts (ed.), *The Wild Animal Story* (Philadelphia: Temple University Press, [1998] 2001), pp. 129–43, p. 137. Burroughs's article was originally published in 1903 in the *Atlantic Monthly*.

[75] Miller, *Empire and the Animal Body*, p. 28.

[76] This point will be developed in Chap. 5, which examines nonsense's combination of flights of fancy with many references to contemporary science.

[77] 'The Deserted Ship' was serialised in the magazine from October 1871 to May 1872.

[78] 'How We Took Our First Slaver' appeared in the July 1871 issue, 'A Night in an African Cruiser' in June 1870 and 'Another Cruise in the "Planet"' in November 1871.

[79] 'Going Up the Hooghly' appeared in the April 1871 issue.

eating kangaroos, opossums, bandicoots, kangaroo-rats, flying-mice, flying-foxes, lizards and iguanas.[80]

The recurrent presence of Australian species is notable in many adventure stories published in the magazine. This is due, in part, to the high number of stories penned by Richard Rowe, often under the pseudonyms of Edward Howe and Charles Camden. In some cases, a number of Rowe's Australian stories were published under different names in the same issue. Edward Howe's *The Boy in the Bush*, serialised in the magazine from December 1868 to October 1869, mentions cockatoos, parrots and lories, iguanas, kangaroos, wallabies, water-moles (platypuses), dingoes and snakes and also refers in passing to koalas and wombats. The piece relates the story of one of the characters, Donald, born 'in the colony' (p. 136),[81] who wanders in the wild with his cousin, like the real and fictional explorers they have read about, from 'Leichhardt, the brave Australian explorer' (p. 136) to Robinson Crusoe. As they set out on 'a secret exploring expedition' (p. 136), the hunting and shooting of animals becomes their main activity:

> One broiling December day – there is no frost or snow, you know, in Australia at Christmas time ... They had a gun with them, and caps, and powder, and shot, and colonial matches in brown paper boxes, and some tea, and sugar, and flour, and three parts of a huge damper (that's a great flat round cake of bread without any yeast in it), and a box of sardines and a can of preserved salmon, that Sydney had given them out of the store, and some salt, and two pannikins, and a Jack Shea (that's a great pot) to boil their tea in, and a blanket to cover them by night, and to hoist now and then as a sail by day. ... of course, as explorers, they wanted to go where they had not been before.
>
> ... When supper was over, the moon had risen, and the boys went down with their gun to the creek to see if they could shoot a duck. The dark water was plated in patches with ribbed and circling silver, and, just in the middle of one of the patches, up came a black something like a bottle. 'Hush, it's a water-mole', whispered Harry; but before he could point his gun at it the queer duck-billed thing had gone under again. ... Donald saw a bandicoot run out of one of the tufts. Up went the gun to his shoulder, and in a second Mr. Bandicoot had rolled over dead upon his back. A bandicoot is a very big brown kind of rat – nicer to eat than any rabbit. The boys soon made a fire, and baked the bandicoot in the ashes, in his skin; and they relished him ten times more than the preserved salmon. (pp. 136–8)

---

[80] 'A Paddle in a Canoe' was serialised between October and November 1871.

[81] Edward Howe [Richard Rowe], 'The Boy in the Bush', part II—'Up a Sunny Creek', *Good Words for the Young* (1 Jan. 1871): 136–40, p. 136. All further references to this instalment will be given parenthetically in the text.

In the bush, as the story shows, snakes bite, kangaroos fight and boys kill wild fauna to survive or to protect birds' nests from venomous snakes. The description of the animals, often limited to how they taste when cooked ('nicer to eat than any rabbit'), foregrounds 'the prominence of hunting as a point of colonial contact',[82] as Miller puts it, explaining that in adventure fiction the 'hero's encounter with animals is generally structured around violence', as part of 'the colonial appropriation'.[83]

Similarly, Howe's 'The Iguana's Eyes. An Australian Story', published in November 1870, emphasises the Australian setting from the title onwards, the opening of the story describing an iguana which mesmerises a young girl and captures her. The description of the 'curious beastie' focuses on its ugliness and wickedness, which is built up by textual repetitions, the better to construct the animal as a foil to the coloniser:

As they rode down the ridge Eppie saw what she had never seen before – an iguana. A lizard, ten feet long, with scales the colour of brown mud, and a tail like a great whiplash. A wasted, wrinkled-cheeked, big headed lizard, with eyes as wearily wicked as the Wandering Jew's. It looked thoroughly tired out with witnessing for centuries all the wickedness that it remembered, and all that it had forgotten; but thoroughly wicked still, only with a wickedness that got no pleasure out of what it did, but kept on being languidly wicked for mere wickedness's sake. It was noiselessly clambering up a gum-tree, putting out its hideous lantern-jawed head, now on this side of the tree, and now on that. Presently it stopped, and suddenly stared at Eppie, mesmerising her with its wearily-wicked old eyes. They were the bush's eyes for her; and, at last, their horrible attraction had become too powerful for her to resist. Silently she dropped from the horse, and wandered away into the still sunny, barren solitude.[84]

[82] Miller, *Empire and the Animal Body*, p. 3.

[83] Miller, *Empire and the Animal Body*, pp. 31–2. Miller's study of Victorian adventure fiction draws upon John MacKenzie's division of hunting into three stages: as commercial, as subsidy (to supply food) and as a 'ritualised demonstration of mastery of colonial space' (through collections of natural history specimens); John M. MacKenzie, 'Chivalry, Social Darwinism and Ritualised Killing: The Hunting Ethos in Central Africa up to 1914', in David Anderson and Richard Grove (eds), *Conservation in Africa: Peoples, Policies and Practice* (Cambridge: Cambridge University Press, 1987), pp. 41–61, pp. 41–57, qtd. in Miller, *Empire and the Animal Body*, p. 11. See also John M. MacKenzie, *The Empire of Nature: Hunting, Conservation and British Imperialism* (Manchester: Manchester University Press, 1988).

[84] Edward Howe [Richard Rowe], 'The Iguana's Eyes', *Good Words for the Young* (1 Nov. 1870): 25–7, p. 25.

More shooting of wallabies and kangaroos can be found in Howe's 'Bush Neighbours',[85] also set in Australia, and published in December 1869, while in 'Running Away to See',[86] the journey enables a little boy who has run away from school to go to sea to see porpoises, dolphins, bonito or sharks in Madeira. All are 'harpooned', 'hooked' or 'hauled on board' (p. 42), and as he samples their flesh, he is given a sense of his own virility: '[A]s he ate, he thought what a much more heroic personage he was' (p. 42). In the South Atlantic, the hero also encounters whales, 'like monstrously magnified pigs in a monstrously magnified strawyard' (p. 42), albatrosses, 'like tame ducks' (p. 42) and later observes zebra-striped Cape pigeons. The description of the unknown animals conjures up creatures which are distinctly odd: the Cape pigeons seem a bizarre mix of bird and beast and of two unfamiliar species; repetition (epanaphora) increases the size of the whales, whilst their comparison to pigs makes them shrink, just like the comparison of the albatrosses to ducks tempers their wildness. The existence of such tensions whenever 'curious beasties' are mentioned is a common feature and reflects the general attempt at domesticating wild and exotic species as so many food sources, hence their comparisons with farm animals here. Either depicted as 'horrid' (p. 42) or as extremely tame, the animals are striking in their wildness; a wildness which is often metaphorised through ugliness and is either let loose and dangerous or safely mastered by humans. Beasts, the stories repeatedly suggest, must be shot or kept in captivity, their spirits broken and they are hardly ever characterised in ways that do not reflect this dichotomy.

Interestingly, the management of wild animals (and sometimes much more familiar creatures) in menageries reproduced part of the imperial adventure, enabling colonial engagement with 'curious beasties' in Victorian England. This is the case in Rowe's stories[87] such as in *The Boys of Axleford*, written under the pseudonym of Charles Camden and serialised in the magazine from January to July 1869. The story presents

[85] Edward Howe [Richard Rowe], 'Bush Neighbours', *Good Words for the Young* (1 Dec. 1869): 93–9.

[86] Richard Rowe, 'Running Away to See', *Good Words for the Young* (1 Nov. 1869): 38–9.

[87] Richard Rowe often pictured boys making collections. 'When I Was Young', written under the pseudonym of Charles Camden and serialised in the magazine from April to September 1871, features a tame monkey which lives in a house, and the story also shows cockle-hunters and boys bird-nesting while the narrator hunts lizard, bringing back home a shed snake's skin; Charles Camden [Richard Rowe], 'When I Was Young – V – Patty Thomas and Her Children', *Good Words for the Young* (1 April 1871): 324–28, pp. 327–8.

100 L. TALAIRACH

boys going bird-nesting and keeping animals they have captured or purchased in England, and which they try to tame:

> One of the stables in the farmyard was our menagerie. We kept guinea-pigs and ferrets there, and they kept away the rats. And we kept rabbits, and a hedgehog, and a young fox, and three squirrels, and two ravens; pigeons and doves, two or three magpies and starlings (whose tongues we had split with a thin sixpence, and yet they wouldn't talk); linnets, and goldfinches, and bullfinches, and canaries; blackbirds, and thrushes, and larks; half-a-dozen puppies, and white mice, and field mice, and dormice, and a mole; a newt in a porter-bottle, and a bowl of goldfish, that were always dying; and some snakes, like whip-thongs made of different coloured ribbons, that were always dying too, or else crawling away, just when we thought that we had tamed them; and I don't know how many newspaper trayfuls of silkworms, and ever so many more things that I can't remember now.[88]

This profusion of animals is portrayed through a play on repetition, suggesting the boys' endless attempt at collecting 'beasties', which they (vainly) try to tame or keep alive. *The Travelling Menagerie*, serialised in the magazine from November 1871 to May 1872, and also written under the pseudonym of Charles Camden, is another example, more sensational in style, which foregrounds the central part played by the menagerie from the title onwards. The story opens with the Rycester fair, offering waxwork shows, peep-shows and roulette-booths. Among the exhibits are the fat lady, the giant, the dancing dogs, the embalmed head of a New Zealand chief, while Jollyman's menagerie displays its 'curious beasties' to a 'large … body of spectators'[89] on the last day of the fair. The menagerie comprises 'an elephant, ridden, and a camel and a dromedary, led, by men in Eastern costumes' (p. 8), as well as the Lion King, 'in laurel wreath, tights, and spangled scarlet doublet, mounted on a mincingly high-stepping piebald horse' (p. 9). The presentation of M. Sohier as the 'Lion King'—in fact, the big cats' tamer—blurs the boundary between the animals and their keepers. This blurring of boundaries between humans and wild animals is maintained throughout the narrative since one of the

---

[88] Charles Camden [Richard Rowe], 'The Boys of Axleford – I – Fibbing Bill', *Good Words for the Young* (1 Jan. 1869): 145–8, p. 146.

[89] Charles Camden [Richard Rowe], *The Travelling Menagerie* (London: Henry S. King and Co., 1873), p. 7. All subsequent references are to this book edition and will be given parenthetically in the text.

menagerie's horse-keepers 'ought to have been exhibited in a bear's cage' (p. 13). 'Little Jollyman' develops a relationship with all animals and later tames many of them: '[H]e was a merry little chap, and had been born in the menagerie, like a lion's cub' (p. 16).

At the heart of *The Travelling Menagerie* lies the management of these 'curious beasties'. The exhibited animals' wild traits are emphasised from the first in the book edition published the following year. The frontispiece represents an attack on the lion tamer during his performance with the leopards, thus introducing the story with the sensational accident involving the Lion King and his failure to master the wild beasts:

> All his subjects were on him. He clutched the crowbar left leaning, as a weapon of last resort, in a corner of the cage. But it was too late. He was down on his face, with a sea of undulating spotted skin surging over him. The women and children in the show made the menagerie ring with their shrieks. A wild stampede of men, women, and children took place.
>
> ... The Lion King had been rolled over towards the front of the cage. One leopard, snarling greedily at the others, was prone upon him; its tail wagging outside the bars. Little Jollyman dived under the safety-rope which ran in front of the cages, and tugged at the tail with all his might. The leopard turned upon him with a savage growl, thrusting both forelegs, with the claws out, between the bars. Fortunately, little Jollyman had fallen backwards beyond its reach.
>
> Meantime, the Lion King, fearfully bitten, had staggered to the door of the cage. There Jollyman and two brother keepers met him. They beat back the infuriated beasts, and backing themselves, dragged him out, almost drenched in blood. When the door was shut-to again like a flash of lightning, a leopard whose paw had been nipped gave a hideous yell. (pp. 21–2)

The outbreak of violence in this passage is paradoxically mapped out by the metaphor of the sea, which annihilates the animals' bodies, suggesting the oozing out or discharge of savagery, as does the metonymy 'undulating spotted skin', mimicking waves on the sea. The beasts' wildness spreads, unconstrained—contaminating men, women and children who become 'wild'. The description of this attack confuses the roles between the 'King' and his 'subjects', the better to render its fierceness and to lay bare the dominant discourses informing the representation of Jollyman Grimstone's uncontrolled animals. The 'infuriated' bloodthirsty predators, with their 'snarl', 'hideous yell' and 'savage growl', serve here to critically map out and subvert, if only briefly, imperial control. The metaphor of the sea, as a trope hinting at colonial exploration and expansion, objectifies the animals by reducing them to their spotted skins, in the same way as collected objects and beasts. As 'an important prop of empire's ideological framework and a significant element of hunting discourse'[90] and a miniaturised part of the British empire, the skins provide in this way a sense of containment. The possibly ambivalent readings of the 'sea of undulating spotted skin' construct the beasts as collected hides whilst simultaneously suggesting oozing and unmappable violence, thus embodying the tension at the heart of the narrative: the story opens and closes with the Lion King's failure to master his big cats, which eventually kill him. Thus, even though Little Jollyman grows up and becomes an expert at collecting and taming 'curious beasties', and even if the illustrations constantly present readers with tamed, caged or captured beasts, the narrative nonetheless closes with the refusal of Little Jollyman to become a tamer of big cats, thus ending the novel on an equivocal note.

Indeed, throughout the story, the narrative revolves around the relationship of humans and wild animals. The presentation of Little Jollyman's little menagerie is followed by his father's menagerie, comprising bigger yet tame creatures, constructing therefore the young boy as a natural tamer and suggesting he will in time replace the Lion King. The animals are, moreover, often presented as 'pets' (p. 81), as if to tone down their wildness (Fig. 3.1):

> Little Jollyman sometimes kept rabbits; but when they grew big and fat his surly father used to kill them, and order his wife to make a pie of them, or to stew them with onions: so little Jollyman generally preferred pets not likely to tempt his father's appetite.
>
> He had a little menagerie of paper fly-cages, which he used to arrange in an oblong like Jollyman's, and in which he kept wood-lice, beetles, bluebottles, caterpillars, moths, lady-birds, grasshoppers, earwigs, glow-worms, dragon-flies, and death-watches.
>
> At different times he had a goldfinch that could draw water for itself in a little bucket, a starling that could talk, a raven, a couple of blind worms, a lizard, a common snake that would lick his face and eat out of his hand, a bat, a dormouse, a squirrel, a cageful of white-mice, a swarm of field-mice, a tame rat, three frogs, a toad, a mole, a couple of guinea-pigs, and three or four pickle-bottles with minnows and sticklebacks and an eft in them.

[90] Miller, *Empire and the Animal Body*, p. 11. Miller draws here upon MacKenzie's three stages of hunting, detailed above.

Fig. 3.1 Charles Camden, 'The Travelling Menagerie', *Good Words for the Young* (1871). (Courtesy of Armstrong Browning Library. Baylor University, Waco, Texas)

But besides these pets, which were his own property, little Jollyman had special pets in the cages of the menagerie: three little lion's cubs as fat as butter, that rolled over on their backs and patted the dog that romped with them as playfully as babies; a silver opossum that would curl its tail round his arm, and grab at the bit of bread, or potato, or loaf-sugar he held in his other hand; a deer-eyed wallaby with which he used to wrestle and hop races (always getting beaten in the races) when it was let out of its cage to have a little exercise during the absence of visitors to the show; a cockatoo that rubbed its head against him, when he gave it its maize or sopped biscuit, and lifted up its wing to be tickled when he said, 'Show your blanket, cocky', a

104 L. TALAIRACH

tiny negro marmoset that would sit on his shoulder and munch apples; the ratels that flung clumsy somersaults expressly for his amusement; the six-banded armadillo that trotted about almost as funnily on the tips of its toes; and the pretty little shy-eyed antelope that licked his hand when he gave it its crust, or, rather, often, *his* crust. (pp. 31–3)

Little Jollyman's creatures are products of Victorian museum culture: the animals are presented and constructed as objects to be displayed ('which he used to arrange in an oblong like Jollyman's'). The significance of food is also to be noted, since the caged animals' tameness shows up in the way they eat, recalling that the feeding of the carnivores in the London Zoological Gardens was often promoted in shows for visitors to visualise their fierceness.[91] Moreover, the play on scale—the series of menageries of different sizes, like so many Chinese boxes—also serves to contain the wildness of the animals. The series of collections creates an ironic distancing of the beasts—one which is further highlighted by the narrator's emphasis on the links between the exhibition of exotic animals and the understanding of the natural world. As a matter of fact, several speeches on the 'beasties' in the collection present the creatures as in a popular science book. Zoological truth articulates imperial power, thus equating scientific authority with colonial control:

> Jollyman's had more than two lions, to say nothing of lionesses and cubs born in various parts of the United Kingdom …
> 'The Lion, ladies and gentlemen, all authorities agree in pronouncing to be the king of beasts. … The present specimens are of the African breed, whose manes vary in colour, from black to pale fulvous. The colour of the bodies of lions runs from chestnut-brown to silver-grey. In South Africa a race of white lions, white as white-wash, is said to exist. The female has no mane, and may further be distinguished from her lord by her smaller head, slighter build, and lighter colour. Several of the lions in the present collection were born in the British Isles …' (pp. 24–6)

---

[91] Diana Donald notes that early promotional prints of the London Zoological Gardens ignored the fiercest carnivores, especially their feeding with live prey so as to evoke 'an elysium where benign humans and animals encountered one another with mutual pleasure'; Diana Donald, *Picturing Animals in Britain, 1750–1850* (New Haven and London: Yale University Press, 2009), p. 182.

3 VICTORIAN MENAGERIES   105

Throughout the novel, the many passages explaining the characteristics and behaviour of lions, monkeys (p. 37), tigers (p. 41), camels (p. 59), rhinoceroses, zebras, quaggas (p. 79), bears (p. 100), seals, walruses and manatees (pp.120–25) march hand in hand with the narrator's frequent references to real beasts currently being exhibited at the London Zoological Gardens and studied by naturalists. Grimstone's menagerie functions as a double of Edward Cross's Exeter Change menagerie, as the many comparisons of his animals with the beasts exhibited at the London Zoological Gardens suggest, especially when his collection is sold by auction at the end of the novel.[92] The mention of 'Old Chuny' (p. 88) confirms the parallel, while the escape of the tiger from the menagerie recalls an episode from 1857 involving Jamrach's menagerie, when a Bengal tiger escaped from its cage and picked up a nine-year-old boy. In addition, 'Old Tom', the seal 'that lived in the Regent's Park Gardens from 1852 to 1856' (p. 34), the elands and the first hippopotamus 'that ever came to England' (p. 76) are all references to real animals kept at the London Zoological Gardens, thus enhancing the realism of the narrative. As with travel writing and adventure stories, the stress upon the factual is crucial since the blurring of fact and fiction helps the narrative perform its 'pro-imperial didacticism',[93] to borrow Miller's terms again.

It is therefore no coincidence that as Little Jollyman collects more beasts, studies their habits, and learns to feed and look after them, the young boy grows up into a naturalist. He becomes an expert at bird-catching, doing his 'birds'-nesting in the scientific-naturalist style' (p. 145) so as not to damage the nests, and sells his collected specimens to shops and collectors. His 'happy family' presents several species of familiar animals living together harmoniously: '[A]t last he had persuaded to dwell together in tolerable unity a number of small birds, an owl, a cat, some mice and a rat, a snake, a frog, and a toad, a wild rabbit, a ferret, a

---

[92] As explained in Chap. 2, when Exeter Change was demolished in 1829, Cross sold some of his animals to the London Zoological Gardens (which opened in 1828) and moved the rest to the Surrey Zoological Gardens in 1831, of which he became superintendent. After Cross's retirement in 1844, the business declined. Cross died in 1854 and all the animals were sold by auction in 1856.

[93] Miller, *Empire and the Animal Body*, p. 23.

106   L. TALAIRACH

sparrow-hawk, a pigeon, a mongrel-pup with bull-terrier blood in him, a rook, a jackdaw, and a magpie' (p. 149). His collections of specimens, and more broadly the connections between the capturing and caging of animals and their display, are therefore justified by the scientific enterprise which underlies the exploitation of animals both in England and abroad:

> Sometimes he felt as if it was not quite right to carry off the merry bright-eyed little things of which he was so fond in captivity, but he comforted himself with the thought that It was his *business* to do so, and, somehow, in spite of his qualms of conscience, he found it a very pleasant business. (pp. 143–4)

As the story comes to a close, moreover, Charles Camden [Richard Rowe] makes clear the links between collecting, exhibiting and studying animals. Little Jollyman, who becomes the 'exhibitor of the menagerie's snakes' (p. 161), shows and keeps 'beasties' he is 'very fond of looking' at, such as the brindled gnu 'because it was so curious' (p. 71). Thus, Jollyman's collection of 'curious beasties' reveals the 'intricately involved nexus of knowledge, power and violence'[94] that Miller reads into the scientific specimen, whilst Jollyman's 'qualms of conscience' illustrate 'the moral and ideological complexity'[95] that informs colonial natural history (Fig. 3.2).[96]

Like the fictional stories in which, as we have seen, the factual was recurrently stressed to perform better their didactic agenda, the non-fictional pieces published in *Good Words for the Young* foregrounded 'curious beasties', and natural history more generally, also using both scientific discourse and caged animals to tame the less familiar creatures. Popular science articles were regularly included in the instalments, mixing familiar and less familiar animals. Charles Kingsley's *Madam How and Lady Why; or, First Lessons in Earth Lore for Children*, published, as already mentioned, in the first issue of the magazine in November 1868, paved the way for many other popular science narratives, such as articles by

---

[94] Miller, *Empire and the Animal Body*, p. 55.

[95] Miller, *Empire and the Animal Body*, p. 85.

[96] The term is borrowed from Alan Bewell's *Natures in Translation: Romanticism and Colonial Natural History* (Baltimore: Johns Hopkins University Press, 2017).

Fig. 3.2 Charles Camden, 'The Travelling Menagerie', *Good Words for the Young* (1872). (Courtesy of Armstrong Browning Library. Baylor University, Waco, Texas)

B. J. Johns on entomology[97] and Hugh Macmillan on ferns and geology.[98] The monthly contributions of Henry Baker Tristram (1822–1906), an English ornithologist and traveller to Bermuda, the Sahara desert and Palestine,[99] included pieces such as 'The Spider and Its Webs', 'Ants and Ant-Hills', 'Rooks and Their Relations', 'About a Caterpillar', 'Bees and

---

[97] B. J. Johns, 'Among the Butterflies', *Good Words for the Young* (1 Jan. 1870): 132–40.
[98] Hugh Macmillan, 'A Lump of Coal' (1 Dec. 1868): 102–5.
[99] Henry Baker Tristram was also an advocate of Darwin, who wanted to reconcile evolution with creation. However, unlike Kingsley, he rejected Darwinism after 1860. As an ornithologist, he accumulated an extensive collection of bird skins which he sold to the World Museum Liverpool.

108   L. TALAIRACH

Beehives' and 'About a Fly'.[100] In similar fashion, Norman Macleod's pieces took its readers to the 'Indian Empire' to encounter elephants.[101] Alfred Wilks Drayson (1827–1901), an army officer, also contributed many articles to the periodical, many of them relating his adventures in South Africa and referring to wild creatures, including rhinoceroses, hippopotamuses, lions, crocodiles and parrot-fish. In 'On the Trail of the Wild Elephant in Africa', for instance, Drayson, on the South African coast, observes many curious species, such as pangolins:

> Amidst this wild, luxuriant foliage are numberless monkeys, and birds of brilliant plumage and quaint forms. On the ground may be seen various creatures, curious and rare in their way. Here we may see a strange sort of mat leaves, which on touching we find to possess life: it is the *manis*, a creature somewhat of the armadillo species, and most curious in its ways.[102]

Drayson's articles also regularly celebrated the hunting and shooting of wild animals, constructing masculine power through violence. In 'Unusual Fishing', for example, Drayson relates how he fished for sharks in the Canary Islands, and for keel-back, rock-cod and parrot-fish on the coast of South Africa. In each case, he compares fishing with hunting and shooting, and praises the 'manly qualities' involved in the activity.[103] The army officer also describes how he hunted for squirrels in the New Forest[104] and shot

[100] Henry Baker Tristram, 'The Spider and its Webs', *Good Words for the Young* (1 Feb. 1869): 171–6; Henry Baker Tristram, 'Ants and Ant-Hills', *Good Words for the Young* (1 March 1869): 242–52; Henry Baker Tristram, 'Rooks and their Relations', *Good Words for the Young* (1 June 1869): 391–6; Henry Baker Tristram, 'About a Caterpillar', *Good Words for the Young* (1 Nov. 1869): 45–7; Henry Baker Tristram, 'Bees and Beehives', *Good Words for the Young* (1 Jan. 1870): 161–8; Henry Baker Tristram, 'About a Fly', *Good Words for the Young* (1 May 1870): 347–54.

[101] Norman Macleod, 'Our Holiday in the West Highlands', *Good Words for the Young* (1 Nov. 1870): 20–24; Norman Macleod, 'Talks with the Boys about India', *Good Words for the Young* (1 Dec. 1870): 116–19.

[102] A. W. Drayson, 'On the Trail of the Wild Elephant in Africa', *Good Words for the Young* (1 Sept. 1871): 644–8, p. 645.

[103] A. W. Drayson, 'Unusual Fishing', *Good Words for the Young* (1 July 1871): 469–72, p. 469.

[104] A. W. Drayson, 'A Ramble in the New Forest', *Good Words for the Young* (1 Aug. 1871): 559–62.

elephants in Africa,[105] systematically presenting game hunting as playful and irresistible—and as part and parcel of the scientific observation of 'curious beasties':

> On more than one occasion we passed several hours of the night concealed among the branches of a tree, in order to have the opportunity of watching the habits of some of those animals whose nature it was to lie hidden during the day. ...
>
> A walk by night in a country where there are such animals as lions, leopards, hyænas, snakes, elephants, buffalo, and other formidable creatures, requires considerable caution; for it would be excessively awkward to run against one of these animals, or to turn a corner of a bush, and suddenly find yourself face to face with one of them. There are, too, some creatures which make a practice of following men at night – the hyæna in particular, as we experienced on more than one occasion. ...
>
> After passing a night in a tree without firing a shot, it was impossible to resist the temptation offered by the proximity of these elephants, so that as soon as there was sufficient daylight for us to see our way, we entered the dense forest, and endeavoured to get a shot at one of them.[106]

Drayson's articles on snakes take his readers from Natal, where he describes shooting pythons, to the London Zoological Gardens.[107] The parallel paves the way for a natural history lesson, as Drayson records the size of his catches and compares the animals to those kept at the London Zoo:

> The particular species to which I shall here refer is called the Natal Rock Snake, or Port Natal Python.
>
> This serpent was not uncommon in the Natal district during my residence there many years ago. ... during my residence of two or three years at Natal, I shot seven pythons, each of which exceeded sixteen feet in length. I killed eight or nine varying from seven to twelve feet in length, and I 'let off' at least a dozen others, whose habits I wished to study, or who escaped because

---

[105] A. W. Drayson, 'A Night in an African Tree', *Good Words for the Young* (1 Dec. 1871): 174–6.

[106] Drayson, 'A Night in an African Tree', pp. 174–6.

[107] A. W. Drayson, 'Venomous Serpents and their habits', *Good Words for the Young* (1 June 1871): 422–6; A. W. Drayson, 'At Home with the Python', *Good Words for the Young* (1 Feb. 1871): 211–13.

110 L. TALAIRACH

I would not alarm the animals in the vicinity by firing off a gun at such unprofitable game.

… I take this opportunity of recording the dimensions of some of the pythons I shot. The measurements were made immediately after death, and when consequently the skin had neither stretched nor contracted.

… In the Zoological Gardens there are two or three very fair specimens of the python, though they are small compared to those one sees in their native wilderness.[108]

Drayson's articles, referring as they do to the animals kept at the London Zoo, are no isolated case. As will be seen, the natural history articles in *Good Words for the Young* regularly invited child readers to compare the animals described to the live beasties exhibited in England, and this play on spaces served to collapse the boundary between home and the colonial world.

Aunt Judy's Magazine similarly featured tales narrating voyages to far-away countries or recounted expeditions to unknown lands in both the fictional and non-fictional pieces published in the periodical. In 'The Cousins and Their Friends', the child characters mentally travel to the South Pole, Australia and South America.[109] The characters also read travel narratives, as in 'Home with a Hooping-Cough'.[110] Some (whether human characters, fairies or birds[111]) travel the world, encountering kangaroos, opossums, iguanas and parrots in Australia.[112] However, more often than not, the 'curious beasties' which pepper the texts are those which were kept in British menageries. The children go to the Zoological Gardens, as in Mrs Robert O'Reilly's 'London Daisies'[113]; Rachel waits for her dad 'to

---

[108] Drayson, 'At Home with the Python', pp. 211–13.

[109] [Annie Keary], 'The Cousins and their Friends', pp. 74, 77.

[110] [Anon.], 'Home with a Hooping-Cough', *Aunt Judy's Magazine for Young People* (London: Bell and Daldy, 1866), pp. 193–204; 257–69; 321–31, p. 323.

[111] [Anon.], 'The Little Bird Who Told Stories', *Aunt Judy's Magazine for Young People* (London: Bell and Daldy, 1866), pp. 332–5, p. 335.

[112] See [Anon.], 'Rather a Long Walk', which takes place in Australia, and Alfred Gatty and Juliana Horatia Ewing's 'Cousin Peregrine's Wonder Stories', some of whose instalments take place in South Australia. Alfred Gatty and Juliana Horatia Ewing, 'Cousin Peregrine's Wonder Stories. Waves of the Great South Seas. Founded on fact', *Aunt Judy's Christmas Volume for Young People* (London: Bell and Daldy, 1874), pp. 415–31.

[113] Mrs Robert O'Reilly, 'London Daisies', *Aunt Judy's Christmas Volume for Young People* (London: Bell and Daldy, 1874), pp. 350–58, p. 357.

take her to the Zoological Gardens',[114] in M. R. Carey's serialised story, 'Rachel's visit to Devonshire', and the zoo is even promoted for children to be able to play with their Noah's Ark toys, as in 'Involuntary Contributions', where one of the child characters has 'learned the names of all the beasts in the Zoological Gardens' so as to be able to identify the animals in his own wooden menagerie.[115]

The Zoological Gardens were not, however, solely constructed as sites of knowledge for children's practical education in natural history. Many pieces conveyed above all British fantasies of imperial power, as in 'Greenhouse forbidden', a story in which the narrator enters a greenhouse, the hot atmosphere inspiring in him a reverie in which he turns into a tiger that ends its life at London Zoo. Ironically, the merging of the narrator and the big cat through the dream reinforces the divide between the human and the animal:

> With great difficulty I managed to push up one of the slides of the frame, and to drop myself gently down; then the frame had to be closed again for fear of the cold air killing the plants, and that was a dreadful struggle; but when safely over I picked up some pots I had knocked down, and settling myself on a piece of matting, became very warm, and comfortable, and happy. I ate up (in imagination) a villageful of Sepoys who had been murdering English men and women, thereby making myself a useful and praiseworthy tiger, and giving a moral to the story; and after a narrow escape of being shot by the white race that I had avenged, I was finally snared in a pit, preserved by my captors, when the story of my good deed was related by the only Sepoy that had escaped from the village, and sent out of gratitude to the Zoological Gardens, to be stared at by the British public as the noble avenger of their murdered countrymen.[116]

The beast, caged as a reward for having eaten up murderous Sepoys (an obvious reference to the Indian Mutiny of 1857), embodies imperial ideology. The narrative tames the beast twice: first by confining the animal

---

[114] M. R. Carey, 'Rachel's visit to Devonshire', *Aunt Judy's Christmas Volume for Young People* (London: Bell and Daldy, 1874), pp. 235–42, p. 225.

[115] [Anon.], 'Involuntary Contributions', *Aunt Judy's Magazine for Young People. The Christmas Volume for 1866* (London: Bell and Daldy, 1866), pp. 12–17, p. 15.

[116] [Anon.], 'Greenhouse forbidden', *Aunt Judy's Christmas Volume for 1875* (London: Bell and Daldy, 1875), pp. 586–91, p. 587.

112 L. TALAIRACH

in the narrator's imagination and second by eventually confining it behind bars in Britain. Used here as a means of controlling the dangerous Sepoys, the tiger is also employed in the text as a political weapon to teach children how 'the white race' successfully subjugates the wildness of its colonised territories. The greenhouse itself, which artificially grows exotic plants, thus defies space and time, framing the narrator's wild dream and ensuring its safe containment.

John Miller's analysis of John Buchan's *The Half-Hearted* (1900), although the latter was published over two decades later, is significant in this context. Like Buchan's adventure novel, 'Greenhouse Forbidden' confirms 'the centrality of the tiger as a metaphor of insurrection in the British mythology of power'.[117] Moreover, 'Greenhouse Forbidden' reverses many of the tales of wild beasties brought to England which escaped from their cages, such as the aforementioned example of the Bengal tiger which escaped from Jamrach's warehouse in October 1857, a few months after the Indian Mutiny.[118] As it releases the symbolical power of the predatory big cat, 'Greenhouse forbidden' thus fuses the native and the animal the better to show how Victorian menageries and sites of exhibition of live 'beasties' symbolised control and management and could serve, especially in children's literature, to indoctrinate young readers.

Another significant example of texts mentioning the Zoological Gardens are Catherine C. Hopley's popular science pieces. Her articles on mongooses ('the largest species of all that are in the Zoological gardens'[119]) and snakes invited young readers to look for other examples of 'curious beasties' at the 'Zoo' whilst giving them directions to the cages to be sought out in particular, thus framing her articles as museum guides for

---

[117] Miller, *Empire and the Animal Body*, p. 40.

[118] The animal, let loose in the heart of London, had seized nine-year-old John Wade, then playing in the street. The story was recounted in several newspapers ('Extraordinary Escape of a Tiger in Ratcliff-Highway', *London Evening Standard* [27 Oct. 1857], 'A Runaway Tiger in Ratcliff Highway', *Reynold's Newspaper* [18 Feb. 1866]). The tiger was later bought by another menagerie and exhibited as 'The tiger that swallowed the child in Ratcliff-Highway' and even appeared in the *Boy's Own Paper* in February 1879, related by Charles Jamrach himself, who explained how he managed to secure the beast and save the boy (Charles Jamrach, 'My Struggle with a Tiger', *Boy's Own Paper*, 1.3 (1 Feb. 1879): 1–2).

[119] Catherine C. Hopley, 'Snake Destroyers', *Aunt Judy's Christmas Volume for Young People* (London: Bell and Daldy, 1874), pp. 164–70, p. 167.

young readers: 'Examples of each may be seen in the Zoological Gardens. ... You may see a rat-snake at the "Zoo". It is in one of the end cases opposite you as you enter, and is so tame that the keeper can do anything he likes with it ...',[120] and so on. Hopley even advised readers of the magazine to interact with the staff at the Zoological Gardens. Seeing things, Hopley suggested, was an integral part of the natural history lesson, the children's periodical becoming inseparable from visits to the zoo:

> The keeper at the reptile-house in the Zoological Gardens occasionally finds a fang in the cage; when he thoroughly cleanses and preserves it as a curiosity. You would be surprised to see what a delicate little ivory thing it is, like a tiny white shell. Perhaps when you go there he will show you one, if you ask him. I am sure he will if you tell him you have been reading about it in *AJM*. And it will help you to remember the many curious things there are connected with snakes' teeth.[121]

Hopley's articles also advised children to observe animals at feeding time, a moment when the 'true disposition and habits of the animal are particularly displayed'.[122] Thus, the Zoological Gardens enabled the readers of the magazine to 'practically learn' the natural history lesson:

> Many young people who go to the Zoological Gardens like to see the animals fed, and run eagerly from one place to another at the hours when it is announced that feeding time has arrived. Indeed it is a sight which (though in many cases unpleasant, and even painful) is not without its uses. The true disposition and habits of the animals are particularly displayed at feeding time, and it is then that children practically learn the meaning of

---

[120] Catherine C. Hopley, 'How Snakes Feed', *Aunt Judy's Christmas Volume for Young People* (London: Bell and Daldy, 1874), pp. 561–6, p. 563. Other examples may be found, such as when she describes the neck-toothed snake in another article: '[O]thers are stripped, and of these latter you may see several in the reptile-house at the Zoological Gardens, in a case at the right-hand side, furnished with evergreens for them to climb and glide along'; Catherine C. Hopley, 'The Deirodon; or, Neck-Toothed Snake', *Aunt Judy's Christmas Volume for Young People* (London: Bell and Daldy, 1874), pp. 626–30, p. 628. See also W. B. H., 'An Adventure with a Boa Constrictor', *Aunt Judy's May-Day Volume for Young People* (London: Bell and Daldy, 1868), pp. 117–19, p. 117.

[121] Hopley, 'The Deirodon; or, Neck-Toothed Snake', p. 629.

[122] Hopley, 'How Snakes Feed', p. 561.

114   L. TALAIRACH

those perplexing words, *carnivorous*, flesh-eating; *herbivorous*, grass-eating; *graminivorous*, grain-eating; &c., &c.

... but of all the sights of the 'Zoo' that excite general curiosity and yet often baffle the comprehension of the beholder, one of the most noteworthy is the manner in which the snakes feed.[123]

Hopley's articles made explicit how the sites where animals could be seen, whether the London Zoo[124] or the British Museum,[125] furnished the materials for natural history lessons, in conjunction with books: 'I hope, however, what you have now read regarding the very wonderful formation of this class of God's creatures will impart new interest to your natural-history books, as well as enable you to derive additional satisfaction from your visits to the Zoo'.[126] Strongly emphasising the role of the visual, and even sometimes of hands-on experience with animals (or animal parts), these articles and stories worked in tandem with the exhibited live 'beasties', constructing the places where live animals were displayed as referential landmarks that fully participated in the construction of reality.[127] In doing so, these pieces used natural history to 'naturalise colonial structures of domination',[128] as Miller puts it, whether they contained the 'curious beasties' through cages, technologies of representation (such as taxidermy) or captured them through signifiers, as in the case of abecedaria.

Images of global domination were also reinforced by texts which established links between the countries their authors were writing from or about and the specimens then being exhibited in Britain. A case in point is Lady Barker, or Mary Anne Barker (1831–1911), who was born in Jamaica but educated in England. Barker married her second husband, Frederick

---

[123] Hopley, 'How Snakes Feed', p. 561.

[124] See also Catherine C. Hopley, 'How a Snake Walks', *Aunt Judy's Christmas Volume for Young People* (London: Bell and Daldy, 1874), pp. 658–62.

[125] 'You can also see skeletons of snakes at the British Museum, and you will be surprised at their numerous ribs, so long, fine, and close together, and curved so as almost to meet underneath'; Hopley, 'How a Snake Walks', p. 659.

[126] Hopley, 'How a Snake Walks', p. 662.

[127] Plants and trees kept and exhibited at Kew Gardens were regularly mentioned as well: 'A palm-tree thus overgrown is to be seen at the Museum at Kew'; Helen Zimmern, 'What the Green Lizard Told Me', *Good Words for the Young* (1 March 1870): 265–7, p. 267. Likewise, in C. Crockford's 'About Philip', the narrator mentions Kew Gardens: 'I do not know exactly why these trees grew there ... perhaps they grew up so that they might be a Kew Gardens or a Bushey Park for the little black sparrows that hopped about there'; C. Crockford, 'About Philip', *Good Words for the Young* (1 April 1870): 328–31, p. 328.

[128] Miller, *Empire and the Animal Body*, p. 55.

Napier Broome (1842–96) in 1865. The latter was a British colonial administrator, whom she followed to New Zealand, South Africa, Mauritius, Australia and Trinidad. Both husband and wife became journalists, contributing to many periodicals, including *The Times*. Barker also published several books recounting her life in New Zealand, such as *Station Life in New Zealand* (1870), which contained her letters home, and its sequel, *Station Amusements in New Zealand* (1873), alongside other stories describing her travels and colonial experiences (*Stories About:–* [1870]). There were also novels aimed at teenage boys and girls (*Sibyl's Book* [1872] and *Boys* [1874]).

Good Words for the Young published several stories by Mary Anne Barker during her time in New Zealand ('Aunt Annie's Stories About Horses',[129] 'A Chapter of Accidents'[130] etc.), in Bengal ('Four Months in Camp',[131] 'More Adventures'[132] etc.), in Jamaica ('A Chapter of Accidents'[133]) and in India ('The Marble Cross'[134]). Although constantly travelling, Barker often turned to the animals of the London Zoological Gardens to help children visualise the creatures she rode, as in 'Four Months in Camp', published in October 1870: 'The elephant I rode was larger than those you see in the Zoological Gardens, but quite as tame'.[135] Likewise, having travelled to and from England, Barker describes herself as a regular visitor to Regent's Park, enjoying the contrast between the creatures in the wild she is familiar with and the caged 'beasties': 'As soon as I came back to England, I went to the Zoological Gardens in the Regent's Park to see the cheetahs there, and found they were very much smaller than those belonging to the Maharajah of his territory, who lent them to our Commander-in-Chief'.[136] Throughout Barker's articles, numerous wild and exotic animals appear, many of them as a result of a personal encounter with the author. In her article on monkeys ('About Monkeys'[137]), for example, the traveller begins by explaining that her knowledge of monkeys

[129] Lady Barker, 'Aunt Annie's Stories about Horses', *Good Words for the Young* (1 Sept. 1870): 600–605.

[130] Lady Barker, 'A Chapter of Accidents', *Good Words for the Young* (1 June 1871): 438–45.

[131] Lady Barker, 'Four Months in Camp', *Good Words for the Young* (1 Oct. 1870): 645–57.

[132] Lady Barker, 'More Adventures', *Good Words for the Young* (1 March 1871): 267–76.

[133] Lady Barker, 'A Chapter of Accidents'.

[134] Lady Barker, 'The Marble Cross', *Good Words for the Young* (1 Aug. 1871): 530–36.

[135] Lady Barker, 'Four Months in Camp', p. 648.

[136] Lady Barker, 'Four Months in Camp', p. 654.

[137] Lady Barker, 'About Monkeys', *Good Words for the Young* (1 Feb. 1870): 196–201.

is 'not out of books'.[138] She then relates her various attempts at purchasing monkeys, thus constructing the animals as consumer goods:

> A year or so afterwards, I was on my way back to England, and the ship stopped at Ceylon for a few days. I had been told that most beautiful little monkeys were to be bought there, and all the time we were at Point-de-Galle I tried very hard to find one to purchase, but it was too early in the spring for the young ones to be taken from their mothers, so there were none in the market.[139]

Because of its title, Barker's article reads first like a popular science article on monkeys. Yet, in this article and many others published in the children's magazine, the woman traveller aims first and foremost at taming monkeys so as to keep them as 'pets'; the creatures she presents are mainly characterised by their ability to be tamed, nursed or groomed.[140] In 'Aunt Annie's Story About Jamaica', for instance, Barker conflates the making of a zoological collection and the keeping of exotic creatures as pets. The pets symbolise the captives' domestication, as in Jane Loudon's *The Young Naturalist's Journey; or, the Travels of Agnes Merton and Her Mamma* and Charles Camden's [Richard Rowe] *The Travelling Menagerie*. Furthermore, her 'love of pets' and 'taste' for animals illustrate her 'hobby', thus associating the 'curious beasties' with leisure while the pets function as a marker of social status:

> Jessie and I had one very decided taste in common, and that was our great love of pets of all kinds, especially of birds. Whilst we lived in England we never could sufficiently indulge this hobby, for the school-room maid rebelled against taking care of more than one cage of canaries, so we were obliged to be satisfied with that; but when we returned to our beautiful summer home in the mountains of Jamaica, we collected a little zoological garden around us in a few months, and it is about these pets I am now going to tell you.[141]

---

[138] Lady Barker, 'About Monkeys', p. 196.

[139] Lady Barker, 'About Monkeys', p. 197.

[140] The wild creatures she mentions are also defined through their tameness: 'Now I think perhaps you would like to hear a little of monkeys who were not pets by any means. Whilst I was in India we went up to [Shimla] very early in the season, before the houses were filled with people; and as the place is nearly deserted in the cold winter months, the monkeys, bears, and panthers get very tame, and I am told almost live in the verandas of the empty houses'; Lady Barker, 'About Monkeys', p. 200.

[141] Lady Barker, 'Aunt Annie's Story about Jamaica. Part II', *Good Words for the Young* (1 July 1870): 518–23, p. 518.

3 VICTORIAN MENAGERIES   117

Throughout these articles, Barker describes the different birds she keeps (owls, hawks, egrets and parrots) as well as some tortoises and records how she looked after each one. The collection also serves as a natural history lesson—with the pursuit of birds and the pursuit of knowledge seen as inextricably bound together ('hunting through the picture-books'):

> Then we had owls and hawks, and once we had a beautiful pair of Egrets given to us. We did not know what these birds were at first, as no one had ever seen any like them, and it was only by hunting through the picture-books about birds that we discovered their likeness under that name. They were flying overhead when some tiresome person who happened to have a gun in his hand shot at them, wounding the female, who fluttered to the ground, and her mate would not desert her, and was easily captured. They were exquisitely beautiful, though with fierce, wild natures. Their legs were bright red and rather long, but their plumage was very peculiar – milk-white, and the feathers which composed their tails and their large crests or top-knots were fluffy, like marabout or the down of the eiderduck. We fed them on raw meat at the risk of having our eyes pecked out, and our fingers were soon covered with wounds, but we bravely persevered, and tried all the surgical art we possessed to heal the poor broken wing of the female, but she only lingered a few days, and then died in great pain, I fear. Her mate became still more fierce and untamable, and we were afraid to let him out of his large cage lest he should share his wife's fate. He ate well and seemed healthy, but very restless and miserable, and we could only keep him alive for three or four months.[142]

Barker's experience with wild and exotic creatures in cages seems to offer a feminine counterpart to the stories by Richard Rowe discussed earlier, which focused on death and cruelty more than on the nursing of animals and the fear of the keepers when faced with wildness. Recording the capacities of animals to suffer recalls previous Romantic representations of sympathy for animals, hinting at the links between pet-keeping and a natural maternal instinct or motherly love typically found in Georgian children's literature, as seen in Chap. 2. However, the 'beasties' in Barker's articles do not really match expectations as to what 'pets' should be like: the wild and exotic beasties she purchases and keeps behind bars do not live in the house, nor are they completely tame—or offer companionship. On the contrary, they are sometimes called 'pets' in a derogatory and

---

[142] Lady Barker, 'Aunt Annie's Story about Jamaica. Part II', p. 522.

118   L. TALAIRACH

patronising sense, as in 'Four Months in a Camp', where Barker highlights a social hierarchy through her references to working-class attendants:

> I must not forget to tell you about the pets, who had a servant all to themselves. There was a beautiful parrot, a small cage full of Java Parrots (who, by the way, are the stupidest littles creatures in the world), and another of parroquets from Ceylon. These were sent on with their attendant, but I kept my latest favourite to travel with me. It was a most beautiful Persian cat, as white as snow, with long silky hair instead of short fur ...[143]

The word 'pet', used here to define 'the stupidest little creatures in the world', reflects the way the creatures are belittled through their caging— they experience a form of enslavement and subjugation to British dominion and the colonial gaze. The 'petting' of the birds suggests paradoxically the absence of any emotional ties, reinforcing British superiority by evoking class consciousness as much as consumer culture ('my latest favourite'). Barker's discourse is similar in 'About Monkeys', where the monkeys are once more associated with feminine 'fondness' and constructed as luxury goods ('in all sorts and sizes'), but also linked with slavery, thus collapsing the distinction between exotic animals and exotic peoples:

> We have just a few minutes more before nurse comes, so I will end with a short story my father told me, of something which happened many years ago in Jamaica. A new Governor was expected to arrive by the following steamer: he was coming from Demerara, where monkeys abound; and his wife was very fond of these pets, and had sent on before her a large iron cage full of monkeys of all sorts and sizes, in charge of a servant. These animals created great excitement on their arrival in Spanish Town, especially among the negroes, who had never seen a monkey, and fancied they were inferior beings of the same class as themselves.[144]

As images of extreme domestication, pets are, according to Ingrid H. Tague, 'the quintessential representative[s] of human culture',[145] hence their frequent comparison with slaves. However, Barker's use of the term 'pet' in this passage seems poles apart from the condemnation of animal

---

[143] Lady Barker, 'Four Months in Camp', p. 646.
[144] Lady Barker, 'About Monkeys', p. 201.
[145] Ingrid H. Tague, *Animal Companions: Pets and Social Change in Eighteenth-Century Britain* (University Park, Pennsylvania: The Pennsylvania State University Press, 2015), p. 64.

cruelty typical of early children's literature. The humanity that informed earlier children's literature is replaced here by a lack of sentimentality; Barker's reports propose an instrumental view of the 'curious beasties' she collects abroad, which are merely purchased for her entertainment. This idea is seen in stark form in 'Four Months in a Camp', where the caged 'pets' are parrots, perroquets, porcupines, antelopes, elephants, whilst the peacocks, generally kept in cages in England, are shot and eaten by the author on her travels in India. Despite some feelings of shame, Barker in fact promotes the eating of exotic birds:

> One afternoon we went to shoot peacocks in the dense patch of tall sugar-canes; but although the birds really were perfectly wild and very strong on the wing I felt more ashamed of myself than ever, as I could not dismiss from my mind the memory of many tame peacocks who had fed out of my hand in England, and it seemed very ruthless to return home with five splendid young birds dangling at our elephant's huge side. They were excellent eating, and I remembered too well that when I was a child, and read historical accounts of great feasts in olden time, I had been filled with indignation at the idea of our ancestors including peacocks in their bill of fare. I little thought I should ever eat them myself, and so will you perhaps, one of these fine days.[146]

Barker's discourse on exotic pets kept in cages, defined as luxury goods which convey a whiff of exoticism and danger (albeit safely contained behind bars in cages managed by servants), completely rewrites the gendering of the love for animals found in Georgian children's literature. Her articles offer a striking example of the colonialist discourse which permeated children's periodicals in the second half of the nineteenth century, illustrating the shifting sensibilities and increased commodification of animals in that period. The contrast between Barker's discourse on the 'curious beasties' found in the British colonies and that of Bagnold, examined earlier—which regularly warned young readers about the cruelty of keeping animals in cages—is telling. It reveals the existence of contrasting perspectives on the natural world and its 'curious beasties' in Victorian periodicals, highlighting the responsibility of both writers and readers. Bagnold creates, for instance, a bond with her readers by addressing them directly, as when she relates the story of Jacko and stresses the monkey's feelings:

---

[146] Lady Barker, 'Four Months in Camp', p. 656.

Reader, never make a pet of a living creature capable of feeling affection and suffering, and then give or sell it to form one of any collection, however well managed. Do not let it know and teach it to value a happier life, and afterwards leave it to pine and weary, a prisoner in a show.[147]

Bagnold's contributions to the magazine reflect a growing interest in animal welfare and at the same time signal a shift of focus towards more exotic animals, rather than restricting concern (as had often been the case up to that point) to Britain's domestic species. Her discourse on caged birds similarly drew upon children's sympathy the better to collapse boundaries between the wild and the domestic, as when she underlined the need for the birds to be regularly taken out of their cages:

It is cruel to keep them always shut up, they are so very fond of a little freedom and change, and space to stretch their wings and give a shake now and then to their feathers is so essential to their comfort; yet if not well watched they will be sure to bite the furniture, tear the carpets and curtains, or pull something to pieces.[148]

Significantly, in both examples, Bagnold hardly describes the animals, preferring to focus instead on the experience of friendship between the birds and monkeys and their owners. Encouraging the moral development of her child readers, she emphasises the 'love and sympathy' (p. 508) shown by the birds to their 'masters', how they 'will become absorbed in the interests of and learn to sympathize with human associates',[149] and provides advice on how to take care of parrots and feed them or cover their cages at night.[150] Looking after animals, keeping pets or making pets of exotic or wild creatures establishes a framework for her natural history lesson—and, ironically, for the subjection of the creatures to their masters. Yet, Bagnold's articles aim more at making the children sensitive towards animals than developing pet-keeping as a desirable activity for children. The type of advice provided by authors such as Bagnold was regularly found in fiction as well. In Camden's [Richard Rowe] 'Boys of Axleford',

---

[147] Bagnold, 'Friends and Acquaintances – Part II', p. 565.

[148] Bagnold, 'Friends and Acquaintances – Part I', p. 504.

[149] Bagnold, 'Friends and Acquaintances – Part V', p. 730.

[150] 'In covering a bird's cage care must of course be taken to provide for the admission of a little air'; Bagnold, 'Friends and Acquaintances – Part V', p. 730. On how to feed parrots, see Bagnold, 'Friends and Acquaintances – Part I', pp. 501–508.

cited above, there is frequent advice on how to keep animals, though it is interwoven in the narrative and serves to characterise the boys:

> There are three birds that ought not to be put into an aviary: the blackbird, and the robin, and the wren; the blackbird is too fond of cocking over little birds, and the robin is too fond of fighting, and the poor little wren is too timid ever to be comfortable in a crowd.[151]

Such advice was as much a part of the natural history lesson as warnings against cruelty to animals, the word 'pet' aiming in this instance at evoking sympathy towards inferior animals, as in 'Aunt Mary's Pets':

> I have at present six pet *Lizards*, and as they belong to a much misunderstood and persecuted race, I should like to say a good few words for them, in hopes of inducing the young readers of 'AJM', if not to follow my example in making pets of them, at least to refrain from ill-using them, and pelting them with stones, as I fear many boys do whenever they see one.[152]

The readers of the children's periodical here are once again not so much encouraged to keep pets as enjoined to 'refrain from ill-using' animals, especially the not particularly attractive ones.

Even if this type of article was more likely to be found in *Aunt Judy's Magazine* than in *Good Words for the Young* (which explains, perhaps, why the periodical was often considered as aimed more at girls than boys, despite Margaret Gatty's refusal to seek out a specifically gendered audience), Strahan's periodical differed from children's magazines such as *Boys of the Empire* or *The Boy's Own Paper*, whose aim was often, as Barbara Black argues, to 'turn[−] boys into men who fulfill conventional prescriptions for masculine virtue (action, aggression, patriotism)'.[153] In *Aunt Judy's Magazine*, animal welfare groups were also regularly advertised, such as the Society for the Prevention of Cruelty to Animals, whose publications were announced in the review section.[154] Recently

---

[151] Richard Camden [Richard Rowe], 'The Boys of Axleford – IV – Shy Dick', *Good Words for the Young* (1 April 1869): 292–4, p. 292.

[152] [Anon.], 'Aunt Mary's Pets', *Aunt Judy's Christmas Volume for Young People* (London: Bell and Daldy, 1870), pp. 299–305, p. 299.

[153] Barbara J. Black, *On Exhibit: Victorians and their Museums* (Charlottesville and London: University Press of Virginia, 2000), p. 152.

[154] 'The Society for the Prevention of Cruelty to Animals is issuing a paper likely to have a considerable influence for good – "The Animal World, a Monthly Advocate of Humanity"

published books advocating the rights of animals were positively reviewed as well and used as a further means of educating children to protect the natural world: 'We beg all good boys and girls to be kind to their little brothers and sisters as well as to the furred and feathered pets of the household'.[155] The articles which presented live collections warned children about the conditions in which the animals were held captive, as in Juliana Horatia Ewing's presentation of the Crystal Palace aquarium and its creatures, safe from visitors' teasing: '[W]hether they hide from custom or caprice, they are quite safe from interference. Much happier, in this respect, than the beasts in the Zoological Gardens. One may disturb the big elephant's repose with umbrella-points, or throw buns at the brown bear, but the "sea-gentlemen" are safe in their caves, and humanity flattens its nose against the glass wall of separation in vain'.[156]

However, both magazines published texts (both fiction and non-fiction) which foregrounded 'the museum as civilization's triumph over the native', to borrow Black's terms.[157] The constant parallels made between fauna observed in the wild and creatures confined in cages, as in the London Zoological Gardens (as thrilling at home as abroad, in some examples), did not merely collapse space so as to build up a miniature world for children to see. The constant change of scale, closing with views

(S. W. Partridge, 9, Paternoster Row). Its sixteen large well-printed pages of literature and art all bear directly or indirectly on the one great object, to make both young and old people kind to animals by dint of making them better acquainted with them'; [Margaret Gatty], 'Book Notices', *Aunt Judy's Christmas Volume for Young People* (London: Bell and Daldy, 1870), pp. 252–3, pp. 252–3.

[155] The magazine's editors are here reviewing *The Humanity Series of School Books* by Rev. Francis Orpen Morris, stressing in particular the books' focus on developing kindness to animals. The parallel between animals and human beings, emphasised by early children's writers in their plea against slavery, is obvious: 'They are not exclusively about the humane treatment of the lower animals, but are collections of tales, verses, &c. all teaching lessons of general kindness. ... The story is quite as closely adapted to the Humanity Series as many others therein included. Moreover we think Mr. Morris fully justified in including tales of humaneness to human beings, as well as to dogs and donkeys. History makes us acquainted with some strange characters, whose humanity towards the lower animals was marked, and whose inhumanity towards the higher race to which they themselves belonged was equally conspicuous', [J. H. Ewing and H. K. F. Gatty], 'Book Notices', *Aunt Judy's Christmas Volume for Young People* (London: Bell and Daldy, 1874), pp. 508–10, p. 508.

[156] Juliana Horatia Ewing, 'Among the Merrows. A Sketch of a Great Aquarium', *Aunt Judy's Christmas Volume for 1873* (London: George Bell and Sons, 1873), pp. 44–57, pp. 49–50.

[157] Black, *On Exhibit*, p. 153.

of the wild animals displayed in cages, built up a fantasy of control and mastery which reflected the British sphere of influence both in Britain and in the colonies. This image of widespread domestication was also widely featured to encourage children to make collections of 'beasties' themselves and look after them, even if the animals in question were often far less exotic, wild and sensational than the ones they read about, as we will see in the next chapter.

CHAPTER 4

Young Collectors

'Oh, I beg your pardon!' she exclaimed in a tone of great dismay and began picking them up again as quickly as she could, for the accident of the goldfish kept running in her head, and she had a vague sort of idea that they must be collected at once and put back into the jury-box, or they would die.[1]

If some one in each country village would but gather the children together as often as possible, and talk to them pleasantly and kindly about the right treatment of horses, asses, dogs, cats, birds, &c., the little lads who will grow up to be grooms, ostlers, and carters, would be likely to remember the teaching they receive, and carry it out in humane treatment of the animals under their charge. I dare not repeat what I know of the cruelties practised by young boys upon birds and their nestlings in the breeding season, but will at any rate try and show some of the motives that lead young people to persecute birds and destroy their eggs. I would classify these motives thus: first, wanton mischief; second, ignorance; and third, collecting mania. The two first mostly influence the poor, and the last the richer classes.[2]

---

[1] Lewis Carroll, *Alice's Adventures in Wonderland* [1865], in *The Annotated Alice*, ed. by Martin Gardner (London: Penguin, 2001), p. 123.
[2] Elizabeth Brightwen, *Inmates of my House and Garden* (New York and London: Macmillan and Co., 1895), p. 130.

© The Author(s), under exclusive license to Springer Nature   125
Switzerland AG 2021
L. Talairach, *Animals, Museum Culture and Children's Literature in Nineteenth-Century Britain*, Palgrave Studies in Animals and Literature, https://doi.org/10.1007/978-3-030-72527-3_4

126   L. TALAIRACH

As seen in the previous chapters, 'curious beasties', especially when on display behind bars, were likely to attract children's attention—attention which ensured their commercial potential. Observing wild and exotic creatures in menageries, moreover, was directly in line with Lockean pedagogical principles. As Samuel F. Pickering has argued, Lockean educators believed that 'a child could be shaped or educated' by 'controlling the objects which impressed the child's senses'.[3] Thus, the observation of animals enabled Georgian and Victorian publishers and writers for children to both rejuvenate fables, as in the case of Aesop's classical tales where the animals served as vehicles for instruction, and to draw upon new pedagogical methods based upon the visual. In addition, although some of the most influential pedagogues of early children's literature, like Sarah Trimmer, used 'curious beasties' and scientific topics more generally to teach moral lessons in order to educate children to be responsible citizens, later children's writers increasingly considered that children could participate in the sciences in a more active way, and suggested that collecting was part and parcel of the pursuit of knowledge.

As Aileen Fyfe has shown, in early children's literature, such as Anna Laetitia Barbauld and John Aikin's *Evenings at Home; or, the juvenile budget opened*, children were invited to 'participate in the sciences' and 'were allowed to learn about science by doing things'. Indeed, unlike Trimmer's children, who 'gathered flowers, but … did so to marvel at their variety and beauty, not to form collections, nor to dissect them', the children represented in Barbauld and Aikin's texts 'made collections of dried plants, collected flowers to dissect, and tried some practical chemistry at home'.[4] This chapter will argue that children's interest in natural history collecting developed noticeably at mid-century and—to borrow Kate Hill's words— that it was 'inculcated largely through books written by middle-class women introducing the techniques of collecting and preserving, the skills of close observation, and the basics of classification and Latin nomenclature'.[5] It was particularly promoted by children's periodicals which invited (middle-class) children to collect the world even more, especially so in *Aunt Judy's Magazine*, a magazine which was edited by

---

[3] Samuel F. Pickering, *John Locke and Children's Books in Eighteenth-Century England* (Knoxville: The University of Tennessee Press, 1981), p. 8.

[4] Aileen Fyfe, 'Reading Children's Books in Late Eighteenth-Century Dissenting Families', *The Historical Journal*, 43.2 (Jun. 2000): 453–73, p. 471.

[5] Kate Hill, *Women and Museums 1850–1914: Modernity and the Gendering of Knowledge* (Manchester: Manchester University Press, 2016), p. 19.

women who embodied 'feminine natural history expertise'.[6] The magazine's emphasis upon collecting helped editor and reader alike play with scale and gauge what Paula Finden terms the 'limitlessness of nature', highlighting the 'paradox of the museum ... in its attempt to confine knowledge while expanding its parameters'.[7] Moreover, through their collections, young readers were brought into contact with animals, whether domestic or more 'curious', living or dead—an indication of shifting attitudes towards animals which was gathering speed at the end of the nineteenth century. Indeed, according to many Victorian women writers, collecting was by no means incompatible with preserving the natural world, if only children were taught about 'the right treatment' from an early age, as Elizabeth Brightwen has it.

Aunt Judy's Magazine offers a good example of the way in which Victorian women naturalists trained children both to collect and to care for the natural world. The periodical illustrates the shift from publications merely inviting children to observe the world in order to marvel at God's creations, in the vein of Trimmer's children's literature, to new formats which made room for children's active participation in the construction of natural history and the understanding of its curiosities. This was due, in part, to the fact that Margaret Gatty, the magazine's founder and editor from 1866 to 1873, was herself a naturalist. Her double role as naturalist and children's writer explains why *Aunt Judy's Magazine* regularly published advice on the techniques of collecting and conserving. The example of Gatty not only illustrates 'feminine natural history expertise' but also emphasises its association with children's education, which Hill characterises as 'domestically based, essentially "hobbyist" in nature, centred on what we might call "transferable skills" of observation, perseverance and self-discipline, and above all suited to popular communication'.[8]

---

[6] Hill, *Women and Museums 1850–1914*, p. 19.

[7] Paula Findlen, *Possessing Nature: Museums, Collecting and Scientific Culture in Early Modern Italy* (Berkeley, Los Angeles, London: University of California Press, 1996), p. 96.

[8] Hill, *Women and Museums 1850–1914*, p. 19.

128   L. TALAIRACH

A 'Taste for Collecting'

Collecting is classification lived, experienced in three dimensions. The history of collecting is thus the narrative of how human beings have striven to accommodate, to appropriate and to extend the taxonomies and systems of knowledge they have inherited.[9]

As scholars have often underlined, collecting and narrative are inseparable.[10] Thomas Richards's construction of the British empire in the second half of the nineteenth century as 'a fiction',[11] created through the 'collect[ing] and collat[ing] [of] information',[12] referred to in the introduction, is a good illustration of the Victorian 'obsess[ion] with the control of knowledge',[13] and the role played by caged and classified 'beasties' and/or objects. As we have seen in the previous chapters, although the exhibition of live animals in eighteenth- and nineteenth-century Britain aimed at entertaining and instructing the lay public, caged animals also enabled naturalists to extend their knowledge of the natural world. Such collections of exotic animals were closely tied to the trade in wild creatures, often sold by animal dealers, and more generally to the traffic in specimens, whether sent or exchanged by professionals or amateurs attached to scientific expeditions, or dispatched back to Britain from the colonies. Such a taste for collecting permeated the children's literature of the period: child characters mimic their parents, like Little Jollyman in Charles Camden's [Richard Rowe] *The Travelling Menagerie* (1871–72), and Victorian travellers, like Mary Anne Barker (1831–1911), collected wild 'beasties' in the British colonies, as we saw in the previous chapter. Barker also explains in one of her articles that when she was a child living in Jamaica, she used to collect coral and shells, species which may have looked less exotic, but were nonetheless 'strange curious creatures':

[9] John Elsner and Roger Cardinal, 'Introduction', in John Elsner and Roger Cardinal (eds), *The Cultures of Collecting* (Cambridge, Mass.: Harvard University Press, 1994), pp. 1–6, p. 2.

[10] For the link between collecting and narrative, see in particular Mieke Ball, 'Telling Objects: A Narrative Perspective on Collecting', in Elsner and Cardinal (eds), *The Cultures of Collecting*, pp. 97–115.

[11] Thomas Richards, *The Imperial Archive: Knowledge and the Fantasy of Empire* (London and New York: Verso, 1993), p. 1.

[12] Richards, *The Imperial Archive*, p. 3.

[13] Richards, *The Imperial Archive*, pp. 4–5.

4   YOUNG COLLECTORS   129

Jessie and I used to go down to the beach the moment the sun sank low enough to allow us to do so in safety, and even venture a little among the rocks to search for the empty crab-shells, out of which their owners had walked. We picked up beautiful bits of coral and lovely shells, looking as if they had been wafted to our feet straight from some sea-palace beneath the sparkling blue waves which rippled away before us, and in whose shallow pools among the rocks we could see fairy treasures of waving sea-grass, sponges, and strange curious creatures without end or number. This was all very delightful to us girls, who had just come out from England, and to whom it was new and charming...[14]

Such collections of smaller, and sometimes more familiar, creatures may be traced back to the beginning of children's literature. Collections lay at the heart of the reading experience, as in seminal works like *Gulliver's Travels* and *Robinson Crusoe* (which were often revisited or imitated in the course of the nineteenth century and adapted for young audiences) or genuine travel narratives, such as James Cook's voyages between 1768 and 1780, sometimes turned into fiction for young readers.[15] As Barbara Black observes, in many of these texts 'journeys lead to exhibition – whether it be of a great goatskin cap, an umbrella, a parrot, or a Brobdingnagian insect tiger'.[16]

The emphasis on the natural sciences in Victorian children's periodicals explains the pervasive presence of collections both in fiction and non-fiction articles. Just like their parents, child readers were invited to practise natural history themselves. In 1876, Margaret Gatty's daughter, Horatia Katherine Frances Gatty, by now editor of *Aunt Judy's Magazine*, wrote:

A taste for 'collecting' is one of the strongest inclinations planted in human hearts. Most people are 'collectors' of some kind, and have their own par-

[14] Lady Barker, 'Adventures', *Good Words for the Young* (1 Feb. 1871): 191–8, pp. 192–3.

[15] See Mary V. Jackson, *Engines of Instruction, Mischief and Magic: Children's Literature in England from its Beginnings to 1839* (Lincoln: University of Nebraska Press, 1989), pp. 117–18, for eighteenth-century examples of 'Robinsonnades' or rewritings of *Gulliver's Travels* and Cook's voyages for children, such as *The Adventures of Philip Quarll* (1727), originally published as *The Hermit* and attributed to Peter Longueville or Alexander Bicknell, Robert Paltock's *The Life and Adventures of Peter Wilkins, A Cornish Man* (1750) and Rudolf Erich Raspe's *Baron Munchausen's Narrative of his Marvellous Travels and Campaigns in Russia* (1785).

[16] Barbara J. Black, *On Exhibit: Victorians and their Museums* (Charlottesville and London: University Press of Virginia, 2000), p. 153.

130   L. TALAIRACH

ticular hoard of treasures. Some collect what is perfectly useless, merely for the enjoyment of amassing things together, and to this class of collectors misers belong, whose heaps of gold, – so long as they *remain* heaps, – are as useless as many of the collections of old postage-stamps, steel-pen-nibs, booksellers' labels, and postal marks, that children delight in. Others collect with a second object in view – not for the pleasure of collecting alone: to these belong all students of Natural History, who desire to see and learn the beauty and order of GOD's works. These, also, are those whose collections are intended for the benefit of their fellow-beings.

This instinct, then, being so strong, and universal, let collectors look well and see which class they belong to, so that their treasures may have some value beyond the miserly gratification of adding heap upon heap.

Children quickly display the taste.[17]

Although the emphasis on collecting appeared to clash with the periodical's admonishments regarding bird-nesting and the cruelty of collecting live specimens, *Aunt Judy's Magazine* regularly offered practical advice on what to collect and why. In so doing, the children's magazine developed both a moral and a scientific approach to collecting. In the above-mentioned article penned by H. K. F. Gatty, the author explains, for instance, why readers of the magazine should collect oyster-shells: the shells are the homes of polyzoa, which can consequently also be collected, creating a collection within a collection. This embedded collection becomes a means for Gatty to invite readers to collect specimens without involving 'any wanton destruction of life',[18] for the shells may be collected by the sea and kept alive. H. K. F. Gatty's emphasis on keeping curious creatures alive and observing 'the beauty and order of God's work's' illustrates the moral considerations which informed children's literature, especially when aimed at a young (female) audience. As we will see, the editors of, and contributors to, *Aunt Judy's Magazine* fit Diana Donald's definition of conservative female moralists who 'saw the world in terms of co-operation and interdependence within a stratified society, which embraced even the "poor dumb animals" as objects of paternalistic concern'.[19] Such magazine articles also reveal how the 'collector's responsibility towards

[17] Horatia Katherine Frances Gatty, 'A New Collection of Old Oyster-Shells', *Aunt Judy's Christmas Volume for 1876* (London: George Bell and Sons, 1876), pp. 157–65, pp. 157–8.
[18] H. K. F. Gatty, 'A New Collection of Old Oyster-Shells', p. 160.
[19] Diana Donald, *Women against Cruelty: Protection of Animals in Nineteenth-Century Britain* (Manchester: Manchester University Press, 2020), p. 149.

4 YOUNG COLLECTORS   131

the welfare of live collections'[20] drew on the prevailing colonialist discourse and reflected Victorian pedagogues' desire to inculcate the duty of care in young readers—an idea which permeates Lewis Carroll's (satirical) rewriting of earlier children's literature in *Alice's Adventures in Wonderland* (1865), as shown in the introduction to this chapter.

H. K. F. Gatty's article reflects the marine aquarium mania which ran from 1850 to 1868,[21] especially among the middle classes. After the repeal of the duty on glass in 1845, scientific equipment made of glass became cheaper. As a result, the craze for natural history, visible in Victorian collections of insects, ferns or seaweeds, boomed in the 1860s, spurred by the development of aquaria, vivaria and Warden sealed cases—equipment which enabled amateur naturalists to keep their finds alive under glass. In 1856, the magazine *The Family Friend* highlighted the extent to which the aquarium, with its 'curious creatures', offered a form of scientific recreation which typified bourgeois taste and education:

> The novel experiment of attempting the domestication of marine and freshwater animals has proved so successful, that vivaria have already taken a place among the indoor recreations of persons of taste; and the stupid old gold-

---

[20] Francis O' Gorman, '"More interesting than all the books, save one": Charles Kingsley's Construction of Natural History', in Juliet John and Alice Jenkins (eds), *Rethinking Victorian Culture* (Basingstoke: Macmillan, 2000), pp. 146–61, p. 158.

[21] The Rev. J. G. Wood describes the craze for aquariums in these terms: 'Some years ago, a complete aquarium mania ran through the country. Every one must needs have an aquarium, either of sea or fresh water, the former being preferred ... The fashionable lady had magnificent plate-glass aquaria in her drawing-room, and the schoolboy managed to keep an aquarium of lesser pretensions in his study. The odd corners of newspapers were filled with notes on aquaria, and a multitude of shops were opened for the simple purpose of supplying aquaria and their contents. The feeling, however, was like a hothouse plant, very luxuriant under artificial conditions, but failing when deprived of external assistance ... So, in due course of time, nine out of every ten aquaria were abandoned; many of the shops were given up, because there was no longer any custom; and to all appearance the aquarium fever had run its course, never again to appear, like hundreds of similar epidemics'; The Rev. J. G Wood, *The Fresh and Salt-Water Aquarium* (London, 1868), pp. 3–6, qtd. in Lyn Barber, *The Heyday of Natural History: 1820–1870* (Garden City, New York: Doubleday and Company, Inc., 1980), pp. 121–2. Also relevant in this context is the way in which the following journalist refers to the craze in 1856 in the *Titan Magazine*: '[T]he mania is raging now at fever point ... in West End squares, in trim suburban villas, in crowded city thoroughfares, in the demure houses of little, unfrequented back streets, and inside the flat, sill-less windows of wretched Spitalfields and Bethnal Green, everywhere you see the aquarium in one form or another'; qtd. in Rebecca Stott, *Theatres of Glass: The Woman who Brought the Sea to the City* (London: Short Books, 2003), p. 132.

fish globe has been abolished to make room for the more scientific tank, with its rare display of waterweeds and curious creatures. Although so short a time has elapsed since the public were admitted to a peep at the bottom of the sea, as exhibited in the tanks of the Regent's Park Gardens, the vivarium has already become a common household ornament, to share with the geraniums and the pet-birds in adding to the resources for tasteful recreation and study.[22]

The opening of the first public aquarium at the London Zoological Gardens on 22 May 1853 was undoubtedly a key date in the history of aquaria. The fish house of the Zoological Gardens was part of a movement that originated, in fact, in 1851, when the Crystal Palace at Hyde Park attracted tens of thousands of visitors to the Great Exhibition. On 28 February 1854, the Zoological Society of London Secretary's Report compared the number of visitors to the Gardens to those of the Great Exhibition.[23] The latter's giant glass and iron building, its shape recalling the Amazonian water-lily (*Victoria regia*), echoed the hothouses and glass cases up and down the country in which Victorians grew their own exotic species. The opening of the Regent's Park fish house was followed by other public aquaria in Dublin, New York, Boston, Hamburg, Hanover, Paris (with a total of three in 1867), Le Havre, Cologne, Brussels, Berlin and Boulogne.[24]

The popular study of living marine and fresh-water animals and plants became a veritable mania—aquaria 'complete, with glass, water, plants, and animals, [being] ... hawked in the streets of London, and sold at a very small price'.[25] The aquarist W. Alford Lloyd (1826–80), who started his business on 14 July 1855, opened the first aquarium shop—'The

[22] [Anon.], 'Parlour Aquaria', *Family Friend*, 2 (1856): 192–7.

[23] [Anon.], 'Zoological Society of London Secretary's Report', *The Zoologist*, 12 (28 February 1854): 4277.

[24] [Anon.], 'The Aquatic Vivarium at the Zoological Gardens, Regent's Park', *The Illustrated London News* (28 May 1853): 420. These places were also research institutes where naturalists could study marine creatures. At the end of her 1873 article on the Sydenham Crystal Palace Aquarium, Juliana Horatia Ewing mentions the recent creation of another 'Great Aquarium' in Naples by the naturalist Dr Dohrn, a 'paradise [for] naturalists' which included tables 'to work at, furnished with necessary appurtenances, including tanks supplied with a constant stream of sea-water'; Juliana Horatia Ewing, 'Among the Merrows. A Sketch of a Great Aquarium', *Aunt Judy's Christmas Volume for 1873* (London: George Bell and Sons, 1873), pp. 44–57, p. 57.

[25] [Anon.], 'The Aquatic Vivarium at the Zoological Gardens, Regent's Park', p. 420.

4 YOUNG COLLECTORS   133

Aquarium Warehouse'—on Portland Road in April 1856, 'Dealing in Living Marine Animals, Seaweeds, Artificial Sea-Water and Marine Fresh-Water Aquaria'. His Aquarium Warehouse sold seawater alongside glass jars containing colourful marine creatures, from anemones to sea-snails and sea-squirts, and books on the natural history of the seashore.[26] As early as the middle of the eighteenth century, the popularity of marine natural history had been triggered by John Ellis's first collections of sea-weeds in 1751 and subsequent collections. The popularity of works like Robert Kaye Greville's *Algae Britannicae* (1830), Mary Wyatt's *Algae Danmonienses* (1833), Isabella Gifford's *The Marine Botanist* (1840) and W. H. Harvey's *Manual of British Algae* (1841) indicates the growing interest in marine life in the early decades of the nineteenth century. In the second half of the century, the Victorians' interest in marine life was boosted by the publication of many popular science books, such as W. H. Harvey's *The Sea-Side Book* (1849), Anne Pratt's *Chapters on Common Things of the Sea-Side* (1850), David Landsborough's *A Popular History of British Sea-Weeds* (1849) and *A Popular History of British Zoophytes and Corallines* (1852), Charles Kingsley's *Glaucus, or the Wonders of the Shore* (1855), J. G. Wood's *Common Objects of the Sea-Shore* (1857), G. H. Lewes's *Sea-Side Studies* (1858) and, of course, the many publications of Philip Henry Gosse (1810–88).[27] Whilst these publications were often read by amateur and professional naturalists alike, their success was undoubtedly due to the increase in seaside visitors, itself fuelled by a boom in railway travel from the 1840s.[28]

[26] Stott, *Theatres of Glass*, p. 126.

[27] *The Ocean* (completed in 1845, published in 1849), *A Naturalist's Rambles on the Devonshire Coast* (1853), *The Aquarium* (1854), *Tenby: A Sea-Side Holiday* (1856), *Evenings at the Microscope* (1859), *A Manual of British Zoology for the British Isles* (1855–56) and *Actinologia Britannica: A History of the British Sea-Anemones and Corals* (1860).

[28] The sea was first and foremost associated with transformation and metamorphosis, inviting scientists to revise their conception of the origins of life on earth in the course of the nineteenth century. In the 1840s, more and more geologists argued that the Earth had started out as a primordial ocean, and thus that single-celled aquatic life forms had gradually mutated, making marine invertebrates examples of primordial life forms. The work of Edward Forbes, Professor of Botany at King's College, London, on starfish (1841) and of Richard Owen (1804–92) on the nautilus and on the anatomy and physiology of invertebrate animals, as well as the rise of the collection of corals at the British Museum, curated by John Gray, and of course Charles Darwin's preoccupations with the classification of barnacles, are all examples of the ways in which marine life fascinated naturalists in the 1840s and 1850s. The different types of reproduction evidenced by sea creatures, in particular, with

134   L. TALAIRACH

These popular works of seaside natural history were written principally by Anglican clergymen advocating natural theology or by women popularisers inviting their mainly middle-class readers to contemplate the Creation in the beauty of the natural world, thus participating in the construction of a bourgeois world view. Practising natural history implied two key elements of that bourgeois ethic, healthy outdoor activity and intellectual engagement, and the Victorians obsessively collected seaweed and shells, insects and ferns—peering into rock-pools or digging into the soil. In so doing, they were consciously or unconsciously advancing the religious and social values of their class, as suggested by the poem, 'Children on the Shore': 'The shells that we gather are so fair ... There is nothing so terrible as rest'.[29] There was indeed only rest for the wicked (Fig. 4.1).

In H. K. F. Gatty's article, the naturalist's emphasis on the need for developing skills of close observation in children and explaining the basics of classification of polyzoa presents collecting as a valuable scientific recreation. H. K. F. Gatty even provides her readers with advice about where to buy their equipment, mentioning for example James Tennant, a dealer in minerals, and Bryce McMurdo Wright, father (1814–74) and son (1850–95), all of whom specialised in minerals, fossils, shells, corals and ethnographic objects:

often sexually complex species or complex sexual arrangements, intrigued scientists. Marine invertebrates' budding and splitting, their asexual reproduction, or in some cases the monogamous sexual couplings of species unsettled Christian conceptions of nature's reproductive modes, especially after the publication of the Danish zoologist Japetus Steenstrup's *On the Alternation of Generations or, the Propagation and Development of Animals through Alternate Generations: A Peculiar Form of Fostering the Young in the Lower Classes of Animals* ([1842], translated into English in 1845); Stott, *Theatres of Glass*, pp. 92–4. In the 1850s, Darwin's work on barnacles and its popularisation, bringing to the fore sea animals' sexuality and metamorphoses, contributed to the explosion of interest in seaside natural history; Jonathan Smith, *Charles Darwin and Victorian Visual Culture* (Cambridge: Cambridge University Press, 2006), p. 60. The mysteries surrounding the reproduction of barnacles thrilled the Victorians and were recurrently addressed in middle-class magazines, such as in Charles Dickens's *Household Words*. Smith cites the articles on 'Sea-Gardens' published in *Household Words*, 14 (1856): 244, and adds that between 1851 and 1854 at least six articles on or related to marine natural history were published, while six other articles appeared as Dickens's *Little Dorrit* was serialised, and two more three months after the novel's completion, one of them referring to Charles Kingsley's *Glaucus; or, the Wonders of the Shore*; Smith, *Charles Darwin and the Victorian Visual Culture*, pp. 293–4.

[29] M. B. S., 'Children on the Shore', *Aunt Judy's May-Day Volume for Young People* (London: Bell and Daldy, 1867), pp. 379–80, l. 17–34.

Let me now add a few hints as to the arrangement of a collection of Polyzoa. In the first place, the specimens must be carefully and gently washed and dried, before putting them away, to prevent mould from accumulating upon them. After this, they should be placed in cardboard boxes, with glass lids, if possible, the name of each being added on a neatly written label, and the boxes then arranged in a drawer. It is sometimes difficult to carry out this plan, as the zoophytes are apt to grow upon shells of very different sizes; but it must be left to the collector's ingenuity either to break the shells into manageable sizes, or to arrange them as best he can. Very charming glass-topped boxes of various sizes can be bought from *Professor Tennant, 149 Strand, London*, or from *Mr. B. M. Wright, 37 Great Russell Street, Bloomsbury*, London. A cheaper method of getting them may be practised at toy bazaars, where boxes containing tin ducks, or other toys, are often to be bought for *1d*. or *2d*. each. A slip of coloured paper or cotton velvet helps to 'set-off' the specimen, if laid in the bottom of the box.[30]

The article also refers to contemporary scientific publications (the classification of polyzoa by George Busk [1807–86], a zoologist and palaeontologist, probably borrowing from his *Catalogue of Marine Polyzoa in the Collection of the British Museum* [1854]), as did many other articles on natural history published in the magazine—a practice which was quite common in popular science books and articles penned by women popularisers.

Interestingly, however, H. K. F. Gatty compares Busk's study of polyzoa with her mother's: 'Mrs Gatty, the late Editor of *Aunt Judy*, also found it [*L. melolontha*] on oyster-shells sent from the south of England, and she gave the following excellent account of it ...'.[31] As a matter of fact, Margaret Scott Gatty (1809–73), though a clergyman's wife, was also 'a botanical bacchante who trekked along the shore in an unconventional collecting costume she devised for herself'.[32] As she argued in the preface to her popular science work, *British Sea-Weeds* (1863), '[a]ll millinery work, silks, satins, lace, bracelets and other jewellery etc. must, and will be, laid aside by every rational being who attempts to shore-hunt'.[33] Margaret

---

[30] H. K. F. Gatty, 'A New Collection of Old Oyster-Shells', p. 164.

[31] H. K. F. Gatty, 'A New Collection of Old Oyster-Shells', p. 161

[32] Ann B. Shteir, *Cultivating Women, Cultivating Science: Flora's Daughters and Botany in England 1760–1860* (Baltimore and London: The Johns Hopkins University Press, 1996), pp. 185–6.

[33] Margaret Gatty, *British Sea-Weeds* (London: Bell and Sons, 1863), p. ix; qtd. in Shteir, *Cultivating Women, Cultivating Science*, p. 186.

Fig. 4.1 M. B. S., 'Children on the Shore', *Aunt Judy's May-Day Volume for Young People* (London: Bell and Daldy, 1867)

Gatty was a close friend of a number of leading scientific figures, including William Henry Harvey (1811–66) (later to become Professor of Botany at Trinity College Dublin in 1857) and George Johnstone (1797–1855). She was introduced to the world of marine biology through William Harvey's *Phycologia Britannica*, which she read while recovering from a bronchial condition at Hastings in 1848. Her interest in marine biology reflected the boom in marine natural history, which started in the 1820s and was related to the development of the compound microscope. Gatty, who owned a microscope, communicated with Harvey as she collected seaweeds, gradually developing a sizeable collection.

Although her relationship with the publisher George Bell[34] undoubtedly facilitated her entry into popular science writing,[35] Margaret Gatty's fame was confirmed by the publication of *British Sea-Weeds* and her *Parables from Nature* (1855–71; eighteenth edition in 1882).[36] She was not only in contact with several influential scientists of the period but also skilled with her pencil, and did etchings on copper which she sent to the British Museum in 1842 with a view to having them exhibited to the public.[37] Her activities as a naturalist were therefore fully developed through the marine specimens (seaweeds) she collected, those she kept in her aquarium, her drawings, as well as the network of contacts she developed actively with other naturalists. Her letters show, moreover, that she could find and identify new species and knew where others were kept, such as in Kew's Herbarium.[38] In addition, her interest in scientific publications was

---

[34] Margaret Gatty and George Bell had been neighbours when they were younger.

[35] Bernard Lightman, *Victorian Popularizers of Science: Designing Nature for New Audiences* (Chicago and London: The University of Chicago Press, 2007), p. 117.

[36] Lightman, *Victorian Popularizers of Science*, p. 107.

[37] Christabel Maxwell, *Mrs Gatty and Mrs Ewing* (London: Constable & Co., 1949), pp. 83–4.

[38] 'At this very moment I am on the edge of having found either a *quite* new species of Elachista, or one which has been lost for years, and of which one, only one very poor specimen is known – that in the Hooker Herbarium. I am in correspondence with Dr Gray of the British Museum about it, but he is so old and infirm, he cannot use the microscope at will'; qtd. in Maxwell, *Mrs Gatty and Mrs Ewing*, p. 92. Other letters indicate that she was familiar with key scientists of the Victorian era. In a letter from November 1868 she writes that she was invited to dine with Richard Owen, Joseph Hooker and the American botanist Asa Gray: '[W]hile you are vegetating among the oaks and lichens of the forest, I have been sitting by some of these mighty men, Owen, Hooker, and Asa Gray … Joking apart, it was really a bit of old times to be bracing energy up for such people, but the attraction to me of the Asa Grays was, that they and Dr Harvey were such tremendous friends. … I thought your Professor Owen a good deal oldened, so very deaf. He did not know me when we met in the

138   L. TALAIRACH

genuine: she often asked her publisher for science books as payment for her own books.[39] Her collaboration with eminent scientific figures, like George Johnston,[40] whom she visited and with whom she went on seaweed expeditions,[41] and with William Harvey, indicate that she was highly experienced in the field of marine biology.[42] As a woman, however, her scientific role was limited to the publication of popular science works, as exemplified by *British Sea-Weeds*, and the editorship of *Aunt Judy's Magazine*, publications in which she concentrated on practical advice for collectors. However, her correspondence shows that her expertise in algology/phycology went beyond the family circle. In a letter she wrote to Harvey in 1857, for instance, in which she mentioned her writing project, Gatty pointed out that she was frequently consulted for 'explanations':

> I have long contemplated making an attempt at a *Horn Book* of Algology. I do not see that Dr. Landsborough's *Popular History* has in reality *simplified* the study; and people who possess both the *Phycologia* and *it*, have for years written to me for explanations. I have been glad to give them, usually making an illustration of what I meant, and the help it has been, has been, I believe, great. And now having brought out a second series of Parables and a Christmas book of Legendary Tales, I am at work on this proposed *Horn Book*.[43]

That two species were named after her by Harvey (*Gattya pinella*) and Johnston (*Gattia spectabilis*) is no coincidence.[44] Although it reflects the fact that nineteenth-century women naturalists were 'gatherers' rather

---

twilight, but afterwards apologized and said if he had but had a bone of my little finger, of course he should have found me directly. Witty wasn't it?'; qtd. in Maxwell, *Mrs Gatty and Mrs Ewing*, pp. 182–3.

[39] When she published *The Fairy Godmothers* in 1851, Gatty asked for no payment, but asked to be given in exchange George H. Johnston's *A History of British Zoophytes* (1838, 2nd edn 1847). For the second edition of *The Fairy Godmothers*, Bell sent her Johnston's *History of British Sponges and Lithophytes* (1842); Maxwell, *Mrs Gatty and Mrs Ewing*, p. 93.

[40] Johnston is the dedicatee of the original edition of *Parables from Nature*.

[41] Maxwell, *Mrs Gatty and Mrs Ewing*, p. 93.

[42] Gatty was also interested in other scientific fields, such as astronomy and mycology (Maxwell, *Mrs Gatty and Mrs Ewing*, p. 102), as well as medicine, especially the use of chloroform, which she actively promoted.

[43] Qtd. in Maxwell, *Mrs Gatty and Mrs Ewing*, p. 94.

[44] Maxwell, *Mrs Gatty and Mrs Ewing*, p. 99.

than 'namers',[45] as Barbara T. Gates puts it, it does nevertheless indicate her crucial contribution to the field of phycology. Moreover, whilst her *British Sea-Weeds* was a highly popular publication, the role played by *Aunt Judy's Magazine* in developing both her knowledge and fame must not be underestimated. As her daughter, Juliana Horatia Ewing, wrote after her death,

> to a large circle of friends, most of whom have gone before her, she was best known as a naturalist in the special department of phycology. She has left a fine collection of British and foreign sea-weeds and zoophytes. Never permitted the privilege of foreign travel – for which she so often longed – her sea spoils have been gathered from all shores by those who loved her; and there are sea-weeds yet in press sent by 'Aunt Judy' friends from Tasmania, which gave pleasure to the last days of her life. She did so keenly enjoy everything at which she worked that it is difficult to say in which of her hobbies she found most happiness; but I am disposed to give her natural history the palm.
>
> Natural history brought her some of her dearest friends. Dr. Johnston, of Berwick-on-Tweed, to whom she dedicated the first volume of the 'Parables from Nature', was one of these; and with Dr. Harvey (author of the 'Phycologia Britannica', &c.) she corresponded for ten years before they met. Like herself, he combined a playful and poetical fancy with the scientific faculty, and they had sympathy together in the distinctive character of their religious belief, and in the worship of GOD in His works.[46]

The life-long practice of natural history suffuses all of Gatty's publications for children, as already mentioned. In *Parables from Nature*, for example, the natural world she describes is never far from the study of the naturalist who collects and classifies species. 'Knowledge Not the Limit of Belief' is set in a naturalist's library, where folio books of specimens fall down and a zoophyte, a seaweed and a bookworm start discussing species classification and the question of human superiority over other species. The work of the naturalist is described through the eyes of her anthropomorphised animals and plants: his washing, drying, squeezing and

---

[45] Barbara T. Gates, *Kindred Nature: Victorian and Edwardian Women Embrace the Living World* (Chicago and London: University of Chicago Press, 1998), p. 102.

[46] Juliana Horatia Ewing, 'In Memoriam, Margaret Gatty, Daughter of the Rev. Alexander John Scott', *Aunt Judy's Christmas Volume for Young People* (London: Bell and Daldy, 1874), pp. 4–7, pp. 5–6.

140 L. TALAIRACH

gumming are all detailed, as are his experiments and observations through the microscope, designed to further the classification of different species:

> he puts you into his collections, not amongst strange creatures, but near to those you are nearest related to; and he describes you, and makes pictures of you, and gives you a name so that you are known for the same creature, wherever you are found all over the world.[47]

The zoophyte then describes being removed from its natural environment. Interestingly, its words echo the contemporary debate over the morality of specimen collection: 'I am only the skeleton of what I once was! All the merry little creatures that inhabited me are dead and dried up. They died by hundreds at a time soon after I left the sea; and even if they had survived longer, the nasty fresh water we were soaked in by the horrid being who picked us up, would have killed them at once' (p. 41). However, the narrative asserts that humans are superior beings ('man is, without exception, the most wonderful and the most clever of all the creatures upon earth!' [p. 44]) and that the naturalist's ability to observe and classify the natural world means that he occupies a particularly elevated position: 'However many mistakes he may make about *you*, he can correct them all by a little closer or more patient observation. But no observation can make you understand what man is. You are quite within the grasp of *his* powers, but *he* is quite beyond the reach of *yours*' (p. 45). The naturalist's superiority over his collected species therefore tempers the view of the naturalist as an immoral individual collecting live specimens.

Gatty's literary output constantly mirrored her activity as a naturalist: in both fields, knowledge emerges from collecting, ordering and classifying, hence constructing a 'representational understanding of the world' based upon 'the spatial juxtaposition of fragments'.[48] The case was similar for her daughters. Juliana Horatia Ewing is today more renowned as a writer of fairy tales and children's fiction than as a naturalist. However, whether she

---

[47] Margaret Gatty, 'Knowledge Not the Limit of Belief', in *Parables from Nature* [1855–71] (Chapel Hill: Yesterday's Classics, 2006), pp. 40–52, p. 48. All further references are to this edition and will be given parenthetically in the text.

[48] Eugenio Donato, 'The Museum's Furnace: Notes Towards a Contextual Reading of *Bouvard and Pécuchet*', in Josué Harari (ed.), *Textual Strategies: Perspectives in Post-Structuralist Criticism* (Ithaca and New York: Cornell University Press, 1979), pp. 213–38, p. 223, qtd. in Susan Stewart, *On Longing: Narratives of the Miniature, the Gigantic, the Souvenir, the Collection* (Durham and London: Duke University Press, 1993), p. 162.

dealt with hedgehogs or flowers, Ewing 'spared no trouble in trying to ascertain whether Hedgehogs *do* or do not eat pheasants' eggs', consulting *The Field*, or 'naming [flowers] scientifically from Professor Asa Gray's *Manual of the Botany of the Northern United States*'.[49] Likewise, Horatia Katherine Frances Gatty, both editor and regular contributor to *Aunt Judy's Magazine*, was as active as her mother in the field of marine biology, pursuing Margaret Gatty's work after her death. H. K. F. Gatty's correspondence with several naturalists of the period shows that she continued collecting specimens, probably, as her mother had done, through her contacts with correspondents in different parts of the British empire. In the 1870s, she corresponded with George James Allman (1812–98), an Irish botanist and zoologist, and with George Busk.[50] Both were active members of the Linnean Society; Allmann, in fact, was President from 1874 to 1881, at the time when he corresponded with H. K. F. Gatty (from 1875 to 1896). As the letters show, H. K. F. Gatty sent Busk a collection of hydroids (animals now seen as belonging to the cnidaria order) for identification. Busk sent the collection to Allman at the end of 1875, and the latter reached a decision in 1878, making drawings and descriptions of the numerous new forms. Allman's letter of 10 February 1878 mentions his desire to publish his conclusions in the *Journal of the Linnean Society* and encloses tracings for H. K. F. Gatty to see. His remarks illustrate the role that H. K. F. Gatty played not only in providing him with marine species but also by putting him in contact with an artist with whom Allman worked for the preparation of his paper on zoophytes. Moreover, H. K. F. Gatty seems to have advised Allman as to the naming or identification of some of the specimens she sent him.[51]

[49] Horatia Katherine Frances (Gatty) Eden, *Juliana Horatia Ewing and her Books* (London: Society for Promoting of Christian Knowledge, 1896), pp. 69, 81.

[50] Busk was involved with editing several scientific journals, notably the *Quarterly Journal of Microscopical Science* (1853–68), the *Natural History Review* (1861–65). He was also a member of the X-Club and therefore close to scientists such as T. H. Huxley. H. K. F. Gatty's letters are in the collections of the London Natural History Museum.

[51] Once Allman had identified her collection, the correspondence between them became more regular, increasingly revealing H. K. F. Gatty's knowledge on zoophytes and Allman's acknowledgement of her skill at tracking and identifying new species. In his letter dating from 6 June 1878, twenty-one new species are listed, and Allman agrees to follow H. K. F. Gatty's suggestion regarding the naming of a species. The letter also reveals that H. K. F. Gatty had asked to see her collection again while Allman was in France; George James Allman, *Letters to Miss Gatty on coelenterates, from G. J. Allman and G. Busk*, Zoology

142   L. TALAIRACH

Their correspondence over several years indicates that H. K. F. Gatty regularly asked questions about marine biology and the species she had found,[52] that she knew about and read recent publications on the subject, owned a binocular microscope[53] and regularly sent Allman botanical specimens (zoophytes).[54] Even though H. K. F. Gatty as a woman was unable

Manuscripts MSS GAT, Natural History Museum, London, 1875–1896. All further references are to this archive collection.

[52] Although we only have Allman's answers to reconstruct the correspondence, it is easy to infer that H. K. F. Gatty regularly asked him questions. For example, on 26 September 1878, when Allman was offered other specimens, she inquired about the differences between two 'apparently distinct forms' which may however be 'intermediate forms' linked 'together in the same species'. Answering one of her questions regarding the name of a scientist whom he believes to be Millen Coughtrey (1848–1908), Allman advised her to read the latter's descriptions of hydroids in the *Proceedings of the New Zealand Institute* ('Critical Notes on the New Zealand *Hydroida*', *Proceedings of the New Zealand Institute*, 8 [1875]: 298–302). He also pointed out Busk's absence of technical description of *Halicomaria* and mentioned his own description of the 'Hydroid of the Porcupine Expedition', offering to send her his report on the Gulf Stream hydroids. The report failed to arrive, as a letter dating from 31 October 1878 shows, and Allman apologised for his delay in a letter dated 18 June 1879, in which was enclosed one of his scientific articles (another one was to follow). Their correspondence also shows that H. K. F. Gatty met Allman ('George James Allman to Miss Gatty', 8 Sept. 1879) and that she provided enough details of the specimens she possessed to enable Allman to make surmises as to the species of zoophyte she owned ('George James Allman to Miss Gatty', 8 Sept. 1879). Allman mentions that one of his monographs is out of print and may only be found through secondhand booksellers, as H. K. F. Gatty guessed in her letter, suggesting that she read scientific publications ('George James Allman to Miss Gatty', 29 Sept. 1879). On 7 November 1879, Allman promised her to send back a parcel of zoophytes, and in his letter from 11 December 1879, he took them with him to visit her, adding that he had identified one of the last forms she sent him. On the 19th, the parcel was sent, together with drawings, but Allman wrote that the last specimen she had sent him had still not been clearly identified. The correspondence reveals, moreover, that H. K. F. Gatty, like her mother, received many species from New Zealand, Australia and Tasmania through readers of *Aunt Judy's Magazine*. We will develop this point later on in this chapter.

[53] As mentioned in her 3 November 1880 letter.

[54] Such as *Menyanthes trifoliata* in a letter written on 1 May 1880. A collection of aquatic plants was sent to Allman around the same period, including nympheas ('George James Allman to Miss Gatty', 5 May 1880). More plants arrived in August 1880, some of them rare (such as a water-lily, *Nuphar pumila*, on 18 May 1880), and Allman planned to describe the hydroids at a forthcoming session of the Linnean Society. More specimens were sent: zoophytes ('George James Allman to Miss Gatty', 16 March 1882) and hydroids ('George James Allman to Miss Gatty', 12 Aug. 1882) (a new species, so far undescribed), which Allman agrees to name 'Diaphana', suggesting thereby that H. K. F. Gatty was proposing names. This is most probably *Sertularella diaphana* (Allman, 1885).

to attend meetings of the Linnean Society,[55] her correspondence shows that she was a skilled naturalist when it came to identification. When Allman eventually published his paper on her collection, he argued that the 'large number of new forms' described 'cannot but form an important addition to the zoology of the Hydroida', and even offered H. K. F. Gatty the possibility of amending his paper.[56] On 20 June 1885, Allman referred to the proofs that H. K. F. Gatty had received and encouraged her to suggest further changes.[57] He then let her choose any zoophyte which she would like to be named after her. H. K. F. Gatty's 'interesting collection',[58] which represented on twenty plates in Allman's paper, reflected her status as a woman 'of literary and scientific pursuits' who had contributed to science many undescribed species of zoophytes through the specimens she had collected.[59] Similarly, among the letters written by George Busk in 1882 and 1883, several show that the zoologist received gifts of specimens from H. K. F. Gatty—specimens which belonged to species that were not at that time known to science (though closely allied to his *Thuiaria crisioides*, described in the *Voyage of the Rattlesnake* [Vol. 1], under another name).[60] Busk also refers in his letters to his own small collection of zoophytes, some parts of which he stated he had been sent by H. K. F. Gatty.

Thus, although neither mother nor daughters were allowed to fully enter the hallowed halls of Victorian natural science, they were acknowledged experts at collecting and identifying marine species, which they managed to gather from countries as distant as Australia, New Zealand and Tasmania. This, I argue, was made possible by the role played by the Gattys in the field of children's literature, especially through their publications in, and editorship of, *Aunt Judy's Magazine* and their emphasis on the practical activity of natural history—a key feature of children's literature in the mid-Victorian period. The magazine nonetheless continued in

[55] H. K. F. Gatty asked Allman in her 19 March 1880 letter; Allman explains that it is not customary for women to attend meetings of the Linnean Society even though '[t]here is no law against their attending'; 'George James Allman to Miss Gatty', 19 March 1880.

[56] 'George James Allman to Miss Gatty', 18 Oct. 1884.

[57] The article appeared in November 1885: G. J. Allman, 'Description of Australian, Cape and other Hydroids, mostly new, from the collection of Miss H. Gatty', *Journal of the Linnean Society*, 19 (1885): 132–61, pls 7–26.

[58] 'George James Allman to Miss Gatty', 10 February 1878.

[59] In addition to the undescribed species mentioned in 1884, another one was sent to Allman in early 1886: 'Your Aglaophenia is apparently an undescribed species'; 'George James Allman to Miss Gatty', 27 Feb. 1886.

[60] See, in particular, 'George Busk to Miss Gatty', 4 July 1882 and 14 July 1883.

144   L. TALAIRACH

the tradition of early children's literature, such as Sarah Trimmer's *Fabulous Histories: Designed for the Instruction of Children Respecting Their Treatment of Animals*.[61] Several stories are related from the animals' point of view and aim at teaching children how to look after, protect and preserve animals and plants. Many of the articles (fiction, non-fiction and poetry alike) promote kindness towards animals. Bird-nesting, for instance, is frequently foregrounded as an undesirable activity. Yet the pieces rarely condemn children as such: instead, they focus on the poor creatures' untimely deaths, playing on feelings and highlighting sensibility. In so doing, the editors and writers show their indebtedness to earlier children's literature and foster the link between women and animal protection. Juliana H. Ewing's poem, 'Three Little Nest-Birds', is a case in point. It relates the story of children who took away little birds to feed them:

> We meant to be very kind,
> But if ever we find
> Another soft, grey-green, moss-coated, feather-lined nest in a hedge,
> We have taken a pledge –
> Susan, Jemmy, and I – with remorseful tears, at this very minute,
> That if there are eggs or little birds in it –
> Robin or wren, thrush, chaffinch, or linnet –
> We'll leave them there
> To their mother's care.
> …
> The way we really did wrong was this:
> We took them for mother to kiss,
> And she told us to put them back,

[61] See, for instance, a review of a new edition of Trimmer's book published in 1874: 'Do any of our readers want a handsome-looking book for a birthday present to a child? We beg to remind them that Mrs. Trimmer's "History of the Robins" (which *we* knew in spotted calf, pictureless, and uniform with the Bible Lessons and the accompanying keys) may now be had of Messrs. Griffith and Farran, St. Paul's Churchyard, with twenty-four exquisite woodcuts of robin-life, and country scenes, finely printed, and gorgeously bound. Little Robin Redbreast's popularity is never likely to diminish, whilst he continues to cheer the dreariness of the winter view from our windows by the roundness of his form, the brightness of his eye, the redness of his waistcoat, and the cheerful notes of his voice. But Mr. Weir has done his best to give him artistic immortality as well. Lovely as the robins are, however, we think the gem of the book is the woodcut called "The Lark's Song". We are tormented by a desire to frame it struggling with our objections to defacing the volume by cutting it out!'; F. K. H. Gatty, 'Book Notices', *Aunt Judy's Christmas Volume for Young People* (London: Bell and Daldy, 1874), pp. 379–81, p. 380.

Whilst out on the weeping-willow *their* mother was crying 'Alack!'
Both what mother told us to do and the voice of the mother-bird.
But we three – that is Susan and I and Jem –
Thought we knew better than either of them;
And in spite of our mother's command and the poor bird's cry,
We determined to bring up her three little nestlings ourselves on the sly.

We each took one,
It did seem such excellent fun!
Susan fed hers on milk and bread,
Jem got wriggling worms for his instead.
I gave mine meat,
For you know, I thought, 'Poor darling pet! why shouldn't it have roast beef
to eat?'
But, O dear! O dear! O dear! How we cried
When, in spite of milk and bread and worms and roast beef, the little
birds died![62]

The structure of the poem, which starts and ends with the girls' tears and feelings of guilt, debunks traditional constructions of little girls' maternal qualities. Maternal care, represented here through the repeated feeding of the birds ('milk and bread and worms and roast beef'), leads to the creatures' death, whilst the mothers (of the girls and of the birds) are both seen to be wiser and warn against stroking the little birds.

Indeed, pets were more often than not at the heart of the magazine's articles, stories and even readers' contributions, suggesting how much the relationship between the readers and other creatures informed the magazine's didactic agenda. To some extent, Gatty's reluctance to dwell at length on pets in her magazine reflects the periodical's support for utilitarian views of animals charged with Christian ideology: animals were useful to humans who held dominion over them, and pets were 'a perversion of humans' relationship to the natural world'.[63] However, the 'beasties'

---

[62] Juliana Horatia Ewing, 'Three Little Nest-Birds', *Aunt Judy's Christmas Volume for Young People* (London: Bell and Daldy, 1874), pp. 726–8, pp. 726–7. As her sister later explained, Ewing had too much sympathy for birds 'to reconcile herself to putting them in cages'; she 'used to wish, when passing bird-shops, that she could "buy the whole collection and set them all free"', Eden, *Juliana Horatia Ewing and Her Books*, p. 87.

[63] Ingrid H. Tague, *Animal Companions: Pets and Social Change in Eighteenth-Century Britain* (University Park, Pennsylvania: The Pennsylvania State University Press, 2015), p. 94.

146   L. TALAIRACH

collected by children differed sharply from Mary Anne Barker's wild and exotic creatures, seen in the previous chapter, which symbolised both the domestication of captive animals and their commodification. The advice provided by contributors to the magazine taught children to respect the natural world, including plants and flowers, both in their non-fiction articles and in poems and stories, while at the same time presenting pets as useful to budding naturalists: as tools capable of providing them with scientific training. In this way, pets could help young readers negotiate tensions around animals, especially when the creatures were captured from the wild and kept in cages to be tamed/petted. More than denoting the pet-keeper's failure to distinguish between animals and humans, engaging with pets thus also meant engaging with natural history specimens, hence with objects worthy of scientific investigation.

It has already been noted that natural history informed all sorts of pieces in the magazine, especially fictional ones,[64] highlighting the central part played by natural sciences in the shaping of future citizens and constantly

---

[64] Stories like *Mrs Overtheway's Remembrances*, published in the first issues of the magazine, aptly mixed fiction and natural history: botanical metaphors, references to the beauty of the natural world and warnings about the need to protect it are found time and again (primroses are described as being 'plucked with reckless indifference', for instance); Juliana Horatia Ewing, 'Mrs Overtheway's Remembrances', *Aunt Judy's Magazine for Young People. The Christmas Volume for 1866* (London: Bell and Daldy, 1866), pp. 15–28, 82–90, 167–78, p. 21. Indeed, the 'science of horticulture' lies at the heart of the story (Ewing, 'Mrs Overtheway's Remembrances', p. 178) and is emphasised both in the gardens and through the botanical works of the manor library. Likewise, 'Home with a Hooping-Cough' lists many different species of flowers grown by one of the characters ('some very ugly ones with remarkably nasty smells, such as tanby and rue ... plenty of others that had literal fragrance – moss rose, woodbine, eglantine, lily, violet, gillyflower, pink, carnation, sweetbriar, &c.'); [Anne Manning], 'Home with a Hooping-Cough; or, How they made the best of it', *Aunt Judy's Magazine for Young People. The Christmas Volume for 1866* (London: Bell and Daldy, 1866), pp. 193–204; 257–69; 321–31, p. 328. Similarly, 'The Prince of Sleona', written by Juliana's future husband, Alexander Ewing, describes the magical metamorphosis of buds into flowers: 'The next day the flower was fully blown. It was a large flower, star-shaped at the top, and having a deep bell, containing highly ornamented and exquisitely formed anthers and pistils'; Alexander Ewing, 'The Prince of Sleona', *Aunt Judy's Magazine for Young People. The Christmas Volume for 1866* (London: Bell and Daldy, 1866), pp. 28–36, 94–9, 151–66, 216–25, 277–90, 344–54, p. 349. The travels of the Prince of Sleona lead him to encounter nautiluses, marine hydrozoans, sperm whales, albatrosses, cape pigeons and so on. In 'From Bex to St Bernard', other types of flowers or insects are referred to as the character walks along the road; Gwynfryn [Dorothea Jones], 'From Bex to St Bernard', *Aunt Judy's May-Day Volume for Young People* (London: Bell and Daldy, 1867), pp. 38–49, p. 41.

emphasising the material culture of natural history. Natural history collections can be found nesting in every nook and cranny of the fictional stories published in the magazine, especially those penned by members of the Gatty family. In Alexander Ewing's 'The Prince of Sleona', for example, the protagonist, whose illness challenges the knowledge and expertise of doctors, is advised to travel around his island. As he does so, he makes 'collections of plants or other natural productions',[65] including a collection of seaweed. Similarly, in Juliana Horatia Ewing's 'Mrs Overtheway's Remembrances', characters look for botanical specimens,[66] whilst in her poem, 'Our Garden',[67] and her non-fiction article 'May-Day, Old Style and New Style', the texts argue that the natural world must be attended to with care and studied responsibly.[68] In Margaret Gatty's 'Nights at the Round Table', the children press flowers and collect shells from the seaside to send to sick children at the children's hospital in London.[69] The young

[65] Alexander Ewing, 'The Prince of Sleona', p. 31.

[66] J. H. Ewing, 'Mrs Overtheway's Remembrances', p. 83.

[67] Juliana Horatia Ewing, 'Our Garden', *Aunt Judy's Christmas Volume for Young People* (London: Bell and Daldy, 1874), pp. 308–10.

[68] 'But, when the sunny bank under the hedge is pale with primroses, when dog-violets spread a mauve carpet over clearings in the little wood, if cowslips be plentiful though oxslips are few, and rare orchids bless the bogs of our locality, pushing strange insect heads through beds of *Drosera* bathed in perpetual dew – then, dear children, restrain the natural impulse to grub everything up and take the whole flora of the neighbourhood home in your pinafores. In the first place, you can't. In the second place, it would be very hard on other people if you could. Cull skillfully, tenderly, unselfishly, and remember what my mother used to say to me and my brothers and sisters when we were "collecting" anything, from freshwater algæ to violet-roots for our very own gardens, *"Leave some for the Naiads and Dryads"*; Juliana Horatia Ewing, 'May-Day, Old Style and New Style', *Aunt Judy's Christmas Volume for Young People* (London: Bell and Daldy, 1874), pp. 424–39, p. 439.

[69] This story anticipates the natural history specimens sent in from 1868 by the readers of *Aunt Judy's Magazine* to the Cot funded by the magazine for the children of Great Ormond Street Hospital for Sick Children. The contributions sent each month by readers of the magazine appeared on the last page of the issue, at the end of the readers' correspondence section. Objects were also sent to the hospital's young patients, most of them books, scrapbooks, toys, flowers and clothing, but natural history specimens regularly appeared as well: shells were sent in 1870, collections of birds' eggs in 1872 and 1874, as well as ferns in 1874 to fill up a Wardian case. It is interesting to note that Aunt Judy's Cot patients became themselves products (or victims) of the Victorian visual culture: the sick children were photographed and the photographs could be purchased (price 2 s. 6d.). Readers of the magazine actually asked to see the patients ('"E. F. S." … The Cot patient can always be seen', [Margaret Gatty], 'Aunt Judy's Correspondence', *Aunt Judy's Christmas Volume for 1870* [London: Bell and Daldy, 1870], pp. 380–84, p. 381).

collectors also exchange elephants and green parrots, salamanders and red coral.[70] The narrator emphasises that objects are worth collecting for their own sake, such as 'knockle-bones which he cleaned milk-white himself'.[71] In another story, 'The Cousins and their Friends', the narrator mentions the 'naturalist in the "Travels of Orlando", who crossed the great desert, stuck butterflies in his hat, and could not be got on because he would collect such numbers of natural curiosities'.[72] Collections of different kinds are also found woven into the descriptions, such as in 'Involuntary Contributions' where an aquarium, 'well stocked with beautiful seashapes' attracts the narrator's eye, whilst 'a nice collection of shells, fossils, and sea-weeds, filled the drawers of a cabinet'.[73]

Advice could be found as well percolating through articles of all kinds. Emphasis was placed on practical knowledge—the sort of 'things' that children 'know by sight and touch already'.[74] References to objects examined in the magazine were not simply a nod to Victorian visual culture, they clearly constructed natural history in material terms: the magazine's writers recurrently invited children to go and see specimens in aquariums, museums or zoological gardens, as we saw in Chap. 3. In addition to encouraging children to observe the natural world in the British countryside or in the cages of zoos and menageries, many articles—fiction and non-fiction alike—advised children how to plant trees, cut flowers[75] or feed birds. In 'Foundling Birdie', the characters ask a naturalist what to feed their bird:

[70] Margaret Gatty, 'Nights at the Round Table', *Aunt Judy's Magazine for Young People. The Christmas Volume for 1866* (London: Bell and Daldy, 1866), pp. 42–55, 99–111, p. 45.

[71] Margaret Gatty, 'Nights at the Round Table', p. 108.

[72] [Annie Keary], 'The Cousins and their Friends', *Aunt Judy's May-Day Volume for Young People* (London: Bell and Daldy, 1867), pp. 27–37; 79–86; 144–55; 202–10; 278–89; 335–43, p. 84

[73] [Anon.], 'Involuntary Contributions', *Aunt Judy's Magazine for Young People. The Christmas Volume for 1866* (London: Bell and Daldy, 1866), pp. 12–17, pp. 12–13.

[74] Margaret Gatty, 'Coral', *Aunt Judy's Magazine for Young People. The Christmas Volume for 1866* (London: Bell and Daldy, 1866), pp. 38–42, p. 38.

[75] '"Everlasting Flowers" should be cut when they are not too fully blown, tied in bunches, and hung in a dry, airy room or cupboard. Some people recommend that they should be hung with the heads downwards'; [J. H. Ewing and H. K. F. Gatty], 'Aunt Judy's Correspondence', *Aunt Judy's Christmas Volume for Young People* (London: Bell and Daldy, 1874), pp. 758–60, p. 758.

[A] friendly naturalist to whom we had applied in our perplexity, told us that, if we objected to feed Birdie on stewed flies, we ought to give him cooked meat instead of raw. His diet was changed accordingly; we fed him about once in two hours on shreds of cooked meat, and water, which he sipped eagerly.[76]

As the readers were encouraged to nurture, and sometimes tame, various sorts of animals, the magazine not only promoted women's maternal devotion, sympathy and compassion—collecting (often living) animals also enabled readers to literally 'tame' knowledge. The 'curious beasties', presented as scientific specimens, framed the links between knowledge and power, yet eschewed violence. This was particularly clear in the readers' correspondence section. The latter served to give a voice both to the editor and to the readers of the magazine, and was further used to share practices with regard to looking after animals, as suggested by the editor's answer to a reader named 'Elaine':

If your jackdaw has not already died of starvation from having his food stolen, we must give you the somewhat obvious piece of advice to let the bird have his meals in a place which is *not* infested with mice, nor open to the inroads of sparrows.[77]

The readers' correspondence section is interesting for it shows how managing animals was not incompatible with drawing attention to the dangers of bird-nesting or the (potential) cruelty of collecting live specimens in the other sections of the magazine. Readers regularly asked the editor for advice, such as 'Fritz and Jack', asking how to look after tortoises,[78] and 'C. F. T.', who 'would be glad if anyone could recommend

---

[76] [Anon.], 'Foundling Birdie', *Aunt Judy's Yearly Volume for Young People* (London: Bell and Daldy, 1869), pp. 15–18, p. 16.

[77] [J. H. Ewing and H. K. F. Gatty], 'Aunt Judy's Correspondence', *Aunt Judy's Christmas Volume for Young People* (London: Bell and Daldy, 1874), pp. 702–4, p. 702.

[78] '"Fritz and Jack". The wants of your tortoise during the winter can be easily supplied. Let him have a warm quiet corner to sleep in, and he will ask for nothing else until he awakens in spring of his own accord. We used to have one who spent his winters in a cupboard near the kitchen fire; he was a great pet, but his shell is now, alas! empty'; [J. H. Ewing and H. K. F. Gatty], 'Aunt Judy's Correspondence', *Aunt Judy's Christmas Volume for Young People* (London: Bell and Daldy, 1874), pp. 126–8, p. 127.

150   L. TALAIRACH

a cheap book on the management of rabbits'.[79] Similarly, 'Kitty D.' is interested in rearing pet birds;[80] 'Puck is looking for a book on the management of pets, which contains instructions for the keeping of *white mice*';[81] 'Aunt Prue' wants 'directions for the treatment of pigeons';[82] and 'Two Little Doves' and 'An Old Crabcatcher' ask how many times a day silkworms must be fed,[83] and so on.

Although the 'beasties' are seldom exotic, let alone curious, advice regarding how to grow, preserve or 'pet' specimens and build up collections of animals or plants appeared in every issue. Readers asked, for example, about plans for 'skeletonizing leaves',[84] drying flowers or keeping ferns, and they frequently got answers not only from the editor(s) but also from other readers of the magazine. The answers to correspondents suggested that the readers' questions were numerous and varied, and the editor(s) sometimes referred to back issues of the magazine when recipes had already been published:[85]

'Eddie and Maud'.... (2) If your ferns are kept in the house under a glass shade they have perhaps died from want of fresh air. Most ferns require to

[79] [J. H. Ewing and H. K. F. Gatty], 'Aunt Judy's Correspondence', *Aunt Judy's Christmas Volume for Young People* (London: Bell and Daldy, 1874), pp. 758–60, p. 758.

[80] '"Kitty D." ... also wants the title of a book on the management of pet birds. We think Beeton's *British Song and Talking Birds: How to rear and manage them*, price 1 s. (Ward, Lock, and Tyler, Paternoster Row, London), will be useful to her'; [J. H. Ewing and H. K. F. Gatty], 'Aunt Judy's Correspondence', *Aunt Judy's Christmas Volume for Young People* (London: Bell and Daldy, 1874), pp. 759–62, p. 760.

[81] [Margaret Gatty], 'Aunt Judy's Correspondence', *Aunt Judy's Christmas Volume for 1872* (London: Bell and Daldy, 1872), pp. 187–92, p. 188.

[82] [Margaret Gatty], 'Aunt Judy's Correspondence', *Aunt Judy's Christmas Volume for 1872* (London: Bell and Daldy, 1872), pp. 701–4, p. 702.

[83] [Margaret Gatty], 'Aunt Judy's Correspondence', *Aunt Judy's Christmas Volume for 1872* (London: Bell and Daldy, 1872), pp. 701–4, p. 702.

[84] '"Katie". "Leonardo" sends the following simple plan for skeletonizing leaves: "Put them into some liquid chloride of lime for a day, and then into fresh water for a time"'; [J. H. Ewing and H. K. F. Gatty], 'Aunt Judy's Correspondence', *Aunt Judy's Christmas Volume for Young People* (London: Bell and Daldy, 1874), pp. 126–8, p. 126.

[85] [J. H. Ewing and H. K. F. Gatty], 'Aunt Judy's Correspondence', *Aunt Judy's Christmas Volume for Young People* (London: Bell and Daldy, 1874), pp. 381–4, p. 382: '"A. E. B" (1) If flowers are dried between sheets of blotting paper, the best rule for preserving the colour is to press them very lightly, and to change the paper often. A receipt was given in our number for June, 1872, which recommended that specimens should be spread out on a wooden board, and then covered with silver sand. We advise you to try both plans, and see which preserves the colours best'.

have the shade occasionally removed, or they will droop from excess of damp. Do not put them near an open window when exposed, or the sudden change of temperature will hurt them. We fear there is no cheap book on the management of window ferns published, but advise you to try to get some practical hints from a gardener, or any friend who grows them successfully.[86]

After Margaret Gatty's death, the magazine's editors recommended their mother's popular science book and the family's contributions to the magazine.[87] The articles published in the children's periodical regularly detailed the methods and equipment necessary for making collections and preserving different types of specimens. A case in point is 'What to Pick up on the Sea-Shore', mentioned earlier. The article was addressed to children—young girls in particular—who did not have mothers to teach them how to build a collection of seaweed. Readers were advised not to

forget to take several quires of clean blotting paper, a quire of white cartridge paper, some camel's-hair brushes, a porcupine's quill (or other similarly pointed instrument which will not rust in the water), as many yards as can be got of old book muslin, or, better still, plain white grenadine, washed free from stiffening, and torn into pieces the size of the blotting paper, a basket, and, lastly, a simple botanical press, or even a couple of boards twenty-four inches long and eighteen wide, between which the specimens can be laid, and weights of books, &c., added on the top, for there are few lodging-house-keepers whose good nature will stretch far enough to allow their clothes-press to be used for scientific purposes. Then there must be suits of clothes fit to wear on the shore, and strong boots that can be rubbed

---

[86] [J. H. Ewing and H. K. F. Gatty], 'Aunt Judy's Correspondence', *Aunt Judy's Christmas Volume for Young People* (London: Bell and Daldy, 1874), pp. 126–8, p. 127.

[87] "'H. L." *British Seaweeds*, by Mrs Alfred Gatty (Bell and Sons, London), price £2 10s., is far the best book for *beginners* that we know. It is written in the simplest language possible, and any scientific terms that are necessarily used have their meaning explained. The introduction contains full directions for laying out specimens, and descriptions of the best places where to find them. If you cannot either purchase or get this book from a library, you will find some simple rules about collecting and preserving seaweeds in the number of our Magazine for June, 1874, in an article called *What to pick up on the Sea-shore*. There is a cheap book, *Common Seaweeds*, by Mr. L. L. Clarke (Warne and Co., Bedford St., Covent Garden, London), price 1 s., which may be useful to you, but we ourselves have not seen it'; [J. H. Ewing and H. K. F. Gatty], 'Aunt Judy's Correspondence', *Aunt Judy's Christmas Volume for Young People* (London: Bell and Daldy, 1874), pp. 759–62, p. 760.

152   L. TALAIRACH

with neat's-foot oil to keep out the water, and so bear a little wading without damage.

Thus equipped they will be provided with all necessaries requisite for beginning a collection.[88]

The article then goes on to explain how to conserve and classify specimens by using books on the subject. As in Margaret Gatty's short story, 'Knowledge Not the Limit of Belief', classification—not mere accumulation—is put forward as the ultimate purpose of natural history (an idea which would be much derided by Victorian nonsense literature for children, as will be seen in the next chapter), an activity that required the help of popular science books and articles.

The readers' correspondence and the editors' comments, queries and answers to questions highlight how the making of collections and the exhibition of collected specimens formed the nucleus of the kind of natural history education promoted by *Aunt Judy's Magazine*. Articles regularly explained how to obtain and conserve specimens, whether dead or alive. In 'Microscoping Objects', Margaret Gatty insisted once again that children should not kill creatures 'for mere amusement' and explained how to obtain butterflies and plants to observe through the microscope.[89] Some of her remarks on microscopes, moreover, reappeared in the correspondents' section of the following issue:

> 'Zummie' must send to Messrs. Field, Birmingham, or other scientific instrument maker for a list of prices and *powers* of microscopes. Years bring improvements, and even reductions in price. There *is* a student's microscope, price 10s. 6d., but Aunt Judy doubts its being sufficiently powerful.[90]

Margaret Gatty's remark regarding the power of the microscope indicates that the magazine genuinely aimed at training naturalists and not simply at educating children to the wonders of the natural world. Some of her comments also suggest that she may have been addressing a dual audience, targeting both child readers and their mothers. Moreover, the fact that

---

[88] [J. H. Ewing and/or H. K. F. Gatty], 'What to Pick up on the Sea-Shore', *Aunt Judy's Christmas Volume for Young People* (London: Bell and Daldy, 1874), pp. 492–500, pp. 493–4.

[89] Margaret Gatty, 'Microscoping Objects', *Aunt Judy's Christmas Volume for 1870* (London: Bell and Daldy, 1870), pp. 311–16, p. 312.

[90] [Margaret Gatty], 'Aunt Judy's Correspondence', *Aunt Judy's Christmas Volume for 1870* (London: Bell and Daldy, 1870), pp. 380–84, p. 380.

4 YOUNG COLLECTORS   153

Margaret Gatty was addressing a principally female audience is clear when she suggests asking the 'brother who is making a collection' of butterflies for samples to observe under the microscope: '[I]f, happily, you have a brother who is making a collection of Lepideptora, beg some scraps of him; you don't want a whole creature, a torn bit of a wing will do'.[91] The gendering of activities practised by the members of the family was similarly emphasised in popular science articles published in the magazine, such as those by E. Horton, who was a frequent contributor to the periodical. In one case, Horton presented the portrait of the ideal middle-class family actively engaged in the pursuit of natural history by separating men and boys, more interested in hunting insects, birds and rabbits, from girls who studied botany, and younger children, who simply collected shells and pebbles. Women, for their part, were depicted by Horton as essentially idle:

> My children could not have been happier in Robinson Crusoe's island, or with the 'Swiss Family Robinson'. Some people would think it impossible to spend six weeks in such an out-of-the-way place; but we had no difficulty. I was very fond of entomology; my eldest son was as much attached to ornithology and rabbit shooting; one daughter was a botanist; the smaller children could spend endless days picking up shells and pebbles, and dabbling among the lovely natural aquariums on the beach; and my wife was never happier than when sitting on a rock near them, and watching the great waves come in from the sea. Thus on fine days we were never in want of amusement, and on wet days we had insects to set, birds to stuff, plants and seaweeds to spread out, shells to clean and, when driven to it, a few books to fill up the time with.[92]

As in Margaret Gatty's earlier presentation of the naturalist's superiority over his collected species, Horton's article is revealing in the way it portrays how the naturalist paradoxically combines the need to protect the environment with the shooting, stuffing and collecting of specimens. Although the son's killing of birds is not described (unlike that of rabbits, generally regarded at the time as agricultural pests[93]), Horton does

[91] Gatty, 'Microscoping Objects', p. 312. This is one of the many examples which explain why the periodical was often considered as aimed more at girls than boys, despite Margaret Gatty's refusal to seek out a specifically gendered audience, as argued in Chap. 3.

[92] E. Horton, 'Flamborough and Flamborough Head', *Aunt Judy's Yearly Volume for Young People* (London: Bell and Daldy, 1869), pp. 103–110, pp. 104–5.

[93] Donald, *Women against Cruelty*, p. 157. As Donald explains, the case was similar for rooks and sparrows.

mention, for instance, 'an association formed for the protection of sea-birds'.[94] Indeed, we will see that the way these writers' encouraged readers to observe, collect and classify the natural world played a significant role in awakening public feeling and fostering the animal protection cause in the last decades of the century.

COLLECTING THE WORLD

The variety of collections that magazine readers were encouraged to assemble, from insects to seaweeds and moss,[95] reflects the museum culture of the age. But sometimes the comments from the editor and contributors went beyond the simple presentation of natural history as a respectable middle-class pastime. In 'What to Pick up on the Sea-Shore', for example, the presentation of the collecting and arranging of seaweed seems to be aimed at training future professional scientists rather than merely combatting children's idleness during the holidays. As already mentioned, the author (most probably Horatia Katherine Frances Gatty) lists the naturalist's essential equipment and describes painstakingly the whole method of collecting, before moving on to explain the classification of different types of seaweed. Moreover, although the readers of *Aunt Judy's Magazine* could vicariously travel the world, following real and fictional naturalists, the Gattys frequently stressed the fact that their readers were expected to get involved themselves by sending new specimens to the magazine's editor. Without ever travelling herself, Margaret Gatty was, in fact, at the heart of specimen exchanges and used the children's periodical to develop her own collections of seaweed and zoophytes. In 'Coral', for instance, she explained that she gave instructions to travellers to pay

[94] 'Since this was written I am happy to say that the matter has been seriously taken up, and an association formed for the protection of sea-birds; the great result of which has been that a Bill, called "The Sea-birds Protection Bill", has lately passed the second reading in the House of Lords, and will therefore become law very shortly. For this we have to thank the Rev. F. O. Morris, whose letter in the "Times", on the destruction of sea-birds at Flamborough, first called the attention of the Legislature to the subject. Some beautiful lines by another Yorkshire clergyman, Rev. R. Wilton, printed in the "Times" last October, have also done much to awaken public feeling in behalf of the poor birds. The Bill provides for sea-birds the same protection, during the breeding season, as that which has been long enjoyed by birds of games'; Horton, 'Flamborough and Flamborough Head', p. 108.

[95] The author has a herbarium in Mona B. Bickerstaffe, 'A Bit of Moss', *Aunt Judy's Christmas Volume for Young People* (London: Bell and Daldy, 1867), pp. 339–48, p. 339.

particular attention to obtaining specimens of the seaweed and zoophytes which interested her:

> We people who stay at home have sometimes friends who not only travel, but are able and willing, when asked, to use their eyes in the particular direction desired by those they leave behind. And so it comes to pass that I have before me two accounts of the Coral Islands from eye-witnesses. One, a quite unscientific friend, but full of kind spirit and observing intelligence, who lately made an expedition from Rangoon to the Andaman Island in search of the sea-weeds and zoophytes he had been asked to look after; the other scientific, and knowing the meaning of what he saw as he canoed over Coral-island wonders in the Pacific.[96]

Likewise, in 'What to Pick up on the Sea-Shore', the author mentions the promise made by a boy who has 'since gone to the paradise of seaweeds, Australia', before adding: 'We trust he will not forget his new promise to bring seaweeds from thence'.[97] Thus, the Gattys were able to procure specimens from around the world through their readers located throughout the British empire. Indeed, many of those specimens were species still unknown to science, and awaiting identification by naturalists, as seen in the first part of this chapter. The readers' correspondence section provides further evidence that the children's magazine was read in places as distant as Tasmania and New Zealand. The editor(s)' answers to subscribers show that specimens were indeed regularly sent to the magazine. In some cases, the items sent had been specifically requested by the editor(s), in others not:[98]

> 'The Young Maori', 'Waratah', and 'Tasmanian Minnie' are thanked … Tatting is a nice thing, but is not Tasmania a long way to send it from? If our friend would give us real pleasure, she will send half a dozen seaweeds instead. Does she not know that Tasmania is an algologists' paradise? – that on some parts of the coast lonely red plants are washed ashore as large as cabbages? Alas! that such exquisite things should be offered to human hands

---

[96] Margaret Gatty, 'Coral', *Aunt Judy's Magazine for Young People. The Christmas Volume for 1866* (London: Bell and Daldy, 1866), pp. 90–93, p. 90.

[97] [J. H. Ewing and/or H. K. F. Gatty], 'What to Pick up on the Sea-Shore', p. 496.

[98] '"Eccentric" is thanked; but Aunt Judy does not collect any of the treasures she so good-naturally offers. Seaweeds, zoophytes, and corals are her pet specialties'; [Margaret Gatty], 'Aunt Judy's Correspondence', *Aunt Judy's Yearly Volume for Young People* (London: Bell and Daldy, 1869), pp. 318–20, p. 319.

156 L. TALAIRACH

so often in vain! If Tasmanian young ladies did but know what a boundless field of pleasure and interest lies within their reach![99]

'Z.' is quite right in supposing that the Australian seaweeds and zoophytes would be acceptable. Aunt Judy sends her hearty thanks. It is the beauty of a seaweed collection to be world-wide. Their scientific arrangement admits of it, and it is delightful to have them from the four quarters of the world. The Tasmanian friends have responded most kindly.[100]

'Alice M. M. H'. The *white* blossoms of the sea-pink, or thrift, are less common than the pink ones, but the plant is the same, *Armeria vulgaris*. It is described as bearing a head of 'either pink or white flowers' in Hooker and Arnott's *British Flora*, p. 335, and also in Bentham's *British Flora*, p. 432. We should be greatly obliged if you would send us a specimen of the white variety.[101]

Although the periodical targeted principally child readers, in other respects it resembled other popular science magazines of the period[102] and functioned as a real hub for amateur naturalists seeking to exchange specimens. In the correspondents' section, Margaret Gatty often advised readers to write to each other in order to obtain or exchange specimens, as in her article on 'Microscoping Objects':

But perhaps you have no *Deutzia scabra* in your garden. Well, write to your friends in the south of England till someone sends you half a dozen leaves, and meanwhile content yourself with a sprig of the common lavender.[103]

[99] [Margaret Gatty], 'Aunt Judy's Correspondence', *Aunt Judy's Yearly Volume for Young People* (London: Bell and Daldy, 1869), pp. 189–92, p. 190.

[100] [Margaret Gatty], 'Aunt Judy's Correspondence', *Aunt Judy's Christmas Volume for Young People* (London: Bell and Daldy, 1871), pp. 318–20, p. 318.

[101] [J. H. Ewing and H. K. F. Gatty], 'Aunt Judy's Correspondence', *Aunt Judy's Christmas Volume for 1874* (London: Bell and Daldy, 1874), pp. 573–6, p. 574.

[102] Some of the editors' remarks illustrate this point. In 1874, the readers were advised: '*The Entomologist's Monthly Magazine*, price 6*d*. (John Van Voorst, 1, Paternoster Row, London), contains articles and notes on all subjects connected with entomology, and especially on British insects. *Newman's Entomologist*, a monthly illustrated journal of British insects (price 6*d*.), refers particularly to those which devour fruit, vegetables, &c., and is a medium for improving collections by exchanging specimens', [J. H. Ewing and H. K. F. Gatty], 'Aunt Judy's Correspondence', *Aunt Judy's Christmas Volume for 1874* (London: Bell and Daldy, 1874), pp. 253–6, p. 253.

[103] Margaret Gatty, 'Microscoping Objects', p. 314.

4 YOUNG COLLECTORS 157

'Aunt Judy's Answers to Correspondents' appeared for the first time in November 1867, as a result of the high number of letters sent to the magazine, many of them anonymous, 'from across both oceans'[104] and in the British colonies. Gatty (and her daughters, after Margaret Gatty's death) provided answers to readers as if conversing with them, as with 'Kate B.', who very regularly sent letters and inquiries to the editor. In the following answer, for example, the editor explains how to make an aquarium and breed frogs and beetles:

> 'Kate B'. It is certainly possible to 'make an aquarium merely from country ponds and streams'; in other words, a *fresh-water* aquarium, as well as a *sea-water* one, but of course the objects would be quite different. If Kate B.'s ideas are very simple she had better begin at first by keeping three or four tadpoles, to have the pleasure of seeing them *cut* their legs and turn into frogs. This is accomplished by filling a good-sized bowl of any sort three parts full with water, having an erection in the middle of bits of stone or rock, so arranged that it shall be partly out of the water. Stones with weed or moss upon them are desirable if they can be had. This serves for a resting-place for the little tadpoles, when, during the process of transformation, they feel the necessity for air. Many sorts of beetles may be added, as they are found, but the aquarium must on no account be crowded with creatures, otherwise they will die, as the people died in the Black Hole at Calcutta. More than this Aunt Judy cannot say here. The subject in an interminable one, and Kate B. must send for one of the many manuals now published in the subject.[105]

Margaret Gatty's refusal to answer some of the readers' queries indicates that the questions asked to the editor were numerous and varied, notably concerning pets. This highlights the centrality of the relationship between readers and animals, whether the audiences kept domestic pets or tamed wilder ones, such as birds. This latter point seemingly annoyed Gatty, as revealed in one issue in 1870, where she ironically advised one of the readers to buy a laughing jackass:

> 'Dorothea and Jenny Wren'. ... An editor is not necessarily either a dress-maker to choose you a 'becoming' dress, or a posture-mistress to advise

---

[104] Margaret Gatty, 'Editor's Address', *Aunt Judy's May-Day Volume for Young People* (London: Bell and Daldy, 1868), pp. 1–3, p. 3.

[105] [Margaret Gatty], 'Aunt Judy's Correspondence', *Aunt Judy's Christmas Volume for Young People* (London: Bell and Daldy, 1870), pp. 445–8, pp. 445–6.

special gymnastics, or a bird and beast fancier to tell you how to feed your pets. However, Dorothea wants another and amusing pet. Let her try an Australian 'laughing jackass' (she needn't be alarmed, it's not a quadruped), and when the 'wet half-holiday' sets in, let her laugh at the jackass till it laughs back to her, and then there will be a pair of them.[106]

During the period of her involvement with the magazine, Margaret Gatty increasingly refused to deal with the management of pets in the correspondents' section,[107] although she did include advice from other readers. In fact, over time, the section became above all a space where readers could communicate with one another rather than one that simply published Gatty's own answers to readers' endless questions:

'Blanche' recommends the following simple remedy for 'Veritas' pet bird, as being less injurious than the cure given in our last number:—'Place a saucer of water under the perch on which the bird sleeps, after it has gone to roost. The red mites will be found in the water the next morning; this must be repeated until they are extinct'.[108]

The correspondents' section is, indeed, particularly interesting for the way in which it throws light on the numerous exchanges and purchases

---

[106] [Margaret Gatty], 'Aunt Judy's Correspondence', *Aunt Judy's Christmas Volume for Young People* (London: Bell and Daldy, 1870), pp. 508–12, p. 510.

[107] '"E. F. B". Aunt Judy has already declined opening her pages to the management of pets. Such subjects occupy too much space; as before, she recommends Bechstein's "Handbook of Caged Birds"'; [Margaret Gatty], 'Aunt Judy's Correspondence', *Aunt Judy's Christmas Volume for Young People* (London: Bell and Daldy, 1871), pp. 253–6, p. 253.

[108] [Margaret Gatty], 'Aunt Judy's Correspondence', *Aunt Judy's Christmas Volume for 1872* (London: Bell and Daldy, 1872), pp. 61–4, p. 62.

4 YOUNG COLLECTORS   **159**

that were made through the magazine. Silkworm eggs,[109] birds' eggs,[110] fern-roots, butterflies, hens[111] and shells were amongst the objects/animals which circulated thanks to the magazine, together with advice related to specimen collecting (and conserving). Although readers asking for silkworms did not aim at making specific collections, collectors of birds' eggs or shells did try to exchange spare specimens in order to complete their collections, as in the case of 'Miranda', who asked for birds' eggs or 'E. F. J.', who tried to exchange butterflies:

> 'Miranda' has the following birds' eggs: 15 sparrows', 2 blackbirds', 4 starlings', 3 swallows', 2 moorhens', 2 chaffinches', and 1 green linnet's, and she wishes to exchange them for bullfinches', cuckoos', corncrakes', and dabchicks'. Address, *M. Smith, Coppice Green, Shifnal, Shropshire.*[112]

---

[109] '"Eulalie" wishes to purchase some silkworms' eggs, if our readers have any to dispose of. All communications to be addressed to the care of *Miss C. E. Todd, The Parade, Monkstown, County Dublin*'; [J. H. Ewing and H. K. F. Gatty], 'Aunt Judy's Correspondence', *Aunt Judy's Christmas Volume for 1874* (London: Bell and Daldy, 1874), pp. 247–57, p. 248; '"A. H. G." offers mulberry-fed silkworms at 1*d.* per hundred. Address, *The Postern, Tunbridge, Kent*'; [J. H. Ewing and H. K. F. Gatty], 'Aunt Judy's Correspondence', *Aunt Judy's Christmas Volume for 1874* (London: Bell and Daldy, 1874), pp. 759–62, p. 759; '"M." offers to supply silkworms' eggs at 1d. per fifty, on receipt of a stamped envelope, and will give the proceeds to Aunt Judy's Cot. Address, *35, Nottingham Place, London, W*'; [J. H. Ewing and H. K. F. Gatty], 'Aunt Judy's Correspondence', *Aunt Judy's Christmas Volume for 1874* (London: Bell and Daldy, 1874), pp. 444–8, p. 444; '"I. and A. E." offer to sell silkworms' eggs at fifty a-penny to anyone who encloses a stamped and addressed envelope with payment. They will be glad to know where they can dispose of the silk, and at what price? Address, *I. A. E. Rochdale, Albany Road, Southsea, Hants*'; [J. H. Ewing and H. K. F. Gatty], 'Aunt Judy's Correspondence', *Aunt Judy's Christmas Volume for 1874* (London: Bell and Daldy, 1874), pp. 700–4, p. 700.

[110] '"Mr. C. H. Scriven" has a collection of forty-seven different sorts of birds' eggs to dispose of, either in exchange or for sale. Price 10*s.* A list will be sent if required. Address, Castle Ashby, Northampton'; [Margaret Gatty], 'Aunt Judy's Correspondence', *Aunt Judy's Christmas Volume for 1872* (London: Bell and Daldy, 1872), pp. 316–20, p. 317.

[111] '"E. K. B." asks if any of our correspondents have some young hens which they will exchange for books, music, &c.? Address, *Froxfield Rectory, Hungerford, Berks*'; [J. H. Ewing and H. K. F. Gatty], 'Aunt Judy's Correspondence', *Aunt Judy's Christmas Volume for 1874* (London: Bell and Daldy, 1874), pp. 63–4, p. 63.

[112] [J. H. Ewing and H. K. F. Gatty], 'Aunt Judy's Correspondence', *Aunt Judy's Christmas Volume for 1874* (London: Bell and Daldy, 1874), pp. 573–6, p. 574.

'E. F. J.' offers the following butterflies in exchange for any of the Cleavewings: —a Camberwell Beauty, a Pale clouded yellow, and a Wood White. Address, *Winton House, Winchester.*[113]

Other readers used the magazine to supply collectors with specimens which they sold, like 'Ada':

'Ada' offers to supply fern-roots from Devonshire at the following prices: eighteen roots of four different kinds for one shilling, or twelve roots of six kinds at the same price. Maidenhair, spleenworts, and hard-ferns at 4*d.* a root. She does not say anything as to the cost of carriage. Address, *Miss Wilson, 21, High Street Bideford, North Devon.*[114]

Or which they simply wanted to give to other readers:

'G. E.' has some birds' eggs to dispose of, and will forward a list if a half-penny stamp is sent. Address, *Latton Vicarage, Cricklade.*[115]

'Kate B.' offers eggs of the Privet Hawk Moth, should any young reader of 'Aunt Judy's Magazine', who is a moth collector, wish to have them – address Hewish Rectory, Marlborough.[116]

The readers must have responded in large numbers, since specimens were quickly sold, given away or exchanged from one issue to the next: 'Kate B.' sent a new letter to the magazine in the following issue, saying that 'she ha[d] had many applications for her Privet Hawk Moth's eggs,

[113] [J. H. Ewing and H. K. F. Gatty], 'Aunt Judy's Correspondence', *Aunt Judy's Christmas Volume for 1874* (London: Bell and Daldy, 1874), pp. 759–62, p. 759.

[114] [J. H. Ewing and H. K. F. Gatty], 'Aunt Judy's Correspondence', *Aunt Judy's Christmas Volume for 1874* (London: Bell and Daldy, 1874), pp. 510–12, p. 511.

[115] [Margaret Gatty], 'Aunt Judy's Correspondence', *Aunt Judy's Christmas Volume for Young People* (London: Bell and Daldy, 1871), pp. 638–40, p. 638.

[116] [Margaret Gatty], 'Aunt Judy's Correspondence', *Aunt Judy's Christmas Volume for Young People* (London: Bell and Daldy, 1871), pp. 638–40, p. 638. 'F. H. Seagrave' also offered a chrysalis of the Privet Moth in an 1872 issue; [Margaret Gatty], 'Aunt Judy's Correspondence', *Aunt Judy's Christmas Volume for 1872* (London: Bell and Daldy, 1872), pp. 757–9, p. 757. In this example, however, it is interesting to note that the reader uses the moth as a pet: '[B]y the time an answer appears, we shall have one ready, as the caterpillar buried himself about a week ago. We always find *two* every year in our garden, *never more*, and amuse ourselves by feeding them and keeping them till they become moths, when we *let them go*, not having the heart to kill them'.

4 YOUNG COLLECTORS   **161**

and ha[d] given them all away, so that she [could not] supply any other correspondents with them', whilst 'J. E.' would 'be much obliged if anyone [would] give her some eggs of the "Death's-head Moth", "Puss Moth" and "Vapourer Moth"'.[117] Readers also used the magazine's section to exchange pond snails for shells,[118] for instance, or books for birds' eggs.[119] As mentioned above, readers seldom asked for exotic specimens, however; the correspondence shows that the species which were exchanged were typically those native to Britain. Nevertheless, the readers of the magazine did often ask those situated in other parts of the country for specimens which they could not find in their own areas. For example, in 1872, a butterfly collector asked for a Purple Emperor, which was common in southern England, in exchange for swallow-tails (*Papilio machaon*), suggesting that the reader lived in the north of the country (and probably north of the Humber, if the contemporary records of the butterfly's distribution are to be believed).[120] Likewise, 'Josie Hollis', a bird-egg collector from Sussex, asked for 'some *sea*-birds' eggs in exchange for inland specimens. She offers for his part eggs from the lesser wood-pecker, the blackcap and the hawfinch.[121]

Although the gender of the often anonymous correspondents is not always easy to identify, the places they indicated showed that the magazine was distributed and read throughout the country and the English-speaking world. The volume of letters sent to the magazine aimed at exchanging, purchasing or selling natural history specimens, evidences the extent to which museum culture was an integral part of the reading experience for

[117] [Margaret Gatty], 'Aunt Judy's Correspondence', *Aunt Judy's Christmas Volume for Young People* (London: Bell and Daldy, 1871), pp. 702–4, p. 703.

[118] "G. Napier" offers to supply "living specimens of two kinds of *Lymnæus* in exchange for shells suitable for a collection". Address, *Merchistoun, Alderley Edge, Cheshire*'; [J. H. Ewing and H. K. F. Gatty], 'Aunt Judy's Correspondence', *Aunt Judy's Christmas Volume for 1874* (London: Bell and Daldy, 1874), pp. 638–40, p. 638.

[119] "Ecila" offers scraps for screens or books in exchange for birds'-eggs. A list of the required eggs will be sent on application. Address, *Berry Head, Brixham, S. Devon*'; [J. H. Ewing and H. K. F. Gatty], 'Aunt Judy's Correspondence', *Aunt Judy's Christmas Volume for 1874* (London: Bell and Daldy, 1874), pp. 444–8, p. 446.

[120] "The ripest of the four Blackberries of Bramble Hill" asks if any one will supply her with a "Purple Emperor" butterfly. She offers a swallow-tail (*Papilio Machaon*) in exchange, or to pay a reasonable price'; [Margaret Gatty], 'Aunt Judy's Correspondence', *Aunt Judy's Christmas Volume for 1872* (London: Bell and Daldy, 1872), pp. 381–4, p. 381.

[121] [Margaret Gatty], 'Aunt Judy's Correspondence', *Aunt Judy's Christmas Volume for 1872* (London: Bell and Daldy, 1872), pp. 571–6, p. 573.

such young audiences—and their mothers. The encouragement of collecting practised by the periodical did not then simply create networks of young collectors. Rather, the emphasis upon collections framed children's experience; it suggested that the world collected by the magazine's young readers, whether comprised of silkworm eggs, birds' eggs, fern-roots, butterflies or shells, was, in Susan Stewart's terms, 'an autonomous world – a world which is both full and singular, which has banished repetition and achieved authority'.[122] According to Stewart, moreover, the collection makes 'temporality a spatial and material phenomenon'.[123] In the case of natural history specimens, the collection implies that 'nature is nothing more or less than that group of objects which is articulated by the classification system at hand'. As a result, she continues, because the collection defines the objects, the latter do not function 'as extensions of the body into the environment' but, on the contrary, 'serv[e] to subsume the environment to a scenario of the personal'[124]—in other words, they reflect the collector's own identity.

We saw in Chap. 3 that representations of 'curious beasties' in cages built up a fantasy of control and mastery—the capturing of wild fauna reflecting the British sphere of influence at home and abroad. Similarly, the play on scale emphasised by *Aunt Judy's Magazine* through the making of collections (as miniature representations of a wider order) and through the exchange or purchase of specimens from around the country and British empire metaphorised the editors' and readers' search for totality, with each issue/section of the magazine strengthening the fiction that the world could be collected and therefore mastered by the magazine's readers. Under the editorships of Margaret Gatty, Juliana Horatia Ewing and Horatia Katherine Frances Gatty, the close community of (middle-class) collectors created and nurtured by the magazine reflects therefore their readers' deep involvement in museums and in museum culture—an involvement which, Kate Hill argues, 'reinforced family relationships and domestic practices'.[125] According to Hill, 'the domestic and the museum were never clearly opposed', perhaps because children, as 'eminently domestic creatures, were repeatedly urged to form their own museum at

---

[122] Stewart, *On Longing*, p. 152.
[123] Stewart, *On Longing*, p. 153.
[124] Stewart, *On Longing*, p. 162.
[125] Hill, *Women and Museums, 1850–1914*, p. 2.

home'.[126] This idea informs *Aunt Judy's Magazine*, as we have seen, all the more so because the magazine often seemed to address a dual audience, targeting both children and their mothers.[127] Inevitably, the magazine reinforced a conception of natural history as a scientific field associated with 'amateurism, frivolity and a tendency to replicate existing knowledge rather than to innovate'[128] (the numerous references to scientific articles and books which the contributors offer the readers are a case in point)— casting both children and women in the role of 'the scientific community's "invisible technicians"', as Barbara T. Gates puts it, working behind the scenes through collection or observation and 'function[ing] more as interpreters of science and its sites than as taxonomists or originators of scientific theory'.[129] However, this representation of collectors, just as the regular invitations to collect more or less curious specimens, fashioned a new vision of the museum: a place that no longer 'consist[ed] just of the physical building and the people who work within it', but rather 'a set of relationships linking people, objects and institutions, [which] spread out in a network across the world'.[130] Kate Hill's concept of the 'distributed museum' matches perfectly Margaret Gatty's editorial project: her children's periodical created a network of collectors that challenged the very idea of 'museums having an "inside" and an "outside"'.[131] Both literary and material, the myriad 'museums' represented, evoked or built by child readers throughout the various sections of the magazine, illustrate, therefore, the part played by children's literature in natural history collecting— a practice indissociable from the development of knowledge and education in the natural sciences.

[126] Hill, *Women and Museums, 1850–1914*, p. 9.

[127] Margaret Gatty wrote in 1870 in the correspondents' section: '"Blanche Cremorne": ...2nd. The magazine is by no means intended for girls only, but for all young people from six years old upwards. You may carry the *upwards* as far as you please, for we flatter ourselves we contain good for grown-up minds as well as infants. There are plain cakes, and there are cakes with sugar crust'; [Margaret Gatty], 'Aunt Judy's Correspondence', *Aunt Judy's Christmas Volume for Young People* (London: Bell and Daldy, 1870), pp. 189–92, p. 189.

[128] Hill, *Women and Museums, 1850–1914*, p. 18.

[129] Gates, *Kindred Nature*, pp. 67, 102. Gates cites William Shapin, *A Social History of Truth: Civility and Science in Seventeenth-Century England* (Chicago: University of Chicago Press, 1994), ch. 8.

[130] Hill, *Women and Museums, 1850–1914*, p. 47.

[131] Hill, *Women and Museums, 1850–1914*, pp. 47–8.

164   L. TALAIRACH

In addition, the way the magazine intermingled zoological and botanical collections with a discourse on animal protection emphasised that the 'taming' of specimens, through drying, stuffing, skeletonising or even caging and petting, was not incompatible with feeling sympathy towards animals. On the contrary, as later Victorian publications illustrate, the activities represented and encouraged in Margaret Gatty's periodical paved the way for new models of animal biographies in the 1890s, as exemplified by Victorian naturalist Elizabeth Brightwen's publications, such as *Wild Nature Won by Kindness* (1890), *More About Wild Nature* (1892) and *Inmates of My House and Garden* (1895), and Alice Dew-Smith's *Tom Tug and Others. Sketches in a Domestic Menagerie* (1898). Some of Brightwen's animal stories appeared in *The Girl's Own Paper*, a periodical published by the Religious Tract Society, launched in 1880 and aimed at girls and young women. Like *Aunt Judy's Magazine*, *The Girl's Own Paper* foregrounded traditional moral values, even though its contents would become more progressive in the twentieth century. Published in a female version of *The Boy's Own Paper*, Brightwen's stories presented the 'curious beasties' she kept at home and in her museum, and, like Margaret Gatty, a few decades before, they invited readers to follow her lead and develop their own 'home museum'. Brightwen's quest for ever more 'pets' was driven by the desire[132] to possess as many 'live curios' as possible (*Wild Nature Won by Kindness*, p. 88), and her relationship with animals revolved around their training and ability to be tamed.[133] Most of Brightwen's creatures were found around her house, but several of them were sent to her from abroad or purchased from the London animal dealer Jamrach's, such as gerbils, Peruvian guinea pigs, lemurs, mongooses, an Egyptian lizard, an Indian fruit-eating bat and an Indian gazelle. After their death, all these creatures were stuffed or skeletonised in order to be exhibited in her museum. The

---

[132] Many passages express her desire for new creatures, such as when she explains that she 'long[s] for another little friend'; Elizabeth Brightwen, *Wild Nature Won by Kindness*, 5th edn (London: T. Fisher Unwin, [1890] 1909), p. 26. All subsequent references are to this edition and will be given parenthetically in the text.

[133] 'The entire absence of odour, its cleanly habits and docility when carefully trained, and its charming playfulness tend to make the ichneumon a very attractive pet'; 'By nature the little creature is as fierce as a tiger, even my specimen when caught in the jungle at four months old, bit the native who captured him pretty severely when he tried to lay hold of him, and Mungo only became the gentle creature he now is, after months of patient care and kindness'; Elizabeth Brightwen, *More about Wild Nature* (London: T. Fisher Unwin, [1892] 1893), p. 31; pp. 32–3. All subsequent references are to this edition and will be given parenthetically in the text.

latter displayed a varied collection of specimens, found or purchased from dealers in natural history. As for her books, not only did they present her collections and those of the London Natural History Museum, but they also contained advice about how to look after animals, such as making books of feathers or insect houses.

There are numerous parallels between Brightwen's writings in the 1890s and the editorship of *Aunt Judy's Magazine* at mid-century, both similarly foregrounding the popularity of natural histories and the way in which natural history education informed children's literature. Although Brightwen's collections of wild and exotic 'beasties' epitomised the 'collecting mania' she often denounced in children, as quoted at the beginning of this chapter, her presentations of animal lives at the end of the nineteenth century were nonetheless 'not just stories offering animal curiosities or subjects for sentimentality', as Gates contends. They were, above all, 'crafted to show something about the animal's raison d'être, its needs, and its personality'.[134] The link made by Gates between the role played by women in 'cataloguing the empire with their collecting'[135] and the lead they took in animal protection and animal rights from the 1870s, building up to their 'crusading [efforts] to save every type of animal' in the 1890s,[136] are highly significant in this context. On the one hand, as we have seen, '[e]xhortations to reject cruelty accordingly became a standard, almost an obsessive feature of improving books for young children' from the late Georgian period onwards, as Diana Donald argues, 'cement[ing] the association in readers' minds between women and animal protection'.[137] On the other hand, as mid-Victorian publications for children encouraged young readers to build up collections, they increasingly bridged the gap between what Diana Donald terms the 'trophies' and the 'adoptees', and between the wild and the domestic. Indeed, as both Victorian male adventurers and female writers were on a common quest through their collections to tame more or less curious creatures, they were united by 'a desire to *possess* animals, whether as trophies, or as adoptees'.[138]

Hence, whilst 'mother[ing]' (*Wild Nature Won by Kindness*, p. 44) meant taming through food, love and gentleness, rather than whips and

---

[134] Gates, *Kindred Nature*, p. 227.
[135] Gates, *Kindred Nature*, p. 102.
[136] Gates, *Kindred Nature*, p. 114.
[137] Donald, *Women against Cruelty*, p. 26.
[138] Donald, *Women against Cruelty*, p. 149.

ropes, as Brightwen's publications suggested, they nevertheless revealed how mid- and late-Victorian women writers' maternalistic discourse on animals capitalised upon the Victorian gendering of attitudes to animals to further animal protection. The 'taming of the wild' through animal collections—'an oft-told tale in Victorian and Edwardian Britain … that eventually became a social province of women'[139]—also meant that collecting could help protect those animals that were not—yet—covered by animal legislation. This is what Brightwen argued when she condemned the killing of creatures to make collections,[140] whilst simultaneously contending that stimulating children's 'rational interest' in wild creatures would draw them 'to do what they can to aid in their preservation'.[141] Her petting and taming of wild creatures, such as bats, blurred the boundaries between scientific investigation and the possession of wild animals in order to assure their protection:

> I live in hopes that some day a young [noctule] will come into my possession. I feel sure that with gentle kindness, patience, and care it would develop into a most interesting pet, for I learn from those who have kept them that they will become tame enough to fly to one's hand and take an insect from it… I hope I have said enough to win favour and protection for this most curious and useful creature. (*More About Wild Nature*, p. 44)

The narratives of captured, collected and tamed animals in the 1890s, albeit still reflecting hierarchies, thus furthered, to quote Diana Donald's words, the trajectory of the '[b]iographies and imagined autobiographies of animals [which] had been written in profusion since the beginning of the nineteenth-century, and … had their roots in earlier anthropomorphic fictions like those of Jonathan Swift, which aimed to make the sufferings of animals palpable to the reader'.[142] However, as naturalists gained greater insights into the objects of their studies, later Victorian women popularisers of natural history were able to propose 'beasties' that no longer spoke 'to say pleasant or humanlike things', but, in fact, were able to 'talk

---

[139] Gates, *Kindred Nature*, p. 218.

[140] 'I do not think very young children should be allowed to kill any living creature in order to make a collection … *That* is the kind of collecting I wholly condemn as both useless and cruel'; Brightwen, *Inmates of my House and Garden*, p. 136.

[141] Brightwen, *Inmates of my House and Garden*, p. 132.

[142] Donald, *Women Against Cruelty*, pp. 158–9.

back'.[143] As we will see in the following chapters, these developments had already permeated Victorian nonsense literature by the 1850s. The urge to build up taxonomies of species, as shown in *Aunt Judy's Magazine*, was humorously subverted in the nonsense texts of Edward Lear and Lewis Carroll, which provided a counterpoint to contemporary pedagogical theories and practices. By offering a journey through the looking-glass of natural history, Lear and Carroll reworked the imperial narrative; they undermined the construction of the British empire as 'a fiction',[144] created through the 'collect[ing] and collat[ing] [of] information'.[145] In this way, they derided the Victorian 'obsess[ion] with the control of knowledge':[146] Alice's attempts to master what she has learnt to identify, collect and classify become, indeed, entirely useless once she has tumbled down that rabbit hole.

[143] Gates, *Kindred Nature*, p. 229. Gates discusses here Alice Dew-Smith's *Tom Tug and Others. Sketches in a Domestic Menagerie*. Unlike Brightwen, Dew-Smith condemned all sorts of caging: 'People who shudder at the thought of the fox's run for life and rapid death, torn to pieces by the hounds, will gaze unmoved at the same creature condemned to the horrors of perpetual confinement in a small cage in a menagerie'; Alice Dew-Smith, *Tom Tug and Others. Sketches in a Domestic Menagerie* (London: Seeley and Co. Limited, 1898), p. 65. For more on Dew-Smith, see Gates, *Kindred Nature*, pp. 224–9.
[144] Richards, *The Imperial Archive*, p. 1.
[145] Richards, *The Imperial Archive*, p. 3.
[146] Richards, *The Imperial Archive*, pp. 4–5.

CHAPTER 5

Nonsense 'Beasties'

It is precisely in the same way that a naturalist, by constantly observing the peculiarities of animal life, acquires the readiest perception of the differences in the structure and habits of the great variety of living beings; and he perceives in each of them qualities which a less practised observer would entirely overlook. Through this habit of observation, the science of *Zoology*, which comprehends all that relates to the description and classification of animals, has been gradually established. By diligent observation, the peculiar structure of vast numbers of individual animals has been ascertained; their habits have been accurately described; and many ancient errors, which arose from hasty examination, have been exploded. This greater accuracy of description has produced a proportionate accuracy of classification; and though no system which attempts to arrange every variety of individual animals according to generic distinctions can be perfect, because exceptions to the rule are constantly occurring, yet an approach to perfection has been made, through a more complete understanding of the organization of each species.[1]

[It] was as if you had shaken up the Zoological Gardens and the Natural History collection at the British Museum in a bag, until you broke everything into bits, and had then glued the pieces together haphazard in the dark.

In the center a Zecamelobra was playing at cards with a Batonkey and an Armaskullamus, while an Elegoapard was tossing-up with a Bob-tailed

---

[1] [Anon.], *The Menageries, Quadrupeds, Described and Drawn from Living Subjects*, vol. 1, The Library of Entertaining Knowledge (London: Charles Knight, 1829), pp. 5–6.

© The Author(s), under exclusive license to Springer Nature Switzerland AG 2021
L. Talairach, *Animals, Museum Culture and Children's Literature in Nineteenth-Century Britain*, Palgrave Studies in Animals and Literature, https://doi.org/10.1007/978-3-030-72527-3_5

170   L. TALAIRACH

Walboon. A Kangarillo was seated on the head of a Monkape, watching the card party, with a Gazelroo looking over its shoulder. A Pigorselope stood farther off, and just below Frank, a Pumag, with a pair of noble antlers, was talking gravely with a Lepunicord.[2]

Tom Hood's *From Nowhere to the North Pole. A Noah's Ark-Æcological Narrative* (1875) relates the wanderings of a little boy in various new and unfamiliar lands. As he enters one of them—Quadrupemia—the sounds of wild animals remind him of his experience at the London Zoological Gardens: the day he had 'ventured too close to the cage of a big monkey, and it had caught hold of him' (p. 93), and he had luckily been rescued by the keeper. Yet, the encounter that awaits Frank in Quadrupedia is even more sensational: none of the four-footed creatures assembled in the centre of an amphitheatre of rocks resembles any of the creatures from the London Zoo. Quadrupemia is inhabited by Zecamelobras, Batonkeys, Armaskullamuses, Elegoapards and many other curious creatures. It seems, in fact, to be a living natural history museum conceived by some mad taxidermist. The 'curious beasties' themselves complain about the way in which they have been prepared, stuffed and reassembled, whilst 'the British Museum' and 'the Taxidermy' become synonyms of 'the devil' to express the animals' anger:

'Why, the British Museum, couldn't you finish us off properly?'

'What do you mean?' inquired Frank.

'Look at my head', said the creature; 'it's only a skull! And the brains inside have shrivelled up into a little dry lump, and roll about into odd corners, where they can't always be found when wanted. That's why they can't cheat me in scoring at cards!' and he ground his teeth and scowled at the Zecamelobra. The Batonkey laughed.

'Look at my legs', broke in the Elegoapard; 'they are only bones! Why, the Taxidermy, didn't you dream a skin to them?'

'Yes, and look at my tail!' growled the Walroon. 'These knobs are always catching in the forks of trees and clefts of rocks. My backbone aches with the constant jerks it gets'.

'And what's the use of my horn?' said the Lepunicorn, sullenly; 'it's stuck on the top of my head where I can't see it, and so I'm never sure of hitting anything with it'.

---

[2] Tom Hood, *From Nowhere to the North Pole. A Noah's Ark-Æcological Narrative* (London: Chatto & Windus, 1875), p. 95. All further references are to this edition and will be given parenthetically in the text.

5 NONSENSE 'BEASTIES' 171

Of course Frank was conscious that he was not to blame for incongruities arising from a dream; but he could not but remember that, when he had laughed at the animals in Noah's Ark, and fancied he could invent better ones, he had not proceeded to construct them with any regard to the mutual fitness of the various portions he had designed to join together. (pp. 100–102)

Hood's fantasy, mocking as it does collections of 'curious beasties' and the Victorian obsession with taxidermy, sheds an uncomfortable light on several aspects of the practice of natural history (Fig. 5.1). To begin with, the creatures that Frank encounters call to mind the frequent collections of animal body parts that arrived in England daily from foreign lands, especially in a period when preservation techniques did not allow naturalists to bring back animals in one piece. The story also illustrates how taxidermy, derived from the Greek *taxis*, meaning order, and *derma*, meaning skin, was inextricably bound up with classification. Interestingly, moreover, as taxidermy 'achieved its aposotheosis in the Victorian imagination',[3] it also became a source of inspiration for naturalists eager to experiment with it, from the anthropomorphised stuffed or mounted rabbits, rats, birds and other domestic animals of Herman Ploucquet (1816–78) and Walter Potter (1835–1918), to the celebrated fakes of Charles Waterton (1782–1865).[4] As a matter of fact, as Michelle Henning has noted, such

[3] Rachel Poliquin, *The Breathless Zoo: Taxidermy and the Cultures of Longing* (Pennsylvania: The Pennsylvania State University Press, 2012), p. 10.

[4] British naturalist and explorer Charles Waterton published in 1825 his *Wanderings in South America, the North-West of the United States and the Antilles, in the Years 1812, 1816, 1820, & 1824, with Original Instructions for the Perfect Preservation of Birds, &c. for Cabinets of Natural History*. He brought back from his voyages numerous specimens and is famous for his original technique of taxidermy which enabled his works—which comprised many hybrid creatures made from different animals' body parts, including his 'Nondescript'—to stand the test of time. See John Simons, *The Tiger That Swallowed the Boy: Exotic Animals in Victorian England* (Faringdon: Libri Publishing, 2012), p. 172, and Michelle Henning, 'Anthropomorphic Taxidermy and the Death of Nature: The Curious Art of Hermann Ploucquet, Walter Potter, and Charles Waterton', *Victorian Literature and Culture*, 35 (2007): 663–78. Waterton's *Wanderings in South America* relates the four journeys of the bold explorer (including his encounters with a cayman, tigers and cougars), his thirst for experiments (as with vampires) and his interest in preservation and modes of preparations of specimens, the naturalist often complaining about contemporary methods of taxidermy. See in particular *Wanderings in South America, the North-West of the United States and the Antilles, in the Years 1812, 1816, 1820, & 1824, with Original Instructions for the Perfect Preservation of Birds, &c. for Cabinets of Natural History*, 4th edn (London: B. Fellowes, 1839), pp. 148, 195, 230, 278, 287.

Fig. 5.1 Frontispiece, Tom Hood, *From Nowhere to the North Pole. A Noah's Ark-Æcological Narrative* (London: Chatto & Windus, 1875)

'alternative' or 'anti-naturalistic' taxidermy practices shaped 'a cultural world thoroughly populated with animals and a natural world thoroughly mediated by cultural representations such as myths and folktales'.[5] Moreover, although Henning considers that such experiments with taxidermy 'sit outside the development of taxidermy as a form of scientific illustration and for preservation',[6] it could be argued that, on the contrary, they represented a period when naturalists willingly explored and promoted the relationship between science and myth, or knowledge and narrative; a time when scientists were reassessing and challenging as well the human/animal divide. As we will see in this chapter, these issues reverberated in the children's literature of the period, and especially so in Victorian nonsense literature.

From the beginning of children's literature, as we saw in Chap. 2, animal descriptions often borrowed from older bestiaries. F. J. Harvey Darton remarks, in addition, that in Francis Newbery's *Natural History of Birds. By T. Telltruth* (1778), '[p]opular speech preserv[ed] some touches of that antique lore, not only in adages but even in beliefs shyly half-held'.[7] What Darton terms the 'steriliz[ation] of the wonder of man's own childhood' did not die out in the following century. On the contrary, the association of nineteenth-century children's literature with medieval bestiaries revealed contemporary ways of thinking which Victorian children's writers playfully adapted for young audiences.[8] Indeed, although Tom Hood's Zecamelobras, Batonkeys and other curious creatures, just like Charles Waterton's stuffed monsters, were obviously satirical in intention,[9] the scientific context of the nineteenth century and the numerous discoveries of hitherto unknown creatures often led scientists to imagine that the beasts

[5] Henning, 'Anthropomorphic Taxidermy and the Death of Nature', p. 664.
[6] Henning, 'Anthropomorphic Taxidermy and the Death of Nature', p. 676.
[7] F. J. Harvey Darton, *Children's Books in England*, 2nd edn (Cambridge: Cambridge University Press, 1970), p. 30.
[8] It is interesting to note here that Enlightenment science was also informed by 'fabular thinking', as Jane Spencer explains in 'Behn's Beasts: Aesop's *Fables* and Surinam's Wildlife in *Oroonoko*', in Jane Spencer, Derek Ryan and Karen L. Edwards (eds), *Reading Literary Animals: Medieval to Modern Perspectives on the Non-Human in Literature and Culture* (New York: Routledge, 2019), pp. 46–65, p. 49.
[9] The assemblage of different body parts to compose new creatures which could be a mix of animal and human parts is also reminiscent of the game 'Heads, Bodies and Legs', which was popular during the Victorian period; see Jo Elwyn Jones and J. F. Gladstone, *The Alice Companion: A Guide to Lewis Carroll's* Alice *Books* (New York: New York University Press, 1998), p. 98.

174   L. TALAIRACH

that peopled ancient legends might just be extinct creatures. Furthermore, with the advent of evolutionary theory, the idea that legendary creatures, such as the mermaids exhibited in fairs and travelling shows (like the fake Fiji mermaid made up of the torso and head of a monkey sewn onto the bottom half of a fish), might in fact be real animals as yet unknown to science galvanised the Victorian imagination.

Rachel Poliquin has described the natural wonders from the sixteenth- and seventeenth-century collections as 'turbulent, category shattering, awe-inspiring, intoxicating objects'.[10] Victorian anthropomorphic taxidermy and fakes 'shared with the curiosity cabinets of the sixteenth and seventeenth centuries a fascination with the anomalous, the hybrid and the grotesque'.[11] But in the context of evolutionary theory, they tapped into, still more strikingly, the dramatically changing relationship between humans and animals,[12] especially in the decades that preceded and followed the publication of Charles Darwin's *On the Origin of Species* in 1859. Charles Gould's *Mythical Monsters* (1886), for example, claimed:

> The great era of advanced opinion, initiated by Darwin, which has seen, in the course of a few years, a larger progress in knowledge in all departments of science than decades of centuries preceding it, has, among other changes, worked a complete revolution in the estimation of the value of folk-lore; and speculations on it, which in the days of our boyhood would have been considered as puerile, are now admitted to be not merely interesting but necessary to those who endeavour to gather up the skeins of unwritten history, and to trace the antecedents and early migrations from parent sources of nations long since alienated from each other by customs, speech, and space.
>
> I have, therefore, but little hesitation in gravely proposing to submit that many of the so-called mythical animals, which throughout long ages and in all nations have been the fertile subjects of fiction and fable, come legitimately within the scope of plain matter-of-fact Natural History, and that they may be considered, not as the outcome of exuberant fancy, but as creatures which really once existed, and of which, unfortunately, only imperfect

[10] Poliquin, *The Breathless Zoo*, p. 18.

[11] Henning, 'Anthropomorphic Taxidermy and the Death of Nature', p. 673.

[12] Laura White contends that although Victorian society 'held vigorously onto the idea that humans were sovereign over animals', hybrids were popular 'because they joked about anxieties about human-animal identity'; see Laura White, *The* Alice *Books and the Contested Ground of the Natural World* (Abingdon, New York: Routledge, 2017), p. 80.

5 NONSENSE 'BEASTIES'   175

and inaccurate descriptions have filtered down to us, probably very much refracted, through the mists of time.[13]

The idea that mermaids, sea serpents and other mythical creatures—the fake composite animals which were the wonders of cabinets of curiosities and fairs—might really have existed shaped the natural world into a marvellous museum of odd creatures capable of entrancing both children and adults. Sceptical naturalists routinely suspected both New World creatures—which sometimes, as in the case of monotremes, looked like composites—and fossils of being made up. Especially at a time when so many fossil collectors mixed up bones to create ever more sensational creatures,[14] zoological and palaeontological collections were positioned on the border between myth and reality:

> Are the composite creatures of Chaldæn mythology very much more wonderful than the marsupial kangaroo, the duck-billed platypus, and the flying lizard of Malaysia which are, or the pterodactylus, rhamphorunchus, and archæopteryx which have been? Does not geological science, day by day, trace one formation by easy gradation to another, bridge over the gaps which formerly have separated them, carry the proofs of the existence of man constantly further and further back into remote time, and disclose the previous existence of intermediate types (satisfying the requirements of the Darwinian theory) connecting the great division of the animal kingdom, or reptile-like birds and bird-like reptiles? Can we suppose that we have at all exhausted the great museum of nature?[15]

This chapter will show that Victorian children's literature frequently reflected the debate surrounding the reality of different species. Even in the eighteenth century, as Harriet Ritvo points out, 'natural history manuals and encyclopedias contained many entries that ironically foregrounded the agonized doubts of their authors about where to draw the line'.[16] Similarly, nineteenth-century children's literature often demonstrated that 'the troublesome categories and the unmanageable information were

[13] Charles Gould, *Mythical Monsters* (London: W. Allen, 1886), pp. 1–2.

[14] Mary Anning's Ichthyosaurus, or fish lizard, and her Plesiosaurus, which Georges Cuvier believed to be a fake, are significant examples.

[15] Gould, *Mythical Monsters*, pp. 18–19.

[16] Harriet Ritvo, *The Platypus and the Mermaid and Other Figments of the Classifying Imagination* (Cambridge, Mass. and London: Harvard University Press, 1997), p. 87.

176   L. TALAIRACH

artifacts of science rather than of nature', to borrow Ritvo's words.[17] Because '[t]he more that was known about a given groups of animals, the more challenging it became to separate them into plausible species',[18] Victorian children's literature recurrently played with the definition of species. Nonsense literature, such as the works of Edward Lear and Lewis Carroll, just like Hood's rewriting of Carroll's *Alice's Adventures in Wonderland* (1865), mentioned in the introduction to this chapter, not only mocked or reworked earlier forms of fiction for children, but also foregrounded and questioned the issue of species classification. According to Jean-Jacques Lecercle, nonsense is a 'patchwork' of pieces of various origin and materials, of 'odds and ends of forgotten genres, borrowed bits extorted through parody',[19] peopled with many hybrid creatures. The following discussion will argue that the 'curious beasties' of nonsense texts reflected the genuine taxonomical difficulties faced by Victorian naturalists with their cabinets of oddities. In doing so, nonsense literature mirrored taxonomic endeavours through the flights of fancy it proposed. Steven Prickett's definition of fantasy as 'a counter tradition'[20] which did not so much oppose Victorian realism as 'extend[ed] and enrich[ed] ways of perceiving "reality"'[21] informs this chapter, which examines how Lear and Carroll subverted the illusion of realism by playing with and emphasising the factual, thereby deconstructing narratives of knowledge, power and violence.

EDWARD LEAR'S ODD COLLECTIONS AND RECONSTRUCTIONS

As argued above, whilst scientists were facing the daily reality of animals which looked alike but did not belong to the same species, some nineteenth-century artists playfully evoked the links between different species by pointing out their resemblance. The renowned humourist and illustrator Edward Lear (1812–88) is a case in point. His drawings often proposed crosses between various species, almost similar to those which

[17] Ritvo, *The Platypus and the Mermaid*, p. 88.

[18] Ritvo, *The Platypus and the Mermaid*, p. 88.

[19] Jean-Jacques Lecercle, *Philosophy of Nonsense: The Intuitions of Victorian Nonsense Literature* (London and New York: Routledge, [1994] 2002), p. 195.

[20] Stephen Prickett, *Victorian Fantasy*, 2nd edn (Waco: Baylor University Press, [1979] 2005), p. 1.

[21] Prickett, *Victorian Fantasy*, p. 3.

featured recurrently in travelling shows. Lear's work is directly linked to developments in natural history in the early to mid-nineteenth century, not only because he was a zoological illustrator, but also because his writings often undermined contemporary attempts at systematic taxonomy and nomenclature.[22]

Lear was first apprenticed to Prideaux John Selby (1788–1867) when he was working on his *Illustrations of British Ornithology* (1821–34), a book that typified the era's interest in new bird species brought back from scientific voyages. Later, in 1830, he made drawings of the parrots kept at the London Zoological Gardens, a couple of years after the Gardens opened. This led to his *Illustrations of the Family of Psittacidæ, or Parrots* (1832), a work whose importance derives not only from its author, but also from the fact that it was the first book drawn from life, and not from stuffed specimens of birds from museums.[23] After its publication, Lear was elected an associate of the Linnean Society and even had two parrots named after him, a macaw (*Anododhynchus leari (Bonaparte)*) and a cockatoo (*Lpochroa leari (Bonaparte)*).[24] The president of the Zoological Society, Edward Lord Stanley, the future 13th Earl of Derby, then invited him to his estate at Knowsley Park, near Liverpool, to make drawings of the animals he kept in his private menagerie in order to make a catalogue (published in the first volume of *Gleanings for the Menagerie and Aviary at Knowsley Hall* [1846][25]). Lear was employed at Knowsley Park from

[22] Ritvo mentions Lear's 'Nonsense Botany' as an example of subversion of zoological nomenclature, together with Charles Kingsley's *The Water-Babies, or a Fairy Tale for a Land Baby* (1863); Ritvo, *The Platypus and the Mermaid*, p. 66.

[23] Lear made his living from painting birds and Psittacidae, both stuffed (as those held in the Zoological Society Museum) and living (in Regent's Park Zoological Gardens or at Knowsley Park, which also had a museum). It is interesting to compare a book like Lear's *Illustrations of the Family of Psittacidæ, or Parrots*, with illustrations made from living birds, with Thomas Bewick's *History of British Birds* (1797–1804). In both of Bewick's volumes, although many of the birds are represented in a natural environment, they are mostly drawn from stuffed specimens from private collections (some were lent by P. J. Selby, for example) or museums, such as the Wycliffe Museum, the Newcastle Museum, the Museum of Ravensworth Castle, the Edinburgh Museum, the British Museum and so on. Some were also shot especially for the book, such as a shrike; Thomas Bewick, *A History of British Birds, Vol. 1 containing the history and description of land birds* (Newcastle: Edw. Walker, 1826), p. 75.

[24] Edward Lear, *The Complete Nonsense and Other Verse*, ed. by Vivien Noakes (London: Penguin, 2002), p. xxii.

[25] The second volume, published in 1850, was illustrated by Benjamin Waterhouse Hawkins.

1832 to 1837 and lived with Lord Stanley and his family. When catalogued before being auctioned off after Lord Stanley's death in 1851, the menagerie was reported to include 345 mammals, representing a total of 94 species and 1272 birds, while 25,000 specimens were on display in the museum.[26] In the menagerie, deer and antelopes were the most numerous, alongside sheep, goats, lamas, zebras, as well as kangaroos, lemurs and armadillos.[27] Lord Stanley's scientific interest in the breeding of exotic species proved successful: one-third of the animals catalogued after his death had been bred in the park. Some of the animals were also, however, 'at least in theory, candidates for the stewpot',[28] as Ritvo explains, which may explain in part Lear's depiction of the sometimes tragic relation between animals and food in his nonsense poetry.

Lear used lithography and not wood-engraving for his illustrations. His works feature numerous people(s) and creatures from around the world and are peppered with references to natural history. His work as an ornithological artist even permeates his personal letters. In 'Scrawl', addressed to 'Ann' and written on 5 April 1831 at Peppering,[29] while he was making drawings for the fifth volume of Sir William Jardine's *The Naturalist's Library* (1835) on pigeons,[30] Lear described himself drawing and looking at the rooks and mentioned that he had 'never procured a teal or a widgeon' (p. 49), two species of duck. The artist's search for creatures to draw also filters through in a letter to Fanny Jane Dolly Coombe, written on 15 July 1832: 'when I ask for beasts or birds—it is only because I feel more pleasure in drawing from those given me by my intimate friends—than I could do from those otherwise come by—not from not being able to get at specimens' (p. 50). Lear even added that his 'rather Zoological connexion' enabled him to visualise beasts at will. Similarly, in a letter written

---

[26] Ann C. Colley, 'Edward Lear's Anti-Colonial Bestiary', *Victorian Poetry*, 30.2 (Summer 1992): 109–20, p. 113.

[27] Harriet Ritvo, *The Animal Estate: The English and Other Creatures in the Victorian Age* (Cambridge, Mass.: Harvard University Press, 1987), p. 239. See also J. C. Stevens, *Catalogue of the Menagerie and Aviary at Knowsley* (Liverpool: Joshua Walmsley, 1851) and William Pollard, *The Stanleys at Knowsley: A History of that Noble Family* (London: Frederick Warne, 1869), p. 111.

[28] Ritvo, *The Animal Estate*, p. 239. See Louis Fraser, *Catalogue of the Knowsley Collections* (Knowsley: s.n., 1850), iii, and Stevens, *Catalogue of the Menagerie and Aviary at Knowsley*.

[29] This is a reference to Peppering House, the home of the Drewitt family whose father was an amateur naturalist (Noakes, in Lear, *The Complete Nonsense and Other Verse*, p. 474).

[30] Noakes in Lear, *The Complete Nonsense and Other Verse*, p. 479. All subsequent references to Lear's writings are from this edition and will be given in the text.

to Harry Hinde in December 1830, Lear refers to his work on the parrots of the Zoological Society museum ('They are at the museum, – When you come you shall see 'em' [p. 47]). The claim, while factual, must however be analysed alongside the illustration Lear included in his letter, which ironically reverses the gaze of the museum visitor upon the exhibited animals. The parrot in the centre of the image not only looks alive (typifying Lear's regular use of living animals), but it also looks down intently upon the fat man who is observing the bird, his hands in his pockets, looking more mildly curious and idle than bent upon studying nature. Such a reversal of the positions of subject and object (of study)—or human and animal—is found time and again in his nonsense poetry.

The links between Lear's nonsense poetry, his personal writings and works as an ornithological artist are often pronounced. For example, there is a clear connection between 'Pelican Chorus' and Lear's work for the fifth volume of John Gould's *Birds of Europe*.[31] Yet, much of his poetry, dealing as it does with children's engagement with exotic creatures of the kind found in nineteenth-century menageries, takes Lear's readers through the looking-glass of Victorian museum culture. Bears, tortoises, pigs and crocodiles feature time and again (Limericks, *More Nonsense*); rides on zebras are mentioned ('A Was Once an Apple-Pie'); whilst elephants ('The Absolutely Abstemious Ass'), rhinoceroses ('[Creatures playing chequers]' [p. 271]) and monkeys perform feats, as do other exotic animals, such as kangaroos ('The Duck and the Kangaroo') and tigers in his 'Nonsense Botany'. His 'Portraites of the inditchenous beestes of New Olland', probably inspired by Gould's visit to Australia,[32] contain kangaroos, a platypus, a porcupine, a wombat and a bandicoot mixed with more domestic animals. In his limericks from the 1846 and 1855 editions of *A Book of Nonsense*, which Lear wrote for the children of Knowsley Park, the characters purchase a barbary ape and a bear (pp. 78, 113).

As suggested above, many of Lear's pieces explore the boundaries between the human and the animal. Human characters are frequently made to look like the animals they own, such as the old man with an owl (*A Book of Nonsense* [p. 176]), whose face and nose closely resemble the owl perched next to him, or 'the Young Lady in White' (Limericks, *More Nonsense* [p. 344]) and 'the Old Person of Nice' who looks like his geese (Limericks, *More Nonsense* [p. 360]). Lear even portrayed himself as a

---

[31] Noakes in Lear, *The Complete Nonsense and Other Verse*, p. 535.
[32] Noakes in Lear, *The Complete Nonsense and Other Verse*, p. 442.

bird. The physical proximity of humans and animals is also significant, as in the old man '... on whose nose,/Most birds of the air could repose' (*A Book of Nonsense* [p. 178]) or 'the Old Person of Crowle' who lives in the nest of an owl and also looks like one (Limericks, *More Nonsense* [p. 369]). As Kaori Nagai remarks, through Lear's nonsense, naming and classifying become 'a joyful process through which humans and animals form a relationship'.[33] However, Lear's emphasis on taxonomy, she argues, must be 'considered in relation to some of the contemporaneous scientific developments of the mid to late nineteenth century'.[34]

Following Ann Colley's study of Lear's 'anti-colonial bestiary', it could be argued that Lear's nonsense frequently hinges upon his readers' familiarity with museum collections, their display and ordering. By mocking the 'manag[ement], organiz[ation], cag[ing], domesticat[ion], and dominat[ion]' of captives, Lear turned the 'myth of superiority [of ornithologists, zoologists, collectors, keepers, hunters, breeders, and exhibitors] upside down'.[35] In 'Eggstracts from the Roehampton Chronicle', the objects in the house are presented as so many curios. Lear uses sensational prose to depict the commonplace, whilst the objects are even sometimes compared to natural species: the rooms are named after the letters of the Greek alphabet as if they were different countries in a different hemisphere; a portmanteau becomes the object of conjecture, believed by some to be 'the gigantic & fossil remnant of an extinct brute partaking of the nature of the ostrich & the domestic caterpillar, habitually walking on 3 feet, its neck, head, & expansive antennæ at once surprising & objectionable' (p. 182). The author's 'duty' is to give as 'accurate' a portrait as possible, in the 'interests of science' (p. 182). As Daniel Brown argues, '[n]onsense appears to offer a childish respite from the seriousness of science and the adult world it shaped, of intellectual, technological and economic progress encumbered with the various existential challenges that flowed from such ideas as geological time, Darwinian evolution and entropy'.[36] Yet, for all its ability to offer a 'childish' escape from the real world, Lear's nonsense does nevertheless resonate with contemporary issues. His satirical portraits of the animals brought daily to England, to be

[33] Kaori Nagai, 'Counting Animals: Nonhuman Voices in Lear and Carroll', in Spencer, Ryan and Edwards (eds), *Reading Literary Animals*, pp. 124–39, p. 124.

[34] Nagai, 'Counting Animals: Nonhuman Voices in Lear and Carroll', p. 124.

[35] Colley, 'Edward Lear's Anti-Colonial Bestiary', p. 115.

[36] Daniel Brown, *The Poetry of Victorian Scientists: Style, Science and Nonsense* (Cambridge: Cambridge University Press, 2013), p. 11.

bred and consumed, illustrate this point, as if providing a counterpoint to his scientific illustrations.[37] Indeed, emphasising as it does some of the evils of his day, in particular those related to the practice of natural history, as seen in the previous chapters, Lear's nonsense reverberates with the uses and abuses of (exotic) animals, as in 'The History of the Seven Families of the Lake Pipple-Popple'.

Written in February 1865, 'The History of the Seven Families of the Lake Pipple-Popple' is situated in a far-away place inhabited by different species (parrots, storks, geese, owls, guinea pigs, cats and fish) and describes, as in a natural history book, the 'habits' and 'history' of the creatures up until their ultimate extinction. The story focuses on the struggle for food (the creatures all die because they fail to share their natural resources) and also introduces an unknown creature, the Clangle-Wangle, 'a most dangerous and delusive beast':

> [Clange-Wangles] live in the water as well as on the land, using their long tail as a sail when in the former element. Their speed is extreme, but their habits of life are domestic and superfluous, and their general demeanour pensive and pellucid. On summer evenings they may sometimes be observed near the Lake Pipple-Popple, standing on their heads and humming their national melodies: they subsist entirely on vegetables, excepting when they eat veal, or mutton, or pork, or beef, or fish, or saltpetre. (p. 202)

Once dead, some of the animals' body parts (the parrot's feathers and the stork's beaks) are collected by other species. But when the parents of all the species realise that all their offspring have died, they purchase all the ingredients for pickling and bottle themselves so as to 'be presented to the principal museum of the city of Tosh, to be labelled with Parchment or any other anti-congenial succedaneous, and to be placed on a marble table with silver-gilt legs, for the daily inspection and contemplation, and for the perpetual benefit of the pusillanimous public' (pp. 205–6). Nagai reads Lear's story as 'a chilling fable of species extinction', which illuminates the 'close relationship between the act of counting and the taxonomic practice

---

[37] In addition to Lear's scientific illustrations already mentioned, Lear probably contributed illustrations to Darwin's *Voyage of the Beagle* volume (Noakes in Lear, *The Complete Nonsense and Other Verse*, p. 29), and Darwin himself 'consulted Lear's scientific illustrations during the 1840s and 1850s'; Matthew Bevis, 'Edward Lear's Lines of Flight', *Journal of the British Academy*, 1 (2013): 31–69, p. 43.

182   L. TALAIRACH

of killing animals for specimens'.[38] In her view, Lear's 'allegory of natural history' is above all an 'allegory of the modern-day environmental movement: it is only when a species becomes endangered (that is to say, countable) that we pay attention to, and try to protect it'.[39] However, by closing on the profusion of specimens in the museum of the city of Tosh, the narrative seems to counteract their extinction, since the creatures on display are now anonymous, lost among millions of other specimens:

> And if ever you happen to go to Gramble-Blamble, and visit that museum in the city of Tosh, look for them on the Ninety-eighth table in the Four hundred and twenty-seventh room of the right-hand corridor of the left wing of the Central Quadrangle of that magnificent building; for if you do not, you certainly will not see them. (p. 206)

Interestingly, Lear's use of the museum for his tale of extinction shows how nonsense's 'eclectic collection[s] of heterogenous things ... resist[–] the conventions of counting and classifying animals'.[40] The case is similar with Lear's hybrids, which are a cross between a human and an animal, as in 'The Adventures of Mr Lear & the Polly [& the] Pusseybite on their way to Ritertitle Mountains'. The title hints at adventurous explorations of unknown lands recounted by Lear and his fellow animals. As the characters are quietly walking in the rain, nature suddenly becomes violent, unleashing a series of natural catastrophes: the characters first fall into 'the raging river' (p. 216) before being 'dashed [to] atoms' by a 'Cataract' (p. 217). When they are found by two 'Jebusites' who try to 'reconstruct them perfectly as 3 individuals', they create a mix of human and animal species until they 'all tumble into a deep hole' (p. 218) and are never seen again, like extinct species from deep time whose remains, when found, are hard to identify.[41] Anna Henchman's study of Lear's constant 'play with parts and wholes of words [and] experimentation with the parts and

---

[38] Nagai, 'Counting Animals: Nonhuman Voices in Lear and Carroll', p. 132.

[39] Nagai, 'Counting Animals: Nonhuman Voices in Lear and Carroll', p. 129.

[40] Nagai, 'Counting Animals: Nonhuman Voices in Lear and Carroll', p. 128.

[41] As will be seen in Chap. 6, the remains of Megatherium and Glyptodon were often found together around Buenos Aires and believed to belong to the same species. Although it would be hard to establish links between debates around the Glyptodon and Lear's nonsense here, it is worth noting that T. H. Huxley's work on the Glyptodon, whose remains had long been confused with those of the Megatherium, dates from the 1860s.

wholes of bodies'[42] is significant in this context. For Henchman, Lear's 'new creations expose the peculiar exchangeability of homologous parts', thus reflecting 'three interrelated Victorian preoccupations: comparative anatomy, the evolution of species, and the evolution of language'.[43] If Lear's 'practice of bodily rearrangement often retains the logic of homology'[44] (such as arms replacing wings, for instance), his nonsense repeatedly undermines the stability of identity by collapsing frontiers between species. Reconstruction, or the odd assemblage of different parts—typifying Lear's 'relish for improbable combination'[45]—is at the root of all his nonsense, such as his nonsense trees, presented, like in a work of natural history, as the result of the association of brushes and trees. Likewise, in 'The Scroobious Pip', various species (including exotic ones such as a kangaroo and a lion) gather around the Scroobious Pip in an attempt to classify him: 'we can't make out in the least/If you're Fish or Insect, or Bird or Beast' (p. 387).

'The Story of the Four Little Children Who Went Round the World' is another tale of exploration which parodies travel books, describing what the children discover in the places they visit, the people they meet and the food they eat. As in 'The History of the Seven Families of the Lake Pipple-Popple', food—and natural resources more generally—remain central to the adventure throughout, with the children finding food or offering it to the different species they encounter. The piece relates the story of Violet, Slingsby, Guy and Lionel who want to see the world and collect objects from the lands they discover, such as veal-cutlets and chocolate-drops. They see birds (red parrots), with whose feathers Violet makes herself a head-dress, and which might recall, as Noakes suggests, 'the feathered head-dresses of Tahiti, which Cook brought back and gave to the British Museum, and … the head-dress given to Cortez which was plucked from the South American bird, the Ketzal'.[46] Among the creatures encountered are blue-bottle-flies, which live in bottles, yellow-nosed apes, a rhinoceros, kangaroos and cranes, as well as an 'enormous Seeze Pyder', 'an aquatic and ferocious creature truly dreadful to behold, and happily only met with

---

[42] Anna Henchman, 'Fragments out of Place: Homology and the Logic of Nonsense', in James Williams and Matthew Bevis (eds), *Edward Lear and the Play of Poetry* (Oxford: Oxford University Press, 2016), pp. 183–201, p. 183.

[43] Henchman, 'Fragments out of Place', p. 183.

[44] Henchman, 'Fragments out of Place', p. 183.

[45] Bevis, 'Edward Lear's Lines of Flight', p. 33.

[46] Noakes in Lear, *The Complete Nonsense and Other Verse*, p. 508.

in these excessive longitudes' (p. 230), which gobbles up the boat they are travelling in. They eventually climb on the back of the rhinoceros to travel back home and have the beast 'killed and stuffed directly' 'in token of their grateful adherence', before setting it up 'outside the door of their father's house as a Diaphanous Doorscraper' (p. 232). Significantly, the illustration ultimately shows the rhinoceros as a stuffed specimen. As Ann C. Colley explains, 'The Story of Four Little Children Who Went Round the World' 'baldly exposes the fate of numerous grand, exotic mammals whose bodies were mounted and exhibited by hunters and scientific institutions'. Lear's rhinoceros thus falls victim to 'colonial authority' at a time when, Colley notes, amateur naturalists were 'encouraged by taxidermy manuals to capture and stuff something small or ordinary, like a squirrel, in order to demonstrate their bravery and command'.[47]

Furthermore, while exotic animals are frequently purchased, ridden upon, stuffed or placed in zoological collections[48] throughout Lear's nonsense poetry, the artist's view of naturalists' activities and research is also manifest in the way in which his characters are depicted as 'creatures that, far from being free-willed agents, are subject to the fixed laws of nature, both their own and that of their physical environment'.[49] Indeed, although Colley contends that Lear's animals were often released 'from the objectifying and classifying gaze of the colonial collector',[50] his creatures remain nonetheless subject to the forces of evolution. As Daniel Brown observes, '[i]n the period stretching from Robert Chambers's *Vestiges of the Natural History of Creation* (1844) through to Darwin's *Descent of Man* (1871) and its aftermath, the odd parallel universe of Lear's verses and drawings offered a surreal articulation of popular Victorian preoccupations with "Man's Place in Nature"'.[51] By depicting the exploitation of animals by

---

[47] Ann C. Colley, *Wild Animal Skins in Victorian Britain: Zoos, Collections, Portraits and Maps* (Farnham: Ashgate, 2014), p. 90.

[48] In 'The Pobble Who Has No Toes', the princess asks the Pobble for his toes, 'To place in [her] Pa's Museum collection' (p. 396).

[49] Brown, *The Poetry of Victorian Scientists*, p. 21.

[50] Ann Colley, 'Edward Lear and Victorian Animal Portraiture', in Raffaella Antinucci and Anna Enrichetta Soccio (eds), *Edward Lear in the Third Millennium: Explorations into his Art and Writing, Rivista di Studi Vittoriani*, 2.1 (2012–13): 11–24, 14, qtd. in Raffaella Antinucci, '"Sensational Nonsense": Edward Lear and the (Im)purity of Nonsense Writing', *English Literature*, 2.3 (Dec. 2015): 291–311, p. 304.

[51] Brown, *The Poetry of Victorian Scientists*, pp. 23–4.

humans, Lear's nonsense resonates with evolutionary conceptions of the natural world, a world ruled by the survival of the fittest and chance.

Thus, Lear's use of nonsense illuminates the links between contemporary views of the natural world and Victorian museum culture—between nineteenth-century science and its objects. Although, as Colley remarks, Lear, as revealed by his correspondence and journals written abroad, 'subscribed to the colonial system and its prejudices against the non-English' and strongly depended 'upon the privileges of imperialism',[52] his rendering of imperial practice and his stance upon museum culture reveal, on the contrary, his 'identification with and sensitivity to animals'.[53] In Colley's view, Lear's animals—most particularly the 'curious beasties' imported into England—carry 'the burden of political symbolism',[54] like earlier animal fables, as seen in Chap. 2:

> The amassing of vast collections of wild animal trophies and exotic bird skins, the sudden sprouting of city zoos (in London, Surrey, Birmingham, Liverpool, and Manchester), the creating of private menageries, the interest of natural history books, the producing of animal furniture (elephant foot stools and African antelope hall stands), the popularity of traveling wild animal displays, the breeding of hybrids like 'tigons', and the consuming of exotic flesh (elephant trunk soup, roast giraffe, kangaroo hams, and Chinese sheep) were all, in a sense, 'patriotic' acts that represented the suppression of the peoples of Africa and Asia (to a lesser extent, those of Australia and Canada). Each act symbolized possession or control.[55]

For Jean-Jacques Lecercle, whether nonsense authors had a special interest in the subject or not, 'the discourse of natural history is present in the genre as a whole', since the genre 'borrows [two themes] from natural history: exploration and taxonomy'.[56] Indeed, nonsense reflects its era, 'a great age of taxonomy, ... of collections ... [and] museums'[57]—most par-

---

[52] Colley, 'Edward Lear's Anti-Colonial Bestiary', p. 110.

[53] Colley, 'Edward Lear's Anti-Colonial Bestiary', p. 112.

[54] Colley, 'Edward Lear's Anti-Colonial Bestiary', p. 112.

[55] Colley, 'Edward Lear's Anti-Colonial Bestiary', p. 112. Colley draws upon John Berger's vision of public zoos as 'an enforcement of modern colonial power'; John Berger, *About Looking* (New York: Pantheon Books, 1980), p. 19.

[56] Lecercle, *Philosophy of Nonsense*, p. 202.

[57] Lecercle, *Philosophy of Nonsense*, p. 203. Lecercle notes that the contemporaneity of the emergence of the genre and of the development of museums was first noted by H. Reichert in *Lewis Carroll, Studien zum literarischen Unsinn* (Munich: s.n., 1970).

186   L. TALAIRACH

ticularly through its 'interminable descriptions, especially if they take the form of lists'.[58] For Nagai, moreover, nonsense's 'close engagement with language' becomes a means of 'resist[ing] the use of natural history to kill and silence creatures' by 'breaking down the classificating system itself'.[59] Nagai's study of Lear's use of binomial Latin-sounding names in his 'Nonsense Botany' shows that while Lear 'parodies the taxonomic conventions of naming', he 'also breaks down the classificating system itself, as he disregards the distinction between species and genus ... and freely mixes plants with other entities'. Hence, in Lear's nonsense, creatures 'are not type specimens, but a celebration of all that evades being captured in the net of taxonomic meaning-making'.[60] As a consequence, through its play with taxonomy, nonsense serves to denounce 'the rage to classify and, thus, to control the natural world'—Lear achieving, as Colley argues, 'a kind of revenge, on the animals' behalf'.[61]

As will be seen in the next section, Lewis Carroll followed in the footsteps of Edward Lear. Indeed, the violence of museum culture percolates through the *Alice* books in similar ways: 'the safe asylum of nonsense'[62] serves to convey Carroll's political commentary. Like Lear, who contributed to many scientific articles[63] and worked for many naturalists, Carroll's relationship with Henry Acland (1815–1900), who initiated the project to establish Oxford's Museum of Natural History, informs his knowledge of many of the 'curious beasties' then exhibited at Oxford.[64] However, seen through the eyes of a little girl fallen into a strange museum of odd creatures, Carroll's wonderland collapses even more radically the boundaries between species, and between reality and myth.

Rewriting Charles Kingsley's Specimens: Lewis Carroll's Mad Museology

... Pray, Dr Hughball,

[58] Lecercle, *Philosophy of Nonsense*, p. 63.

[59] Nagai, 'Counting Animals: Nonhuman Voices in Lear and Carroll', p. 127.

[60] Nagai, 'Counting Animals: Nonhuman Voices in Lear and Carroll', p. 127.

[61] Colley, 'Edward Lear's Anti-Colonial Bestiary', pp. 115.

[62] Colley, 'Edward Lear's Anti-Colonial Bestiary', pp. 118.

[63] See Colley, 'Edward Lear's Anti-Colonial Bestiary', pp. 112–13.

[64] As recorded in his diary, Carroll took six photographs of the museum specimens for Henry Acland on 16 June 1857, as well as photographs of the museum and the people involved, including Benjamin Woodward (1816–61), one of the architects.

5 NONSENSE 'BEASTIES'   187

Play the man-midwife and deliver him
Of his huge tympany of news – of monsters,
Pygmies and giants, apes and elephants,
Gryphons and crocodiles ...[65]

... an ornamental alligator with a fetching parasol, a little clay doll, a cruet, and a remarkable zoological monstrosity (shown in the lower right-hand corner of this page), which may be one of the animals which Alice saw in Wonderland.[66]

In June 1866, *Aunt Judy's Magazine* published a review of *Alice's Adventures in Wonderland*, promoting Carroll's fairy tale for its young readers. The reviewer claimed that 'never was the mystery made to feel so beautifully natural before', and went on to describe how 'the exquisitely wild, fantastic, impossible, yet most natural history of "Alice in Wonderland"'[67] evoked the wild yet possible natural history specimens which were visible all over the country in Victorian shows and exhibitions. Even though the reviewer (probably Margaret Gatty) warned parents not to look for 'knowledge in disguise'[68] in Carroll's narrative, she neverthe-less linked the fantasy closely to the reality of Victorian natural history. Her reasons for defining Carroll's fantasy in such terms may relate to her own passion for natural history and her expertise in the field, as seen in the previous chapter, but *Alice's Adventures in Wonderland* nevertheless often reads like natural history through a humorous looking-glass, with some of its odd creatures standing out as potential exhibits in a cabinet of curiosities.

Both *Alice* books are, indeed, permeated by allusions to contemporary natural history. In *Alice's Adventures in Wonderland*, Alice's fall down the

---

[65] Richard Brome, *The Antipodes* [c. 1640], in *Three Renaissance Travel Plays*, ed. by Anthony Parr (Manchester: Manchester University Press, 1995), 1.1.176–81.

[66] Laura B. Starr, 'Found in Uncle Sam's Mails. [From photographs taken expressly for George Newnes, Ltd.]', *The Strand Magazine*, 16.92 (Sept. 1898): 148–52, p. 149.

[67] [Margaret Gatty], [Review of *Alice's Adventures in Wonderland*], *Aunt Judy's Magazine for Young People. The Christmas Volume for 1866* (London: Bell and Daldy, 1866), p. 123. For more on Lewis Carroll and natural history see White, *The Alice Books and the Contested Ground of the Natural World*. White argues that Carroll owned Gatty's *British Seaweeds* (1863), together with other natural histories, such as J. G. Wood's *Common Objects of the Sea Shore* (1857), Charles Kingsley's *Glaucus, or the Wonders of the Shore* (1855) and Philip Henry Gosse's *A Year at the Shore* (1865); White, *The Alice Books and the Contested Ground of the Natural World*, p. 152.

[68] [Gatty], [Review of *Alice's Adventures in Wonderland*], p. 123.

188   L. TALAIRACH

rabbit hole enables the little girl to discover mysterious worlds in which her size constantly changes. She undergoes metamorphosis, like insects, becoming as tiny as caterpillars or flowers, and is terrified at the idea that she might be gobbled up by a puppy or stung by a giant insect. Moreover, Alice's obsession with size throughout her adventures points to the narrative's preoccupation with questions of scale. At first, Alice believes she is falling through the earth and will end up on the other side of the globe, and the tale recurrently calls to mind the period's urge to travel and to see ever farther, to peer into unknown worlds—or even scrutinise the skies, as the motif of the telescope suggests.[69] The tale can be read in fact as a humorous rewriting of a travel narrative,[70] with the young naturalist entering a strange and wild land where codes of conduct and signs of civilisation have become irrelevant or turned upside down.[71] Defined as a 'fairy tale' by Carroll,[72] *Alice's Adventures in Wonderland*, like its sequel, takes us into a realm inhabited by talking animals and flowers, extinct species, mythical creatures and imaginary beings. But Carroll's tale, though aimed at children, is highly representative of the issues that concerned Victorians of all generations: it echoes debates about representations of nature at a time when such definitions were changing, especially with the popularisation of evolutionary theory. In addition, it alludes to the development of scientific knowledge linked to a new mapping of the world and the discov-

[69] 'Oh, how I wish I could shut up like a telescope!'; Lewis Carroll, *Alice's Adventures in Wonderland* [1865], in *The Annotated Alice*, ed. by Martin Gardner (London: Penguin, 2001), p. 17. All subsequent references are to this edition and will be given parenthetically in the text.

[70] Jean-Jacques Lecercle sees Alice as 'in the same position as Mungo Park – she is in the power of those strange creatures, and must obey their orders, even if she is secretly convinced of the superiority of her own world; they, in turn, treat her as a "fabulous monster", even as the Africans sometimes mistook Mungo Park for a ghost. Alice, therefore, is the nonsensical reincarnation of the traditional figure of the British explorer *qua* traveller, a figure that disappeared with imperialist colonisation'; Lecercle, *Philosophy of Nonsense*, p. 203.

[71] Many imitations of *Alice's Adventures in Wonderland* published in the following decades aimed at popularising science. These rewritings reveal how much Carroll's fantasy was informed by natural history, as in Gertrude P. Dyer's *Elsie's Adventures in Insect-Land* (London: Ward & Co., 1882) or Frank Stevens's *Adventures in Pondland* (London: Hutchinson & Co., 1905).

[72] Carroll called his fantasy a 'fairy tale' when he wrote to a professional illustrator (Tom Taylor) to ask him for advice on the title: 'I should be very glad if you could help me in fixing on a name for my fairy-tale', Charles Dodgson to Tom Taylor, 10 June 1864, qtd. in Morton N. Cohen (ed.), *The Selected Letters of Lewis Carroll* (Basingstoke: Macmillan [1982] 1989), p. 29.

ery of exotic, unknown or unseen species and peoples; and it tackles the diffusion of such knowledge, either through allusions to specimens exhibited at the time in museums (like the dodo in the Oxford University Museum of Natural History[73]) or to works of popular science. Furthermore, the little girl's experience of natural selection and her struggles for life in Wonderland recall the journeys proposed at the time by writers of natural history books, inviting their readers to travel into unknown territories in order to discover new species, and in so doing constantly triggering wonder and curiosity.[74]

That *Alice's Adventures in Wonderland* may be read as a nonsensical natural history for young readers cannot be gainsaid, as already underlined. Carroll was even known to have borrowed a natural history book for his own illustrations for *Alice's Adventures Underground*.[75] Jean-Jacques Lecercle sees nonsense as 'a by-product of the development of the institution of the school'—'the institution that develops the need for meaning and a reflexive attitude towards language'.[76] Moreover, in his view, the 'meaning-preserving activity' of nonsense functions to establish 'the cor-

---

[73] For potential links between the *Alice* books and the Oxford University Museum of Natural History, see Jones and Gladstone, *The Alice Companion* and White, *The* Alice *Books and the Contested Ground of the Natural World*, in particular pp. 206–8. White also analyses some of Carroll's photographs of the Oxford natural history specimens (pp. 59–60) and records the lectures and debates Carroll attended around Darwinian science (p. 62). Although White reads Carroll's relationship with the Oxford specimens, and nineteenth-century science more generally, as systematically reflecting Carroll's anti-Darwinian satirical worldview, I contend that his engagement with Victorian science and museum culture, seen through the prism of nonsense, is not a series of 'jokes', as she argues, but rather a significant lens for young audiences to reflect upon the 'curious beasties' his texts display.

[74] Carroll frequently made references both to voyages of exploration and museum collections in his writings. In *The Rectory Umbrella* (1849–50), written when he was only a boy and shaped as a children's magazine, significant references to the world of collecting and exhibiting animals punctuate the series of articles. The Rectory articles were mostly written by Lewis Carroll himself, although members of his family contributed; Jan Susina, *The Place of Lewis Carroll in Children's Literature* (New York: Routledge, 2010), p. 54. For instance, Lewis Carroll proposed 'Zoological Papers' in which ornithological species ('The One-Winged Dove', 'The Lory'), just like rare species ('The Pixies'), were presented to children, the collection of creatures parodying Carroll's contemporary natural science. *The Hunting of the Snark*, Carroll's mock-heroic epic, is another example, since the Bellman expedition aims to find the enigmatic creature.

[75] Roger Lancelyn Green, *The Diaries of Lewis Carroll*, 2 vols (London: Cassell, 1953), p. 193, qtd. in Rose Lovell-Smith, 'Eggs and Serpents: Natural History Reference in Lewis Carroll's Scene of Alice and the Pigeon', *Children's Literature*, 35 (2007): 27–53, p. 28.

[76] Lecercle, *Philosophy of Nonsense*, p. 4.

respondence between signifier and signified'; its 'matter-of-factness' giving, as will be shown below, 'a pseudo-scientific authority'[77] to the challenge it posed to the reality of species classification. Indeed, it is easy to see, following Lecercle's links between the *Alice* books and the nineteenth-century educational system, that Carroll's choice of animals in *Alice's Adventures in Wonderland* is by no means coincidental. As Rose Lovell-Smith has pointed out, the creatures inhabiting Wonderland resemble far more 'the animals referred to in contemporary scientific discourses' than they do 'fairy-tale or fable creatures'.[78] Wonderland tests Alice's knowledge of animals, such as that of Cheshire cats ('You don't know much', said the Duchess; 'and that's a fact' [p. 63]). She is also regularly presented with the 'history' of some of the animals, such as that of the mouse or the Mock-Turtle, and invited to study these creatures in the accompanying illustrations,[79] in imitation of children's natural history books.

More interestingly, perhaps, some of the animals may have been inspired by the creatures on display in menageries, zoological gardens or pet shops.[80] Laura White notes that Carroll 'liked to take children to exhibits

[77] Brown, *The Poetry of Victorian Scientists*, p. 16. Brown is here rephrasing Jean-Jacques Lecercle's argument on nonsense.

[78] Lovell-Smith, 'Eggs and Serpents: Natural History Reference in Lewis Carroll's Scene of Alice and the Pigeon', p. 28.

[79] And so is the reader, since the narrator advises the reader who does not know what a Gryphon is to look at the illustration on the page (p. 124). Edith Nesbit often invited her readers to do the same in her books, as will be seen in Chap. 6.

[80] In *Sylvie and Bruno*, likewise, whose central story, 'Bruno's Revenge', was written in 1867 at the request of the editor of *Aunt Judy's Magazine* (Margaret Gatty), exotic animals, such as those the children could see in the London Zoological Gardens, pepper the narrative, sometimes indirectly (such as the elephant, the rattlesnake, the hippopotamus, the kangaroo and the bear, which all appear in the gardener's song), alongside an anthropomorphised crocodile. Several characters are 'fond of Natural History' (p. 94) ('you know I am very learned in Natural History (for instance, I can always tell kittens from chickens at one glance)' [pp. 196–7]) or botany, and recent developments in natural history, albeit satirised, are referred to: improbable transformations, such as the mouse which turns into a lion, take place, using Darwin ('"A development worthy of Darwin!" the lady exclaimed enthusiastically. "Only *you* reverse his theory. Instead of developing a mouse into an elephant, you would develop an elephant into a mouse!"' [pp. 64–5]); Lewis Carroll, *Sylvie and Bruno* [1889] (New York: Dover Publications, 1988). *Sylvie and Bruno* (1889) and *Sylvie and Bruno Concluded* (1893) resulted from Carroll's compilation of stories he had written for twenty years. The former 'was to be a serious book, full of philosophical, scientific, and religious implications, quite different in that respect from the *Alices*', writes Morton N. Cohen; Morton N. Cohen and Edward Wakeling (eds), *Lewis Carroll and His Illustrators*.

of animals, living animals as at aquariums and zoos and stuffed or skeletonized ones as at museums of natural history'.[81] It is tempting, for instance, to imagine that the Lobster Quadrille was intended as a variant of the Hippopotamus Quadrille (or Polka), inspired by the Hippomania, a craze ignited by the first hippopotamus seen in Britain, Obaysh, which arrived at the London Zoo on 25 May 1850.[82] Alice believes, moreover, that she sees 'a walrus or hippopotamus' (p. 25) when she is in the pool of tears, but which only turns out to be a mouse. The Lobster Quadrille, meant as an oblique reference to Carroll's own poor dancing technique, was seen by the latter as resembling some of the wild animals at the Zoological Gardens: 'Did you ever see the Rhinoceros, and the Hippopotamus, at the Zoological Gardens, trying to dance a minuet together?'[83] The Lobster is, furthermore, later humorously presented as a species already seen by Alice at 'Dinn' (p. 107). The change from 'dinner' to 'Dinn'—a place where animals may be seen—matches the changing function of the zoo in the second half of the nineteenth century, as Jed Mayer explains, shifting 'from one mode of consumption to another: from the acclimatization of foreign species to be consumed as meat and leather, to the display of exotic animals to be consumed by avid spectators'.[84] In addition, Alice's fear that the flamingo might bite recalls the many accidents which occurred daily in Victorian menageries. Similarly, John Simons argues that the hedgehogs which roll themselves into balls in *Alice's Adventures in Wonderland* may have been inspired by the armadillos found at Jamrach's shop in London.[85] It has also been suggested that Carroll's dormouse may have been modelled on Dante Gabriel Rossetti's wombat—a 'curious beastie' purchased at Jamrach's as well, alongside

Collaborations & Correspondence, 1865–1898 (Ithaca and New York: Cornell University Press, 2003), p. 102.

[81] White, *The* Alice *Books and the Contested Ground of the Natural World*, p. 190.

[82] John Simmons, *The Tiger That Swallowed the Boy: Exotic Animals in Victorian England* (Faringdon: Libri Publishing, 2012), p. 114.

[83] Qtd. by Martin Gardner; Lewis Carroll, *Alice's Adventures in Wonderland*, in *The Annotated Alice*, p. 131.

[84] Jed Mayer, '"Come Buy, Come Buy": Christina Rossetti and the Victorian Animal Market', in Laurence W. Mazzeno and Ronald D. Morrison (eds), *Animals in Victorian Literature and Culture. Contexts for Criticisms* (London: Palgrave Macmillan, 2017), pp. 213–31, p. 215.

[85] John Simons, *Rossetti's Wombat: Pre-Raphaelites and Australian Animals in Victorian London* (Middlesex University Press, 2008), p. 102.

other creatures, like kangaroos, wallabies, armadillos and a zebu.[86] The crocodile, used by Carroll to rework Isaac Watts's 'How does the little busy bee', may likewise have been one of the creatures on display at the London Zoo. According to Mary Elizabeth Leighton and Lisa Surridge, the crocodile 'served generally as a liminal sign, marking the beginning of fantastic or exotic adventures': '[r]epresenting the cannibalistic, the greedy, the unevolved, the unpredictable, and the highly dangerous, the crocodile thus functioned as the quintessential sign of alterity'.[87] As the crocodile rewrites Watts's 'busy bee', the boundary between the wild and the domestic thus collapses, as does the frontier between the familiar (the nursery rhyme) and the unfamiliar (the exotic creature). By revealing Alice's confrontation with the Other, the 'curious beastie', which came 'to stand for generalized imperial anxieties'[88] in the nineteenth century, hints therefore at contemporary colonial ideology, albeit parodied through nonsense.

As Lovell-Smith demonstrates, moreover, the influence of John Tenniel's illustrations of animals throughout *Alice's Adventures in Wonderland* 'offers a visual angle on the text … that evokes the life sciences, natural history, and Darwinian ideas about evolution, ideas closely related by Tenniel to Alice's size changes, and to how these affect the animals she meets'.[89] Carroll had admired Tenniel's illustrations for Reverend Thomas James's 1848 edition of *Aesop's Fables*, a work that had established his reputation, and he considered that Tenniel would be ideal for the creatures in the *Alice* books. Tenniel, for his part, accepted 'because it gave him the opportunity to create more animal drawings, a subject that had fascinated him since he was inspired years before by the work of the

---

[86] This idea is suggested by Martin Gardner, but as Simons explains, *Alice's Adventures in Wonderland* was written in 1864, thus before Rossetti's wombat, Top, had arrived. However, Carroll did visit Rossetti and his family, took photographs of them and Rossetti had several dormice, alongside owls, armadillos, a raccoon, kangaroos, wallabies, jackasses, peacocks, parakeets and a Japanese salamander; Simons, *Rossetti's Wombat*, p. 113.

[87] Mary Elizabeth Leighton and Lisa Surridge, 'The Empire Bites Back: The Racialized Crocodile of the Nineteenth Century', in Deborah Denenholz Morse and Martin A. Danahay (eds), *Victorian Animal Dreams: Representations of Animals in Victorian Literature and Culture* (Aldershot: Ashgate, 2007), pp. 249–70, pp. 254–5.

[88] Leighton and Surridge, 'The Empire Bites Back', p. 255.

[89] Rose Lovell-Smith, 'The Animals of Wonderland: Tenniel as Carroll's Reader', *Criticism*, 45.4 (Fall 2003): 383–415, p. 385.

French illustrator J. J. Grandville'.[90] The sources of Tenniel's creatures were the animals kept in menageries and zoological gardens.[91] Furthermore, his scientific accuracy is evident, notably in some of the backgrounds he chose (such as the meadow setting in the first illustration of the white rabbit, or his use of foxgloves or mushrooms to suggest Alice's size), the anatomical details of many of the creatures (which are never grotesque),[92] or even in his stylistic allusions to natural history illustration, as Lovell-Smith argues.[93] Indeed, Tenniel's 'cross-hatching and fine lines used to suggest light, shade, and solidity of form in the Mock Turtle's shell and flippers, or the crabs' and lobster's claws',[94] are typical conventional techniques of realism. According to Lovell-Smith, Tenniel's illustrations 'pick up on but also extend this Darwinist and natural history field of reference in Carroll's text'.[95] Details such as the addition of a glass dome in the background of the croquet-lawn, for instance, refer possibly to 'the dome at the old Surrey Zoological Gardens [...which] therefore constitutes another reference to the study of animals'.[96]

[90] Susan E. Meyer, *A Treasury of the Great Children's Books Illustrators* (New York: Harry N. Abrams, 1997), p. 66.

[91] William Cosmo Monkhouse, *The Life and Works of Sir John Tenniel* (London: Art Union Monthly Journal, Easter Art Annual, 1901), p. 28, qtd. in Lovell-Smith, 'The Animals of Wonderland: Tenniel as Carroll's Reader', p. 395.

[92] According to Lovell-Smith, '[e]ven if creatures are endowed with personality and facial expressions, his animals, unlike his humans, are never grotesques. In fact, nineteenth-century natural history illustration also delights in endowing the most solidly "realistic" creatures with near-human personality or expressiveness, a quality that Tenniel builds on to good effect, for instance, in his depiction of the lawyer-parrots, which remind one of Edward Lear's magnificent macaws'; Lovell-Smith, 'The Animals of Wonderland: Tenniel as Carroll's Reader', p. 388.

[93] The scientific accuracy of Tenniel's animals must be contrasted with Carroll's opinion on zoological realism. In a letter to Harry Furniss (the illustrator of *Sylvie and Bruno*) written on 29 November 1886, Carroll writes: 'Of course the Herrings must *not* have sharks' heads, and the Badgers must *not* be the dogs I have made of them. But as to the disproportion in size, I fear you have been too much to the Zoo, and have got ideas too accurate for our present purpose. Surely it is better to get the pictures as funny as we can, than to be zoölogically correct? If you really feel that you *must* draw them in right proportion, of course I give in: but, if you *can* strain your standard of zoölogical propriety, I *think* we shall gain in drollery'; qtd. in Cohen and Wakeling (eds), *Lewis Carroll and His Illustrators*, pp. 133–4.

[94] Lovell-Smith, 'The Animals of Wonderland: Tenniel as Carroll's Reader', p. 391.

[95] Lovell-Smith, 'The Animals of Wonderland: Tenniel as Carroll's Reader', p. 388.

[96] Lovell-Smith, 'The Animals of Wonderland: Tenniel as Carroll's Reader', p. 388. Lovell-Smith also analyses the potential connections of some of Tenniel's illustrations with contemporary natural history books. She surmises, for instance, that Carroll and Tenniel were

With its wild and domestic animals, its hookah-smoking creature (recalling imports from the British colonies), and dodo (brought back from the Indian Ocean), *Alice's Adventures in Wonderland* is thus undeniably grounded in Victorian museum culture, its collections and exhibitions. But Carroll's story also owes much to Charles Kingsley's fantasy, *The Water-Babies, or a Fairy Tale for a Land Baby* (1863), published a few years earlier, and which overtly aimed at disseminating evolutionary theory in a narrative adapted to a young audience whilst playfully alluding to contemporary scientific controversies and debates. In both fantasies, the scientific discourse is therefore inextricably bound up with museum culture, the links between science and museum specimens enabling both writers to denounce scientific materialism.

As mentioned in Chap. 2, Kingsley's *The Water-Babies* can be seen as one of the many literary progeniture of the 'Harris Papillonnades' published at the turn of the nineteenth century by John Harris, and imitated by many other publishers,[97] which paved the way for later entertaining works of popular science. The hero of the children's story, Tom, who falls into a river and is transformed into a water-baby, becomes an intriguing specimen likely to 'spoil [scientists'] theories' (p. 161). He thus spends his time trying to avoid getting collected and given 'two long names of which the first would have said a little about Tom, and the second all about himself' (p. 157). As a potential museum exhibit, however, Tom represents the 'wonders of nature' (p. 159) denied by the tenets of scientific materialism. His flight is therefore symbolic, and illustrates Kingsley's critique of material—and museum—culture. The book's presentation of the naturalist as a giant fully equipped to collect and conserve nature in glass jars, and to frame it by taking pictures, typifies Victorian museum culture, whilst

---

familiar with J. G. Wood's *Illustrated Natural History* (in 1862, the second volume included a picture of a dodo, as well as illustrations of numerous more familiar animals that appear in the words and/or pictures of *Alice*, such as the hedgehog, the crab, the lobster, the frog, the dormouse, guinea pigs, flamingos, varieties of fancy pigeon and so on); Lovell-Smith, 'The Animals of Wonderland: Tenniel as Carroll's Reader', pp. 396–7. As Lovell-Smith points out, Michael Hancher notes as well the strong resemblance between a Bewick hedgehog in *General History of Quadrupeds* (1790) and the croquet-ball hedgehog at Alice's feet; Lovell-Smith, 'The Animals of Wonderland: Tenniel as Carroll's Reader', p. 399. See Michael Hancher, *The Tenniel Illustrations to the 'Alice' Books* (Columbus: Ohio University Press, 1985), pp. 29–30.

[97] Tess Cosslett, *Talking Animals in British Children's Fiction, 1786–1914* (Aldershot: Ashgate, 2006), p. 118.

the narrative hints at the period's controlling obsession with collecting, mapping and dissecting the natural world:

> He was made principally of fish bones and parchment, put together with wire and Canada balsam; and smelt strongly of spirits, though he never drank anything but water: but spirits were used somehow there was no denying. He had a great pair of spectacles on his nose, and a butterfly-net in one hand, and a geological hammer in the other; and was hung all over with pockets, full of collecting boxes, bottles, microscopes, telescopes, barometers, ordnance maps, scalpels, forceps, photographic apparatus, and all other tackle for finding out everything about everything, and a little more too.[98]

Although Kingsley does not condemn museums or places of exhibition as such,[99] his fairy tale humorously bridges the gap between natural history and the objects representing that knowledge which the naturalists attempt to capture and pickle in preserving fluid. Tom, as a water-baby, will not be captured, objectified and exhibited in a jar. He thus becomes a visual representation of what cannot be grasped nor represented, what cannot be measured or gauged, what does not fit classification and yet which people must believe in—unlike the bottled specimens displayed on museum shelves.

Kingsley's fairy tale, which follows the journey of a little boy who turns into a museum specimen and who meets all sorts of creatures under water, though fantasy, nonetheless reads very much like a popular science book, as I have argued elsewhere.[100] References to popular science books devised for children, such as Bewick's,[101] or scientific venues, such as the Zoological Gardens, punctuate Tom's adventures in fairyland. But as an amphibious creature, Tom is a natural enigma—a curiosity likely to intrigue naturalists. What is 'contrary to nature' but yet exists—from 'impossible monster[s]' like the elephant and the giraffe, considered 'when first brought to Europe, contrary to the laws of comparative anatomy' (p. 72) to flying dragons (Pterodactyls)—reveals the limitations of

---

[98] Charles Kingsley, *The Water-Babies, a Fairy Tale for a Land Baby* [1863] (London: Penguin, 1995), p. 294. All subsequent references are to this edition and will be given parenthetically in the text.

[99] The Crystal Palace ('Great Exhibition') is, for example, mentioned on page 277.

[100] Laurence Talairach-Vielmas, *Fairy Tales, Natural History and Victorian Culture* (Basingstoke: Palgrave Macmillan, 2014), pp. 38–46.

[101] As already seen, the engraver Thomas Bewick (1753–1828) was most famous for *A General History of Quadrupeds* (1790) and *A History of British Birds* (1797–1804).

empiricism as a scientific method. On the contrary, Kingsley's fairy tale aimed to bring to light how contemporary science depended on scientists' belief in impossible things. Kingsley praises more than he derides evolutionary theory, but he nevertheless condemns the ways in which nature is collected and exhibited. His Professor Ptthmllnsprts, chief professor of *Necrobioneopalæonthydrochthonanthropopithekology*, who humorously combines palaeontologist/biologist T. H. Huxley (1825–95) and anatomist Richard Owen (1804–92), with their violently opposed views on evolution, embodies Kingsley's critique of Victorian scientific materialism. His name, moreover, Ptthmllnsprts (Put-them-all-in-Spirits), seems to suggest that the bottling of specimens sucks all substance out of them, just as the vowels have been removed from the name, shrinking it and reducing it to a series of consonants—an ironically dry name for a compulsive creator of wet specimens.

From the beginning of Tom's journey, the discovery of, and knowledge about, nature works in tandem with the collection and exhibition of natural specimens. When Tom looks at the water and sees 'beetles and sticks; and straws, and worms, and addle-eggs, and wood-lice, and leeches, and odds and ends, and omnium-gatherums', the profusion of natural creatures instantly conjures up naturalists' collecting activities: there is 'enough to fill nine museums' (p. 111). Recurrent references to stolen birds' eggs and creatures caught in nets, to people keeping creatures in tanks and bottles, birds being shot, stuffed and 'put ... into stupid museums' (p. 255), allude repeatedly to the world of Victorian collecting.

It is logical that the museum culture underpinning the narrative and which Kingsley satirises is inseparable from the tale's obsession with classification. A significant instance of this is the scene when the otter tries to define Tom in order to understand where to place him in the food chain. The animal applies here principles of comparative anatomy which are shown to be completely ineffective:

> 'Come, away, children', said the otter in disgust, 'it is not worth eating, after all. It is only a nasty eft, which nothing eats, not even those vulgar pike in the pond'.
>
> 'I am not an eft!' said Tom; 'efts have tails'.
>
> 'You are an eft', said the otter, very positively; 'I see your two hands quite plain, and I know you have a tail'.
>
> 'I tell you I have not', said Tom. 'Look here!' and he turned his pretty little self quite round; and, sure enough, he had no more tail than you.

The otter might have got out of it by saying that Tom was a frog: but, like a great many other people, when she had once said a thing, she stood to it, right or wrong; so she answered:

'I say you are an eft, and therefore you are, and not fit food for gentlefolk like me and my children. You may stay there till the salmon eat you (she knew the salmon would not, but she wanted to frighten poor Tom). Ha! ha! they will eat you, and we will eat them'; and the otter laughed such a wicked cruel laugh – as you may hear them do sometimes; and the first time that you hear it you will probably think it is bogies.

'What are salmon?' asked Tom.

'Fish, you eft, great fish, nice fish to eat'. (pp. 105–6)

The scene does not simply change the order of predation. By deriding the 'anthropocentric assumption that language is what makes us human',[102] the talking otter also suggests that assigning a name and practising classification may be fatal. This idea is most clearly visible in Carroll's rewriting of this scene, which illustrates, as Nagai puts it, that classification is a system that 'silences nonhumans [and] which is nothing but a form of violence'.[103] When Alice, following the caterpillar's advice, eats a part of the magical mushroom and sees her neck grow so much that a pigeon mistakes her for a snake, the idea that she might not be a human silences the little girl who is no longer able to know, and thus to classify, herself:

'But I'm *not* a serpent, I tell you!' said Alice. 'I'm a—I'm a—'

'Well! *What* are you?' said the Pigeon. 'I can see you're trying to invent something!'

'I—I'm a little girl,' said Alice, rather doubtfully, as she remembered the number of changes she had gone through that day.

'A likely story indeed!' said the Pigeon in a tone of the deepest contempt. 'I've seen a good many little girls in my time, but never *one* with such a neck as that! No, no! You're a serpent; and there's no use denying it. I suppose you'll be telling me next that you never tasted an egg!'

'I *have* tasted eggs, certainly,' said Alice, who was a very truthful child; 'but little girls eat eggs quite as much as serpents do, you know.'

'I don't believe it,' said the Pigeon; 'but if they do, why then they're a kind of serpent, that's all I can say.'

This was such a new idea to Alice, that she was quite silent for a minute or two, which gave the Pigeon the opportunity of adding, 'You're looking

---

[102] Nagai, 'Counting Animals: Nonhuman Voices in Lear and Carroll', p. 123.

[103] Nagai, 'Counting Animals: Nonhuman Voices in Lear and Carroll', p. 124.

for eggs, I know *that* well enough; and what does it matter to me whether you're a little girl or a serpent?'

'It matters a good deal to *me*,' said Alice hastily; 'but I'm not looking for eggs, as it happens; and if I was, I shouldn't want *yours*: I don't like them raw'. (pp. 56–7)

Like Kingsley, Carroll rewrites the rules by which we classify the natural world, but his nonsense goes a step further: it shatters Alice's sense of self the better to question the animal-human hierarchy. This matches John Miller's contention that classification 'emerges not just as a series of technologies and practices for signifying non-human others, but also, critically, as a means of signifying the self'.[104] Indeed, whilst both Kingsley's and Carroll's fantasies revolve around the issue of species classification and allude to the debate on evolutionary theory, nonsense enables Carroll to 'disrupt anthropocentric … and political power relationships … as Wonderland's animals refuse to see Alice in a position of power'.[105]

Carroll's scholars have not always agreed on Carroll's stance with regard to Darwinian evolution. Laura White's study of the *Alice* books, for example, presents the fantasies as 'perform[ing] an anti-narrative of the survival of the fittest'.[106] Whether Carroll 'consider[ed] Darwinism sympathetically'[107] or not, it is clear that *Alice's Adventures in Wonderland* constantly foregrounds the human/animal divide being challenged by the evolutionary sciences at that time. The work also destabilises the hierarchy between humans and animals, as many Carrollian critics have shown.[108] As Rose Lovell-Smith argues, indeed, the Pigeon scene may be read 'in the

---

[104] John Miller, *Empire and the Animal Body: Violence, Identity and Ecology in Victorian Adventure Fiction* (London: Anthem Press, 2014), p. 56.

[105] Anna Feuerstein, '*Alice in Wonderland*'s Animal Pedagogy: Governmentality and Alternative Subjectivity in Mid-Victorian Liberal Education', *Victorian Review*, 44.2 (Fall 2018): 233–50, p. 234.

[106] White, *The* Alice *Books and the Contested Ground of the Natural World*, p. 73.

[107] White, *The* Alice *Books and the Contested Ground of the Natural World*, p. 59.

[108] Unlike Laura White and Jessica Straley (*Evolution and Imagination in Victorian Children's Literature* [Cambridge: Cambridge University Press, 2016]), who both stress Carroll's sceptical views of evolution and his attempt to reassert human agency, critics such as Lovell-Smith ('The Animals of Wonderland: Tenniel as Carroll's Reader'; 'Eggs and Serpents'), U. C. Knoepflmacher (*Ventures into Childland: Victorians, Fairy Tales and Femininity* [Chicago and London: Chicago University Press, 1988]) and Ruth Murphy ('Darwin and 1860s Children's Literature: Belief, Myth, or Detritus', *Journal of Literature and Science*, 5.2 [2012]: 5–21) argue, on the contrary, that the *Alice* books challenge human superiority. As already suggested, I am not interested here in Carroll's personal views on

context of contemporary developments in natural history'.[109] Not only does it hinge upon the contemporary natural history topos of what Lovell terms the 'egg-thief' motif, typically found in children's literature advocating kindness to animals (such as in Sarah Trimmer's *Fabulous Histories: Designed for the Instruction of Children Respecting Their Treatment of Animals* [1786], as seen in Chap. 2, as well as in Victorian children's periodicals, as seen in Chap. 4), but Lovell-Smith also suggests that the scene owes a debt to John Audubon's illustration of the mockingbird in *Birds of America* (published in England between 1827 and 1838), with the egg-thief being 'probably best understood as a kind of sub-group of the many Victorian depictions of predation and conflict in the animal world'.[110] Natural history writers such as Reverend J. G. Wood also used the motif, for example in his *Illustrated Natural History* (1853) and in *The Boy's Own Book of Natural History* (1861), where he mentioned a blacksnake attacking a mockingbird nest, and 'perhaps based his description on two of Audubon's plates'.[111] In Wood's *Illustrated Natural History*, moreover, Lovell-Smith adds, a picture of a predatory snake and a nest illustrates the 'Reptiles' volumes.[112] Such echoes between Carroll's illustrated narrative and contemporary scientific writing or works of popular science thus invite readers to 'understand[–] the many animals in Wonderland as fantastic versions of "real" animals like those which were the objects of contemporary scientific observation and description'.[113]

evolution; my point is rather to show how Carroll's fantasies question the animal-human hierarchy, in particular through their allusions to museum culture and scientific materialism.

[109] Lovell-Smith, 'Eggs and Serpents', p. 27.

[110] Lovell-Smith, 'Eggs and Serpents', pp. 29–30. For Lovell-Smith, the scene reflects Carroll's unease with predator-prey relations. Laura White's interpretation of the scene differs from Lovell-Smith's, arguing that the passage 'shows Carroll's black humour' about the cruelty of such relations; White, *The* Alice *Books and the Contested Ground of the Natural World*, p. 195.

[111] Lovell-Smith, 'Eggs and Serpents', p. 33.

[112] Lovell-Smith, 'Eggs and Serpents', p. 33.

[113] Lovell-Smith, 'Eggs and Serpents', p. 35. It is interesting to notice, moreover, that Tenniel did not draw Alice's elongated neck, although Tenniel was 'clearly conscious of natural history references in the *Alice* books and often evoke[d] natural-history models for his animal drawings'. If several explanations may be attributed to this absence, Lovell-Smith surmises that it was to avoid religious connotations in the scene: a 'visual connection via natural-history illustration [Audubon's rattlesnake, coiled round a central tree branch, is visually reminiscent of the serpent seen coiled in various grand Renaissance paintings of the Temptation of Eve] may have resulted in textual reference to the Fall of Man here – and a familiar religious image may lie behind Carroll's endowing Alice with her peculiar snake-like

As Alice's 'sense of self' is constantly challenged and threatened by the animals she encounters (who regularly ask her who she is or mistake her for another species)—and thus '*by nature*'[114]—the fantasy resonates recurrently with the preoccupations of evolutionary science. Several animals encountered by the little girl strongly recall Darwin's own research, such as the pigeon—'a Darwinian bird'.[115] Similarly, as U. C. Knoepflmacher argues, 'the Dodo, the Caterpillar, [and] the Pigeon … are imports from a Darwinian world of aggression, voracity, and sexual selection'.[116] The narrative's numerous references to eating or being eaten add to the evolutionary threat which permeates Alice's adventures. Her constant misclassification aims to rewrite her position as a human being and place her among the animals. For Lovell-Smith, Carroll's fairy tale illuminates 'the dislodging of humanity from its confident "overseeing" of nature'.[117] Lovell-Smith's analysis of the way in which Carroll's fantasy reverses 'the usual direction of the natural history gaze' and 'bring[s] Alice under nature's gaze'[118] may be read in the context of Victorian museum culture: like Lear's depiction of the caged parrots of the museum of the London Zoological Society looking down on visitors in his December 1830 letter to Harry Hinde, or Charles Kingsley's Tom, Alice is herself shaped as a 'curious beastie', subjected to the gaze of Wonderland's captives. The contemporary scientific context that resonates throughout *Alice's Adventures in Wonderland* cannot be separated, therefore, from the construction of Wonderland as a land peopled by 'curious beasties' which spur the girl's interest, just as much as Alice arouses their interest as an undefinable species.

As a matter of fact, the first animal Alice encounters in Wonderland is 'the white rabbit with pink eyes' which speaks to itself and keeps a watch in its waistcoat pocket. The sight of the rabbit instantly arouses her

neck' (Lovell-Smith, 'Eggs and Serpents', p. 38). This idea confirms her assertion that the visual in *Alice* is meant to imply 'a modern, scientific idea of nature replacing a religious one, and thus add[s] to an overall impression that nature in Wonderland disputes the human superiority of the child Alice' (Lovell-Smith, 'Eggs and Serpents', p. 38).

[114] Lovell-Smith, 'Eggs and Serpents', p. 39.

[115] Lovell-Smith, 'Eggs and Serpents', p. 41. Darwin studied the domestic pigeon in *The Origin of Species* ('Variation Under Domestication').

[116] Knoepflmacher, *Ventures into Childland*, p. 167, qtd. in Lovell-Smith, 'Eggs and Serpents', p. 42.

[117] Lovell-Smith, 'Eggs and Serpents', p. 28.

[118] Lovell-Smith, 'Eggs and Serpents', p. 28.

curiosity ('burning with curiosity' [p. 11]), just like the Cheshire cat later on ('It's the most curious thing I ever saw in all my life' [p. 69]; 'looking at the Cat's head with great curiosity' [p. 91]), the March Hare's clock ('Alice had been looking over his shoulder with some curiosity' [p. 74]), the cards painting the roses red ('Alice thought this a very curious thing' [p. 83]) or the 'curious croquet-ground' (p. 88). Curiosity quickly becomes, indeed, the key word guiding the little girl's steps as she encounters the inhabitants of Wonderland:

> Just as she said this, she noticed that one of the trees had a door leading right into it. 'That's very curious!' she thought. 'But everything's curious to-day. I think I may as well go in at once'. And in she went. (p. 81)

Yet, as noted above, Alice also turns into one of the curiosities of Wonderland, just like Kingsley's water-baby. When she sits at the Mad Hatter's table, the latter looks at Alice 'for some time with great curiosity' (p. 73), and when she tries to recite, 'You are old, Father William', her words prick the curiosity of the Gryphon and the Mock Turtle (p. 109). Moreover, her metamorphosis, such as when she grows in the rabbit's house, participates in her reification: Alice is defined through her arm ('Who ever saw one that size? Why, it fills the whole window!' [p. 42]), whilst her monstrosity objectifies her: 'Well, it's got no business there, at any rate: go and take it away!' (p. 43). Likewise, Alice's feeling of alienation resurfaces when she loses her shoulders (p. 56) or her feet (p. 20). Her journey in Wonderland is therefore a journey of alienation and dismemberment; her objectification becomes indissociable from the way in which her body is cut up into its constituent parts like so many different exhibits. Thus, both the subversive reversal of the natural order experienced by Alice ('being ordered about by mice and rabbits' [p. 40]) and her objectification unravel against the backdrop of museum culture. What is 'curious' in 'this sort of life' (p. 40) is Alice's entry into a live cabinet of curiosities, in which visitors and exhibits constantly change place and evolve—even words evolve, becoming 'curiouser and curiouser' (p. 20). In Carroll's live cabinet of curiosities, or disorderly zoological gardens, 'all sorts of little birds and beasts' (p. 114), 'a Duck, a Dodo, a Lory, and an Eaglet, and several other curious creatures' (p. 28) and even the Queen herself, who looks 'like a wild beast' (p. 86)—perhaps just escaped from the Surrey Zoological Gardens whose dome appears in the background of her croquet-ground,[119] as suggested—are jumbled up, mixing zoological periods and species, and reality and myth, as illustrated by the Gryphon.[120]

In addition, the stress on Alice as the object of the curious beasties' gaze lays bare the links between vision and power, thereby subverting the authority of those who generally look on animals, as when the Cheshire Cat—a 'curious' object in itself[121]—looks boldly at the King, or when the Gryphon watches the Queen until she is out of sight before chuckling and deriding her power (p. 99). The objects on display in Wonderland's zoo, be it a little girl, a queen, a king or a Cheshire cat, are thus never fixed; as when, in *Through the Looking-Glass, and What Alice Found There* (1871), Alice is puzzled by the fact that '[t]he shop seemed to be full of all manner of curious things', but 'whenever she looked hard at any shelf, to make out exactly what it had on it, that particular shelf was always quite empty, though the others round it were crowded as full as they could hold'.[122]

However, unlike Charles Kingsley's fairy tale, in which Kingsley's stance upon Victorian biology and morphology is 'a less devoted parody of the Linnaean system',[123] in both Alice books, form is untameable, 'indomitable'; the narratives construct 'a lasting burlesque of the general science of ordered form'.[124] As Thomas Richards observes, indeed,

> Logic, the Linnaean logic of form, is the only monster in the *Alice* books. The shapes the monsters assume there – cards, chessmen, cats – tend toward the domestic and the serene, not the strange and the portentous. Nothing in *Alice in Wonderland* and *Through the Looking-Glass* turns out to be more truly monstrous than the operation of logic itself, which dictates shape and configuration at every instant.[125]

[119] The Surrey pleasure gardens featured flower beds and fountains, whilst domestic and exotic plants and trees welcomed visitors. Its conservatory (300 feet in circumference and more and 6000 square feet) was, however, not only aimed at exhibiting plants (for flower shows) but also contained cages for the exotic animals then kept at the Royal Surrey Gardens, hence merging the animal and the vegetable behind its glass panels in ways very similar to Carroll's fantasy.

[120] See note 78 in Chap. 6 on Archæopteryx, which was sometimes called 'Griffosaurus'.

[121] The king looks at the cat 'with great curiosity' (p. 91).

[122] Lewis Carroll, *Through the Looking-Glass, and What Alice Found There* (1871), in *The Annotated Alice*, ed. by Martin Gardner (London: Penguin, 2001), p. 211. All subsequent references are to this edition and will be given parenthetically in the text.

[123] Thomas Richards, *The Imperial Archive: Knowledge and the Fantasy of Empire* (London and New York: Verso, 1993), p. 55.

[124] Richards, *The Imperial Archive*, p. 53.

[125] Richards, *The Imperial Archive*, p. 53.

In a world where objects, creatures and human beings constantly change shape or direction, where subjects and objects swap roles as much as predators and preys, Alice 'cannot fix the visible world in stable logical categories, and just as importantly, she cannot remember her own name', denying therefore any 'causal explanation of development'.[126] Carroll's play on categorical instability, making 'one feel as if one was asleep in a very fantastical world of animals all topsy-turvy', as Margaret Gatty puts it,[127] thus invites Alice to believe in impossible things, as the White Queen advises her.[128] In Charles Kingsley's *The Water-Babies*, similarly, Professor Ptthmllnsprts ultimately believes in 'unicorns, fire-drakes, manticoras, basilisks, amphisbœnas, griffins, phœnixes, rocs, orcs, dog-headed men, three-headed dogs, [and] three-bodied geryons' (p. 162), and so many other fabulous monsters found in Victorian museums. Carroll's cabinet of curiosities, informed as it is by Victorian museum culture, thus ultimately explores the boundaries between myth and reality, reflecting, in so doing, not so much a realm of fairy and folk tales devised for children but practices typical of Victorian scientific venues and publications. In *Through the Looking-Glass, and What Alice Found There*, moreover, Victorian museum culture invites readers even more explicitly to look at the instability of species categories through the looking-glass. It is Alice, this time, who becomes a 'monster'.

---

[126] Richards, *The Imperial Archive*, pp. 54–5.

[127] As seen in Chap. 4, Gatty was also interested in such experiments, as in her 'Knowledge Not the Limit of Belief' (1855) where 'a seaweed and a zoophyte discuss their own categorical instability—caught between taxonomies of plant and animal'; Straley, *Evolution and Imagination in Victorian Children's Literature*, p. 47.

[128] 'Now I'll give *you* something to believe. I'm just one hundred and one, five months and a day'.

'I can't believe *that!*' said Alice.

'Can't you?' the Queen said in a pitiful tone. 'Try again: draw a long breath, and shut your eyes'.

Alice laughed. 'There's no use trying', she said: 'one *can't* believe impossible things'.

'I daresay you haven't had much practice', said the Queen. 'When I was your age, I always did it for half-an-hour a day. Why, sometimes I've believed as many as six impossible things before breakfast ...' Carroll, *Through the Looking-Glass, and What Alice Found There*, p. 251.

Carroll's 'Fabulous Monsters' Through the Looking-Glass

MORE LAST WORDS
Who killed the sea-serpent?
'I', said PROFESSOR OWEN,
In Zoology so knowing;
'And I killed the sea-serpent!'

Who won't say 'die' to the serpent?
'I', said CAPTAIN McQUHAE,
'I stick to what I say;
And I won't say "die" to the Serpent'.

Why couldn't it be a serpent?
'Cause OWEN's never seen one.
Has there, therefore, never been one?
Why couldn't it be a serpent?

What was't if not a serpent?
We know it wasn't a seal,
And 'twas too big for an eel,
And it certainly *was* a serpent.

There's six of us saw the serpent,
With a mane upon its back,
And a tail and not a track;
We'll all swear to the serpent.[129]

On 6 August 1848, the captain of H. M. S. Daedalus arrived at Plymouth claiming that he had sighted a creature resembling the legendary sea-serpent. His word was instantly contradicted by the authoritative opinion of Richard Owen who believed it only to be a great seal or sea lion.[130] Oscillating as it did between the realms of reality and of legend, this mythical creature was one of the many curiosities that captivated the Victorians,

---

[129] [Anon.], 'More Last Words', *Punch*, 15 (1848): 243.

[130] Richard D. Altick, *The Shows of London* (London and Cambridge, Mass.: Belknap Press of Harvard University Press, 1978), p. 306.

as evidenced by the numerous desiccated specimens purporting to be mermaids that still circulated in shows and fairs.

Popular science works also played recurrently upon the frontier between myth and reality. Gould's *Mythical Monsters*, cited above, is a case in point. The book denied that 'chimeras' could not exist:

> For me the major part of these creatures are not chimeras but objects of rational study. The dragon, in place of being a creature evolved out of the imagination of Aryan man by the contemplation of lightning flashing through the caverns which he tenanted, as is held by some mythologists, is an animal which once lived and dragged its ponderous coils, and perhaps flew; which devastated herds, and on occasions swallowed their shepherd; which, establishing its lair in some cavern overlooking the fertile plain, spread terror and destruction around, and, protected from assault by dread or superstitious feeling, may even have been subsidised by the terror-stricken peasanty, who, failing the power to destroy it, may have preferred tethering offerings of cattle adjacent to its cavern to having it come down to seek supplies from amongst their midst.
>
> To me the specific existence of the unicorn seems not incredible, and, in fact, more probable than that theory which assigns its origin to a lunar myth.
>
> Again, believing as I do in the existence of some great undescribed inhabitant of the ocean depths, the much-derided sea-serpent, whose home seems especially to be adjacent to Norway, I recognise this monster as originating the myths of the midgard serpent which the Norse Elder Eddas have collected, this being the contrary view to that taken by mythologists, who invert the derivation, and suppose the stories current among the Norwegian fishermen to be modified versions of this important element of Norse mythology.[131]

Gould's book traces many monsters and chimeras typically found in legends, myths, folk or fairy tales to a natural origin, lifting the veil of fantasy on many 'curious beasties'. He argued, moreover, that extinct creatures could provide an explanation for dragons and giants:

> [T]he fossil remains of animals discovered from time to time, and now relegated to their true position in the zoological series, were supposed to be the genuine remains of either dragons or giants, according to the bent of the mind of the individual who stumbled on them …

---

[131] Gould, *Mythical Monsters*, pp. 2–3.

> The annexed wood-cut of the skeleton of an Iguanodon, found in a coal-mine at Bernissant [sic], exactly illustrates the semi-erect position which the dragon of fable is reported to have assumed.[132]

As Gould's popular science book illustrates, Darwinian theory not only fascinated the Victorians for suggesting that all types of 'transformations' were possible, it also implied that ever more unknown creatures might one day be found, which could then explain past myths and legends. Gould's exploration of the slippery boundary between science and myth at a time when scientific materialism claimed it could explain everything was far from original. Popular science books, just like museum guides, regularly capitalised on the relationship between science and myth, more especially perhaps when dealing with young audiences whose attention was more likely to be attracted by fairies and monsters than by 'dry facts'. Thus, in E. W. Payne's *More Pleasant Mornings at the British Museum; or, the Handy-Work of Creation. Natural History Department*, the children explore the various meanings and forms of mythical beasts as they look at the collections on display. The children's aunt hints at the impossibility of some legendary creatures, but also stresses the connections between the mythical and the real:

> 'Are not dragons fabulous creatures?', inquired Lucy.
> 'The dragon of ancient fable was described as made up of limbs never found existing in the same animal, but these lizards so far resemble those imaginary monsters, that they are distinguished by the name. Their apparent wings consist only of the skin of the sides spread out upon very long ribs, which acts like a parachute as they leap from tree to tree'.[133]

Surprisingly, however, when the children later come across the remains of a dodo, they are astonished to see merely a head and foot, along with a picture of it, instead of the 'real' bird. Their aunt informs them that:

> They are the remains of a bird called the *dodo*, said to have inhabited Madagascar, but supposed to be now extinct. The painting was done in Holland, and reputed to be from the living bird; but these remains, and the

---

[132] Gould, *Mythical Monsters*, pp. 199–200.

[133] E. W. Payne, *More Pleasant Mornings at the British Museum; or, the Handy-Work of Creation. Natural History Department* (London: The Religious Tract Society, 1856), pp. 102–3.

painting, exhibit so little correspondence of parts, that it is doubtful whether there really was such a creature.[134]

Whether the aunt means that the representation of the dodo is not faithful to the actual bird or that dodos never existed, despite the remains that the children are facing in Case 108, is interestingly hard to tell and may be confusing.[135] Nevertheless, the doubt she expresses highlights the shaky boundaries between myth and reality, making both dragons and dodos potential 'imaginary monsters'.

The way in which children's fiction and non-fiction staged the possibility for 'imaginary monsters' to become 'real monsters' resonated with contemporary scientific research on the reality of species, as already suggested. According to Thomas Richards, the scientific discoveries of the nineteenth century, in particular evolutionary theory, established 'that there was no longer any place for monstrosity within the biology of living matter'.[136] However, at the end of the century, 'a new form of monstrosity arose to outwit Darwin': monsters 'that fell outside the sureties of lineage enshrined in morphology', which 'were threats to the global claim of Darwinism'.[137] Richards starts from the premise that none of the renowned naturalists and explorers of the nineteenth century, from Richard Francis Burton (1821–90) and Henry Morton Stanley (1841–1904) to Charles Darwin, encountered monsters while travelling around the world. He argues, moreover, that Victorian literature is devoid of monsters, apart from some fabricated ones, such as Mary Shelley's creature, for instance. This absence of monsters, Richards claims, where monstrosity remains most of the time merely adjectival—the result of 'the behavioral perversions of the self'[138]—cannot be separated from contemporary studies of the natural world and its species, in particular the interest in form. Victorian

[134] Payne, *More Pleasant Mornings at the British Museum*, p. 134.

[135] Because only a head and a left leg were preserved in Oxford, as well as a leg at the British Museum, doubts as to the existence of dodos were frequent in the nineteenth century. They were dispelled when a Danish zoologist (John Theodor Reinhardt [1816–82]) discovered two other specimens in Copenhagen in seventeenth-century collections unstudied until then. Another skull was later discovered in Prague (in boxes containing seventeenth-century specimens). The discovery of many fossil bones of dodos in Mauritius in 1865—incidentally, the year Carroll's fantasy was published—put an end to beliefs that dodos might have been legendary creatures.

[136] Richards, *The Imperial Archive*, p. 48.

[137] Richards, *The Imperial Archive*, pp. 48–9.

[138] Richards, *The Imperial Archive*, p. 45.

208   L. TALAIRACH

morphology developed from Linnaeus's construction of 'an ideal taxonomy of pure forms'[139] that was capable of 'includ[ing] all known species of plants'.[140] In Richards's words, '[f]orm in Linnaeus meant taxidermy; single specimens in little boxes defined in terms of one generalized rubric, however defined'.[141] Nevertheless, in the course of the nineteenth century, 'the Linnaean metaphysics of the fixed form, the *forma formata*, gave way to the new field of the changing form, the *forma formans*'.[142] In other words, evolutionary theory enabled scientists to account for every form that ever existed or would appear; monsters would either 'disappear forever or mutate themselves into a form which eventually becomes the norm'.[143] As a result, 'all the exceptions that the old taxonomies had once relegated to the category of the monstrous could theoretically be rehabilitated'.[144]

Richards's insightful analysis of Victorian morphology, its links with taxidermy—and hence museum culture—and its influence on Victorian literature are valuable here, for they highlight the significance of Lewis Carroll's *Alice* books at mid-century—books 'about a little girl dropped into a world of monsters'.[145] Indeed, the shift from Linnaeus's *forma formata* to Darwin's *forma formans* recalls vividly Alice's experience in the shop full of curious things but which cannot be apprehended by the gaze, mentioned above, and her encounters with the shape-shifting Cheshire Cat. Like *Alice's Adventures in Wonderland, Through the Looking-Glass, and What Alice Found There* reads like a travel narrative featuring a little girl who wants 'to make a grand survey of the country she [is] going to travel through' (p. 177):

> 'It's something very like learning geography', thought Alice, as she stood on tiptoe in hopes of being able to see a little further. 'Principal rivers – there *are* none. Principal mountains – I am on the only one, but I don't think it's got any name. Principal towns – why, what *are* those creatures, making honey down there? They can't be bees – nobody ever saw bees a mile off, you know –' and for some time she stood silent, watching one of them that

---

[139] Richards, *The Imperial Archive*, p. 47.
[140] Richards, *The Imperial Archive*, pp. 45–6.
[141] Richards, *The Imperial Archive*, p. 47.
[142] Richards, *The Imperial Archive*, p. 47.
[143] Richards, *The Imperial Archive*, p. 56.
[144] Richards, *The Imperial Archive*, p. 47.
[145] Richards, *The Imperial Archive*, p. 50.

was bustling about among the flowers, poking its proboscis into them, 'just as if it was a regular bee', thought Alice.

However, this was anything but a regular bee: in fact in was an elephant – as Alice soon found out... (p. 177)

Although Laura White reads such references to large animals as characteristic of Carroll's 'comedy of scale',[146] it is important to note that Alice's attention is immediately captured by the 'creatures' that inhabit Looking-Glass world, which she attempts to place according to her knowledge of Victorian menageries. Among the curious Looking-Glass 'beasties' she encounters are elephants, making honey, creatures which 'look[–] like kangaroos' (p. 277), and imaginary lions or tigers which sound like steam trains (p. 197).[147] These either do not match the shape and size of the creatures generally exhibited in zoos and museums or come from the realm of nursery rhymes, such as the walrus or the lion, a literary world much wilder than that of Victorian menageries. As in *Alice's Adventures in Wonderland*, the narrative constantly destabilises species typologies and categorisation. The play on dreams-within-dreams, when Alice is told, for example, that she is not 'real', being only a 'thing[–]' (p. 198) in the King's dream whilst she is herself dreaming her Looking-Glass adventure, as well as the embedded texts featuring wild or 'curious beasties', intensify the slippery boundary between reality and fiction—or myth. When Alice encounters the unicorn, for instance, the nursery rhyme (and legendary)

---

[146] White, *The* Alice *Books and the Contested Ground of the Natural World*, p. 132.

[147] 'Here she checked herself in some alarm, at hearing something that sounded to her like the puffing of a large steam-engine in the wood near them, though she feared it was more likely to be a wild beast. "Are there any lions or tigers about here?" she asked timidly' (p. 237). In both Henry Kingsley's *The Boy in Grey* (1871) and Tom Hood's *From Nowhere to the North Pole. A Noah's Ark-Æcological Narrative* (1875), written in imitation of Carroll's *Alice's Adventures in Wonderland*, the sound of wild beasts is associated with the London Zoological Gardens: 'The moment he was out of sight there was such a noise as you never heard. If all the beasts in the Regent's Park were to have nothing to eat for a fortnight, there would be a fine noise, I doubt not; but it would be nothing to the noise which arose the instant the naturalist's back was turned' (Henry Kingsley, *The Boy in Grey and other Stories and Sketches* [New York: Longmans, Green & Co., 1899], p. 53); 'There was a tremendous gnashing of teeth, tossing of horns, and waving of trunks going on, to the accompaniment of various disordered noises, that painfully reminded Frank of the Zoological Gardens at feeding times'; Hood, *From Nowhere to the North Pole*, p. 97. As will be seen in Chap. 6, Edith Nesbit also mentions the Zoological Gardens in *Five Children and It* (1901).

210   L. TALAIRACH

creature, suddenly made 'real', instantly turns the little girl into a 'fabulous monster':

> 'This is a child!' Haigha replied eagerly, coming in front of Alice to introduce her, and spreading out both his hands towards her in an Anglo-Saxon attitude. 'We only found it to-day. It's as large as life, and twice as natural!'
> 'I always thought they were fabulous monsters!' said the Unicorn. 'Is it alive?' (p. 241)

Just like the Gryphon in *Alice's Adventures in Wonderland*,[148] the unicorn shapes *Through the Looking-Glass* as a travel narrative, with Alice becoming a specimen worthy of (scientific) interest, a wonder to be displayed in a cabinet of curiosities. The unicorn forcefully conveys here Carroll's discourse on Victorian museum culture: the museum ironically represents a place where categories are never fixed but always potentially evolving. As Harriet Ritvo argues, by the end of the Victorian period '"unicorn horns" had come, in the eyes of progressive modern curators, to symbolize the dark ages of museology'.[149] But unicorns 'continued to figure in zoological discussion throughout the period',[150] alongside other mythical beasts and monsters. Thus, despite new developments in natural history and the fact that Darwin's theory of evolution disregarded monsters or made them unsensational, it remained difficult throughout the nineteenth century to cast monsters aside and consider them as nothing more than dusty curios belonging to old cabinets of curiosities.

The first reason explaining the enduring presence of 'monsters' in Victorian England is that 'monsters'—freaks—were regularly seen in London and throughout the country, exhibited in travelling shows and fairs. The public wanted 'monsters', and 'monsters' were on public display well into the twentieth century.[151] Dwarfs and giants were exhibited along-

---

[148] Ronald Reichertz, *The Making of the Alice Books: Lewis Carroll's Uses of Earlier Children's Literature* (Montreal and Kingston, London, Ithaca: McGill-Queen's University Press, 1997), p. 47. Reichertz also adds that the Gryphon is 'especially associated with antipodes … Mandeville included a report on griffins in his account of his voyages, griffins so enormous that humans used their talons for drinking cups'. This link with travel literature supports the view of Carroll's narrative as a rewriting of tales of exploration.

[149] Ritvo, *The Platypus and the Mermaid*, p. 176.

[150] Ritvo, *The Platypus and the Mermaid*, p. 176.

[151] According to Ritvo, Kenneth Grahame identified the 'disappearance of freaks and monstrosities' as 'perhaps the greatest change that has taken place in show-life in our generation'; Kenneth Grahame, 'Sanger and his Times' introduction to 'Lord' George Sanger, *Seventy*

5 NONSENSE 'BEASTIES'   211

side wild beasts performing feats and human curiosities from the distant corners of the world, from 'pied Blacks' to Hottentots. 'Monsters' and curiosities ensured audiences and income for travelling shows and museums, just as they combined entertainment and instruction. Unsurprisingly, children were frequently targeted by advertisements. As Ritvo mentions, an 'early-nineteenth-century guide to the Leverian Museum, designed also as a natural history primer for children' promoted '*anatomical preparations* and monsters'.[152] The British Museum even had a department of 'Modern Curiosities' in its early years,[153] which included monstrosities, such as a lamb with two heads and a lizard with two tails, and Ritvo traces such exhibits in the collections of the British Museum in the twentieth century.[154]

However, 'monsters', whether stuffed, pickled or living, were of interest to professional scientists as well.[155] Although teratology—the science of 'monsters'—became a scientific discipline in the nineteenth century, it was still difficult for scientists to explain away some of the natural curiosities that had been the pride of Renaissance cabinets and which enlightened natural philosophers had subsequently disparaged.[156] According to

Years A Showman [1910] (London: J. M. Dent, 1926), pp. 17–18, qtd. in Ritvo, *The Platypus and the Mermaid*, p. 134.

[152] Ritvo, *The Platypus and the Mermaid*, p. 136.

[153] Ritvo, *The Platypus and the Mermaid*, p. 136.

[154] Later examples found in museum guides illustrate this point, as Ritvo underlines: 'Early-twentieth-century accounts of the British Museum (Natural history) holdings noted that among the domesticated animals was a "collection of monstrosities ... of quite recent origin" and that the equine series included several "horned horses", the significance of which was "at present inexplicable"'; R. Lydekker, 'Domesticated Animals, Hybrids, and Abnormalities', in *The History of the Collections Contained in the Natural History Departments of the British Museum*, vol II, *Separate Accounts of the Several Collections Included in the Department of Zoology* (London: British Museum, 1906), p. 73; [R. Lydekker], *Guide to the Specimens of the Horse Family (Equidae) Exhibited in the Department of Zoology, British Museum (Natural History)* (London: British Museum, 1907), pp. 9–10; qtd. in Ritvo, *The Platypus and the Mermaid*, p. 136.

[155] Ritvo mentions the case of Jacob Butler, an anatomy demonstrator at Cambridge who invited touring giants and dwarfs home to dine; R. T. Gunter, *Early Science in Cambridge* (Oxford: Oxford University Press, 1937), p. 307. She adds that preserved monsters occupied the centre of attention at scientific gatherings and that '[b]y the early eighteenth century, anthologies of monsters designed for gentlemen's libraries had become a recognizable publishing genre'; Ritvo, *The Platypus and the Mermaid*, p. 137. See also Dudley Wilson, *Signs and Portents: Monstrous Births from the Middle Ages to the Enlightenment* (London: Routledge, 1993), p. 114.

[156] Ritvo, *The Platypus and the Mermaid*, p. 138.

Ritvo, the explanation can be found in 'the difficulty of developing persuasive replacements',[157] whilst classification often remained speculative.[158] Moreover, teratologists frequently 'attempted to supplement nature's raw material by fabricating their own monsters in the laboratory'.[159] More significantly, perhaps, as she argues, the new discoveries of species coming from distant places were regularly interpreted as if they were composites,[160] continuing thereby 'a Renaissance habit of interpreting American novelties as monstrous recombinations of familiar parts, analogous to the chimaeras and yales of medieval bestiaries'.[161] The platypus, for example—a mix of bird, reptile and mammal—was therefore as much a 'fabulous monster' as Carroll's unicorn or Gryphon. Thus, whilst the general public had a taste for monstrosities and curiosities of sorts exhibited in London and throughout the country, specialists too kept the idea alive, through their interpretations, and sometimes their inventions.[162]

Consequently, by recurrently foregrounding the unreliable or misleading quality of species categories, the *Alice* books function like menageries, hosting species that defy classification. Alice herself represents the narrative's discourse on species when the Lion asks her, playing the traditional guessing game, whether she is 'animal – or vegetable – or mineral' (p. 242). Alice, as a 'fabulous monster' (p. 242), encapsulates, indeed, every possibility. The 'fabulous monster', beyond categorisation, thus cuts off all attempts at classifying or naming, recalling Alice's loss of her own name earlier on in the wood where 'things have no name' (p. 185) and when she

---

[157] Ritvo, *The Platypus and the Mermaid*, p. 140.

[158] Ritvo gives the examples of John Hunter (1728–93), Richard Owen and Johann Friedrich Meckel (1781–1833), whose different definitions of monsters and classificatory systems are a good illustration of the debates around monstrosity; Ritvo, *The Platypus and the Mermaid*, pp. 140–41.

[159] Ritvo, *The Platypus and the Mermaid*, 141. Ritvo cites Marie-Hélène Huet, *Monstrous Imagination* (Cambridge, Mass.: Harvard University Press, 1993), p. 108ff.

[160] Ritvo gives the examples of the opossum, half fox and half ape, the sloth, half bear and half ape or the armadillo, as a pig with a turtle's shell; Ritvo, *The Platypus and the Mermaid*, p. 132.

[161] Ritvo, *The Platypus and the Mermaid*, p. 132.

[162] Victorian palaeontologists' reconstructions of extinct species, often aiming at sensationalism, as typified by Richard Owen's moa birds, might be a case in point. This point will be developed in Chap. 6. This link between species categorisation and fiction/myth is particularly striking in *Alice's Adventures in Wonderland* when the Pigeon accuses Alice of 'trying to invent something', 'A likely story indeed!' (p. 76) which Alice seems to have made up to avoid being taken for a serpent.

fears she might be given another one and will have to track the 'creature' which has been given hers. The word 'creature' itself, applied as much to Alice as to the other characters she encounters, gradually becomes a mere series of uncertain signs, as the King makes explicit:

> ... 'I'll make a memorandum about her, if you like – She's a dear good creature', he repeated softly to himself, as he opened his memorandum-book. 'Do you spell "creature" with a double "e"?' (p. 240)

Thus, Looking-Glass 'creetures' and 'creatures' deceive the ear as much as the eye, as Alice constantly wonders whether she has been deluded by 'a conjuring trick' (p. 241), as she faces creatures which change shape, melt away, like the Goat's beard (p. 181) or turn into other species: the sheep 'gets more and more like a porcupine' (p. 212), the gnat is 'about the size of a chicken' (p. 181) and elephants make honey like bees. The human child, however, observed and objectified through a telescope, microscope and opera-glass and eventually named a 'fabulous monster', epitomises the collapse of species categorisation (the monster) and the annihilation of the boundary between myth and reality ('fabulous'). Alice becomes, therefore, a taxonomic anomaly, exemplifying the failure of the taxonomic system. The genuine nature of the human child is being all the more emphasised by Haigha's phrase, 'as large as life, and twice as natural!'

As argued, according to Nagai, classification is a system that 'silences nonhumans [and] which is nothing but a form of violence'.[163] As Carroll's subversion of taxonomic classification illustrates, moreover, classification is, in fact, a means of fixing the reality of species and 'is useful to the people that name them' (p. 182), not to the species themselves. This idea is exemplified in the various embeddings and reversals that punctuate the narrative. The Rocking-horse fly, which lives on sap and sawdust, the Snap-dragon-fly, which lives on frumenty and mince-pie, or the Bread-and-butter-fly, which lives on weak tea with cream in it (p. 184), which Alice encounters at the beginning of her journey (as Looking-Glass versions of the horsefly, the dragonfly and the butterfly), all embody Carroll's nonsensical discourse on naming; the 'matter-of-factness' of Looking-Glass insects challenges the arbitrariness of the link between signifier and

---

[163] Nagai, 'Counting Animals: Nonhuman Voices in Lear and Carroll', p. 124.

214   L. TALAIRACH

signified.[164] This play on the correspondence between signifier and signi-fied is further magnified by the Jabberwocky poem, where invented words give life to imaginary creatures, which turn the whole natural system on its head, from its very first stanza:

'Twas brillig, and the slithy toves
    Did gyre and gimble in the wabe:
    All mimsy were the borogoves,
    And the mome raths outgrabe. (p. 155)

The poem-within-the-text, first printed in reverse, which needs to be deciphered and which reappears several times in the narrative, like a nurs-ery rhyme refrain, constructs classification as a complex, mysterious—and almost alchemical—activity. The use of portmanteau words, as compounds of different words, just as the imaginary compound creatures, which unsettle types of diet as much as species categories, embody—literally— Carroll's nonsensical discourse on taxonomy. These include toves, a spe-cies of badger with short horns like a stag, 'something like lizards – and ... something like corkscrews' (p. 226), which make nests and live on cheese; borogoves, an extinct kind of parrot without wings, 'something like a live mop' (p. 227) which live on veal; and raths, a species of land turtle with a mouth like a shark, 'a sort of green pig' (p. 227). Thus, as an embedded piece within the main narrative, the Jabberwocky poem does not function as a reflecting device to help Alice and the reader understand the mystery of species classification. On the contrary, it mystifies classification through its use of portmanteau words and imaginary creatures, making it even more puzzling as it is printed in reverse. The text-within-the-text, peopled as it is with 'curious-looking creatures' (p. 226), is thus a literal (reversed) mirror image of Looking-Glass creatures, whose species categories are as insecure as they are changeable. Yet, its meaning must be sought in the main text and decoded by Looking-Glass creatures, such as Humpty Dumpty who sheds light on some of its words or the King who compares the White Queen to a Bandersnatch later on in the narrative (p. 240).

---

[164] Laura White also remarks that their description matches the expectations of natural his-tory, since 'each is described in terms of structure, habitat, means of locomotion, and diet'. In addition, Tenniel's illustrations 'show each insect as a natural history exhibit' since the insects 'are displayed in order of their threat'; White, *The* Alice *Books and the Contested Ground of the Natural World*, p. 197.

As puzzling when printed in mirror-writing as when not, proceeding as it does in reverse (the main text reflecting the embedded one and not the other way around), the Jabberwocky poem hosts however at its heart a 'monster'—the eponymous Jabberwocky. As already suggested, the embedded Jabberwocky, just like Alice, the human child, as a 'fabulous monster', evokes the failure of the classification system to accommodate all species. But as it symbolises mystification, the monstrous Jabberwocky lays bare Carroll's mad cabinet of curiosities as a mere literary creation, an association of signifiers and of literary pieces capable of engendering ever new creatures. In doing so, the monster illuminates what Jean-Jacques Lecercle sees as central to nonsense: 'a process ... of reflexivity'.[165] As a result, Carroll's presentation of classification through an embedded poem in which arbitrarily associated signifiers create 'curious beasties' illustrates the way in which classification, as a narrative—or myth—supersedes the materiality of the exhibited objects themselves. Indeed, the curious creatures begotten by literary creativity, in the main text and in the embedded ones (the unicorn, extracted from a nursery rhyme, or the Jabberwocky in the poem), explore the link between materiality and meaning. Susan Stewart's study of the collection as an 'object[–] generated by means of narrative'[166] or as 'a form involving the reframing of objects within a world of attention and manipulation of context'[167] here matches significantly Carroll's use of nonsense.

Indeed, the metatextual quality of *Through the Looking-Glass*, with Alice reading books or holding pens, and characters whose names are embroidered on their collars (Tweedledum and Tweedledee) or written all over their faces (Humpty Dumpty), as well as the recurrent reflection upon naming ('Must a name mean something?' [p. 219], Alice asks Humpty Dumpty), offers a monstrous image of the making of collections: beings become mere signs whilst signs in turn are capable of producing new creatures. Mastering words, especially those with 'a temper' (p. 224), as Humpty Dumpty explains, might prove more difficult than taming such beasts as flamingos and hedgehogs to play croquet with the Queen. Linked to wax-works (like Tweedledum and Tweedledee [p. 189]) or taken for 'stuffed figure[s]' (p. 218) (like Humpty Dumpty), Looking-Glass

---

[165] Lecercle, *Philosophy of Nonsense*, p. 2.

[166] Susan Stewart, *On Longing: Narratives of the Miniature, the Gigantic, the Souvenir, the Collection* (Durham and London: Duke University Press, 1993), p. xii.

[167] Stewart, *On Longing*, p. 151.

creatures, many of them from the literary world of nursery rhymes, as suggested, make up a veritable museum collection of literary objects, an odd assemblage of creatures much wilder than those found at the London Zoological Gardens. They do not simply challenge the established order of nature, but the very way in which it is ordered.

'Why, the Taxidermy'

Thus, Lear's and Carroll's nonsense, which both draws heavily on and recurrently satirises earlier children's literature, revisits in sensational form the birds and beasts found in eighteenth-century didactic literature, as indeed does Alice when she tries to recite Isaac Watts's 'Against Idleness and Mischief'. Their composite creatures certainly hint at the issue of classification in a period obsessed with evolution and taxonomic practices, turning them upside down to entertain their readers. Literally *substantiating*—that is, giving substance to—what cannot be grasped nor represented, like Charles Kingsley's water-baby, whilst questioning the reality of known species, including human children, Lear's and Carroll's fantasies explore—and shake—the foundations of the museum culture of their day. Whilst Thomas Hood's *From Nowhere to the North Pole*, quoted in the introduction to this chapter, turned 'the Taxidermy' and 'the British Museum' into tropes evoking devilish practices, Lear's and Carroll's play with words and meanings deconstruct, for their part, the discourses shaping museum specimens; reconstructing in this way the story of imperial Britain as a tale of violence and silencing. Nevertheless, neither Lear nor Carroll attempt to use their nonsensical composites to convey 'the nonsensicality of evolution'[168] and promote divine design or natural theology. This is, however, not the case in Hood's fantasy, where Tom's adventures in Quadrupemia rewrite Alice's adventures in Wonderland through the story of a little boy who fears being made a museum specimen, a work that often reads like a defence of natural theology.[169] As suggested above, moreover, Lear's stance on museum culture is often intertwined with the threat of extinction, as in the example of 'The History of the Seven

[168] Straley, *Evolution and Imagination in Victorian Children's Literature*, p. 91.

[169] Jessica Straley argues that 'Hood's translation of *Wonderland*-like scenarios into defenses of natural theology is not unique'; another example is Albert and George Gresswell's *The Wonderland of Evolution* (1884); Straley, *Evolution and Imagination in Victorian Children's Literature*, p. 91.

Families of the Lake Pipple-Popple'. As Laura White argues, Carroll also 'makes a joke of extinction' through his Bread-and-Butter-fly, a creature who 'could never have had a supply of foodstuffs, even though it is *itself* a foodstuff', since 'its diet does not include itself'.[170] The absurdity of the hybrid creature is therefore due to its inability to adapt. In the following chapter, it will be seen that extinction and non-adaptation featured prominently in both Victorian museums and in children's literature in the long nineteenth century. The reconstruction of curious fossilised creatures produced other types of monstrous creations, and the taste for monstrosities[171] on the part of both amateurs and professionals permeated a whole range of scientific disciplines. Even dragons became creatures which could be seen in museums—or kept by professors, as Edith Nesbit suggests in 'The Seven Dragons', when the children return home after their adventure:

> Mother said: 'Oh, my darlings, my darlings, you're safe – you're safe! You naughty children – how could you be so disobedient? Go to bed at once!'
> And their father the doctor said:—
> 'I wish I had known what you were going to do! I should have liked to preserve a specimen. I threw away the one I got out of Effie's eye. I intended to get a more perfect specimen. I did not anticipate this immediate extinction of the species'.
> The professor said nothing, but he rubbed his hands. He had kept his specimen – the one the size of an earwig that he gave Harry half a crown for – and he has it to this day.
> You must get him to show it to you![172]

---

[170] White, *The* Alice *Books and the Contested Ground of the Natural World*, p. 199. White reads *Through the Looking-Glass, and What Alice Found There* as 'a textbook case of animal extinction'; p. 198.

[171] Ritvo, *The Platypus and the Mermaid*, pp. 147–8.

[172] Edith Nesbit, 'The Seven Dragons III. The Deliverers of their Country', *The Strand Magazine*, 17.101 (June 1899): 586–94, p. 594. In the illustration, the dragon captures an elephant from the London Zoo. The references to the creature from the London Zoological Gardens once again strengthen the connection between myth and reality.

CHAPTER 6

Prehistoric 'Beasties'

The Dean of Hereford ... should be glad to see those models multiplied at a price which would enable them to be introduced into village and ordinary schools, as every one could not visit the Crystal Palace, and he therefore hoped that specimens like those before them might be rendered attainable by those in remote and secluded districts, who would not have the advantage of witnessing the splendid and gigantic illustration of the extinct creation of the early ages of the world which would be there exhibited. He would also express a hope that specimens like those might be introduced in connection with the approaching Educational Exhibition of the Society, as it would be of great importance in an educational point of view, and schoolmasters and teachers of the young might thereby have an opportunity of seeing what had been done, and what they might do for themselves, if they could purchase those models at a moderate price. It would be doing for the extinct world what they had not done for the existing one, because, in many of the rural districts the young were taught the nature and habits of elephants, lions, and tigers, and yet had never seen either a living specimen or even an accurate model of those animals.[1]

In 1854, Victorian educationalist and Dean of Hereford, Richard Dawes (1793–1867), recommended the use of models of extinct creatures in

[1] [Anon.], 'Education Exhibition', *Journal of the Society of Arts*, 78.2 (19 May 1854): 443–9, p. 447.

© The Author(s), under exclusive license to Springer Nature Switzerland AG 2021
L. Talairach, *Animals, Museum Culture and Children's Literature in Nineteenth-Century Britain*, Palgrave Studies in Animals and Literature, https://doi.org/10.1007/978-3-030-72527-3_6

219

schools as teaching aids to give children who had never been to the zoological gardens the opportunity to get a sense, at least, of what prehistoric 'beasties' looked like: '[P]ut in a form in which they would be easily attainable by those engaged in education', with 'some lighter and less brittle material … – papier-maché, or some such material',[2] the extinct 'beasties' might then help 'those engaged in education'[3] to present the creatures as a prehistoric equivalent of a Noah's Ark bought from the toyshop.

As palaeontology and geology became increasingly popular during the nineteenth century, the idea that some creatures had become extinct both fascinated and intrigued. The frequent collaborations between palaeontologists and artists ensured the sensationalisation of the emerging scientific disciplines. But the Crystal Palace Gardens and its reconstructions of a number of distinct ecosystems brought prehistory to life even more strikingly, inviting museum-goers to participate by placing themselves in the midst of extinct creatures. When the Crystal Palace was re-opened at Sydenham by Queen Victoria and Prince Albert on 10 June 1854, 40,000 spectators discovered for the first time a collection of full-sized models of dinosaurs and other extinct animals displayed according to geological periods. Commissioned from the London sculptor and illustrator Benjamin Waterhouse Hawkins (1807–94) and supervised by the anatomist and palaeontologist Richard Owen (1804–92), the reconstructions made the Crystal Palace Park highly popular. The dinosaurs were visually impressive, and their relocation within a natural habitat reflected the cultural urge of the period to show people past ecosystems. As this chapter will argue, mid-Victorian and Edwardian children's literature frequently alluded to recent palaeontological finds and the methods used to analyse fossil discoveries. In addition, contemporary books for children referred time and again to three-dimensional models of extinct creatures that were being exhibited in Britain at that time, especially the Sydenham 'beasties'. Poised between fantasy and popular science, these narratives designed for a young audience illuminate the impact of museum culture on the trajectory of children's literature, especially when the exhibits promised their readers a journey both in space and time.

[2] Papier-mâché models, such as those made by the French anatomist Louis Auzoux (1797–1880), were at the time particularly used to teach anatomy.

[3] [Anon.], 'Education Exhibition', p. 448.

The Crystal Palace Gardens and the Popularisation of Extinction

The world before the flood. Yes. Ages, perhaps, before Noah's ark was built, or launched, or thought of. In this shed the Triton known as Waterhouse Hawkins has conspired with the King of Animals, Professor Owen, to bring back those antediluvian days where there were giants in the land. Pre-Adamite, perhaps; pre-Noahite, certainly. Modelled according to some subtle theory, admirably carried out into practice; the marvels of what we may call scientific art – plasticity applied to comparative anatomy – are the great monsters and reptiles of the fossil world. The ichthyosaurus, the plesiosaurus, the megatherium, the mastodon, igua-arneton; gigantic creatures of lizard, toadlike, froglike, beastlike form grin at you, crawl at you, wind their hideous tails round you. … All these antediluvian monsters, which will finally be executed in a composition as hard as stone, will be placed on the shores of two artificial islands in the lake; one exhibiting the secondary, the other the tertiary epochs of the world. There, among reeds and slime, the great fish lizards crawl, and higher up the great Irish elk reposes. All of which is explained to us in a little studio, where sepia sketches of elks and mastodon, and megatheria mingle with clay sketch models and casts of skulls and femurs of fossil mammalian and reptiles.[4]

When Charles Dickens described the Crystal Palace at Sydenham in December 1853, his presentation of Benjamin Waterhouse Hawkins's dinosaur park highlighted the union of science and art that the park epitomised, emphasising the magical power of comparative anatomy—the scientific method which enabled scientists to recreate extinct creatures from bones or fragments of bones. The Crystal Palace was not only a symbol of recent advances in science and technology; it also displayed the idea of progress through exhibits showing visitors the evolution of humankind from ancient times to modern civilisations. In the Natural History Department, plaster models of 'savages' from Africa, India and the Americas were on display within their natural habitat, aligned with exotic animals. The visit continued in the gardens, where visitors could encounter models of creatures from ancient times. The display was thus intended as a journey through space and time, and the place as a whole was advertised as a museum aimed at visitors of all classes. As Harriet Martineau pointed out, 'the People's Palace … is the token of one phase of the

---

[4] Charles Dickens, 'Fairyland in fifty-four', *Household Words*, 193 (3 December 1853): 313–17, p. 317.

people's trouble being past, and of a beneficent change having come over their taste and capacity for amusement'.[5] Visually impressive, the Crystal Palace Gardens were aimed at the 'recreation' of 'the multitude',[6] and exemplified how the popularisation of science in the second half of the nineteenth century merged amusement and instruction. The whole project was designed to turn people 'as ignorant as a Hottentot', as Dickens puts it,[7] into educated citizens, and it is thus no coincidence that this open-air museum appeared in many a book for children,[8] or in caricatures mocking contemporary educational principles, as illustrated by *Punch*'s 'A Visit to the Antediluvian Reptiles at Sydenham—Master Tom Strongly Objects to Having His Mind Improved' (1855). As Dickens's reference to the Hottentot makes explicit, moreover, education goes hand in hand with the march of progress. Indeed, the whole Crystal Palace project offered a striking illustration of human evolution through the ages, with the dinosaur park—as the project's beating heart—representing in striking form how dramatically the earth had been transformed over very long periods of time.

The Crystal Palace dinosaur park illustrated the trend in popular science towards the gigantic and the sensational. For Elizabeth Eastlake, author of a review published in 1855, the prehistoric exhibits matched contemporary methods of popularisation by appealing to the visitors' 'love of the marvellous' whilst at the same time providing a valuable lesson in geology:

> One feature however is novel here, which will be sure to attract even the laziest walker to the uttermost end of this ample domain. It is easy to foresee that the extinct animals which occupy the islands on the lowest lake will be permanent favourites with the public, not only because the rising generation may be better instructed in the science that has established their existence, but because they appeal strongly to that love of the marvellous which mankind, it is to be hoped, will never be too wise to indulge. This is one of the most successful hits of the Company, for ignorance and knowledge will be alike gratified here. And in this instance too, the unfinished state of the

[5] Harriet Martineau, 'The Crystal Palace', *Westminster Review*, 6 (1854): 534–50, p. 535.
[6] Martineau, 'The Crystal Palace', p. 537.
[7] Dickens, 'Fairyland in fifty-four', p. 313.
[8] Books such as George Samuel Measom's *The Crystal Palace Alphabet: A Guide for Good Children* (London: Dean & Son, 1855) appeared quickly after the opening of the Crystal Palace Gardens.

scenery about them adds a temporary effect. Doubtless an appropriate class of vegetation is intended to spring up around, and the time will come when sedges and rushes, and overgrown waters, will increase the probability of these swan-like reptiles and magnified frogs and toads; but, as they now stand, they harmonise singularly with the features in their immediate gaping caverns, and upturned geological strata which suggest the idea of a world that is extinct too.[9]

Eastlake's article underlines the idea that the site was first and foremost designed to entertain, using Benjamin Waterhouse Hawkins's artistic skills—or magic wand—to turn reptiles, frogs and toads into attractive creatures likely to prick the viewers' interest. Moreover, the reasons why such prehistoric creatures had vanished from the earth seemed to matter less than the way in which the exhibits were intended to strike viewers with the reality of extinction itself.[10] Eastlake's remarks may be linked to the way in which the dinosaur reconstructions were embedded within scientific controversies, pointing particularly to the tensions between science and religion which emerged as various theories of evolution were being debated. In the decades that preceded and followed the publication of Charles Darwin's *On the Origin of Species* (1859), Scripture was still used to explain natural history, mostly because many science popularisers were

[9] [Elizabeth Eastlake], 'The Crystal Palace', *Quarterly Review*, 192 (March 1855): 303–55, pp. 352–3.

[10] The life-sized sculptures of extinct animals served of course to feed debates of the period about the evolution of species and the potential causes of their extinction: the creatures on display were accompanied by guidebooks, such as Routledge's, explaining that the monsters had not been drowned in the Noachian Flood but had become extinct at different geological periods: 'What vast changes must have occurred, not only in the climate, but in the character of this country, since these uncouth monsters held undisputed sway over the rank herbage and wild prairies which existed ages before the vast deposits of chalk and oolite were formed! Does the visitor seek to know by what means these vast changes were brought about, and these tremendous extinctions of successive creations effected? Geological science can afford no satisfactory answer; and the stone tablets of creation, the sepulchres of successive formations, afford no guide to the solution of the interesting question. Entombed in the Wealden lies many a great monster, whose roar awoke many an untried echo, and whose tread shook the sedgy and half-formed earth; but you search in vain for the record of the means by which the various giant races became extinct'; G. Routledge, *Routledge's Guide to the Crystal Palace and Park at Sydenham: with descriptions of the principal works of science and art, and of the terraces, fountains, geological formations and restoration of extinct animals therein exhibited* (London: G. Routledge, 1854), p. 196. We will see further on how Richard Owen conceived the site however to counter Darwin's theory of evolution.

Anglican parsons. In addition, the most renowned geologists (or mineralogists) and palaeontologists of the age, men like Adam Sedgwick (1785–1873), William Buckland (1784–1856) and William Whewell (1794–1866), were also clergymen-scientists for whom natural science and Christian theology were in no way incompatible.[11] As a result, although science was becoming professionalised and increasingly naturalistic, with young professionals, such as T. H. Huxley (1825–95) and John Tyndall (1820–93), seeking to establish a clear distinction between science, religious dogma and natural theology, it was nonetheless often filtered, rewritten and offered to the British reading public by Anglican theologians who attempted to reconcile the findings of contemporary science with the religious ideals of the Anglican Church. This was perceptible in many of the reviews of the time, as the writers refused to connect the extinct creatures exhibited in the Crystal Palace Gardens with the history of the earth and of humankind on earth. For example, as Harriet Martineau seems to suggest, whether the creatures had become extinct because of natural processes or because they were 'too large to go into the ark' was not the main purpose of the exhibition. What mattered, rather, was to stress the gap between past ecosystems and the modern civilisations of the present—between the iguanodon and the camelopard or the hylæosaurus and the porcupine:

> [The school-children, the artisans and tradespeople] are disposed to visit the antediluvian region at the bottom of the garden. They have their own conceptions of the Flood, and, therefore, an interest in seeing what sort of creatures lived, as they have been told, before that event. 'What *are* those?' exclaimed a passenger in the railway carriage, as it ran along the embankment above the gardens. 'The antediluvian animals, to be sure', a comrade informed him. 'Why antediluvian?' 'Because they were too large to go into the ark; and so they were all drowned'. Such was the explanation. Whole classes of people go to see those monsters, every one provided with more or less a theory, after the manner of this railway passenger: but how much better to go with an untenable theory than with no ideas whatever! The mere rustic, if he finds his way here, looks at the iguanodons as he did at the camelopard just now: and he stares at the spines of the hylæosaurus as he will presently at those of the great porcupine from Central America. He knows nothing of the vast scenery of ages disclosed to the reverted eye of science:

---

[11] Bernard Lightman, *Victorian Popularizers of Science: Designing Nature for New Audiences* (Chicago and London: The University of Chicago Press, 2007), p. 40.

but the man who separates the world before the flood from the present, however erroneously, has a prodigious advantage over him.[12]

The displays offered to visitors collapsed space and time, as visitors could walk from the exotic exhibits of the Natural History Department to the dinosaur reconstructions. What Martineau implies, therefore, is that by positioning the dinosaur park at the end of the gardens, following the visit of the Natural History Department and its stuffed animals and its plaster natives, the visitor could effortlessly gain a sense of the evolution of life on earth and the connections between exotic contemporary species and the monstrous creatures of the past. In fact, the sensational originality of the Crystal Palace dinosaur park was above all the radical break in museology offered by its reconstructions. Instead of visualising prehistoric monsters through skeletons or in illustrations, the full-sized creatures now enabled visitors to encounter something resembling a prehistoric zoo:

> Hitherto we have been accustomed to wonder at or study these monsters of the old world, either in pictures of a small scale, illustrating the descriptions of writers on the subject; or in museums where a vast fragmentary or an almost complete fossil skeleton, gave us a vague idea of these predecessors of the living family of nature. But at Sydenham we are not to be contented with either pictures or dry bones.
>
> On one of these islands will be placed, in natural attitudes, and amid appropriate vegetation, animals of the secondary, and others of the tertiary period; while opposite to each will be full-proportioned representations of the strata in which the remains of these vast beasts were found. To add to the illusion, the waters of the pool will rise and fall, partially submerging the amphibious inmates from three feet to eight feet alternately, during the playing of the great waters, after the manner of an actual tide. Thus, then we shall see, pausing among the rushes, the Iguanodon, or monstrous lizards, thirty feet high, and a hundred feet from snout to tip of tail. The Megatherium or monster Sloth, will appear in the act of climbing an antediluvian tree; huge Chelonians are to bask upon its banks. The Plesiosaur, with its reptile form and bird-like neck, will wallow in the mud; while the Brobdignagian grandfather of turtles, gaping, shall frighten aldermen with ideas of retribution in its monstrous jaws.[13]

[12] Martineau, 'The Crystal Palace', p. 540.
[13] [Anon.], 'The Crystal Palace at Sydenham', *Illustrated London News*, 23 (31 Dec. 1853): 599.

The reconstruction and exhibition of three-dimensional extinct creatures within their habitat ('amid appropriate vegetation', 'an antediluvian tree')—with the exception of the anachronistic background foliage—and the visualisation of the various geological epochs ('full-proportioned representations of the strata') aimed at helping visitors conceive the ages of the earth, thereby physically experiencing the position of humans in terms of geological time.

This sensational—'Frankensteinic'[14]—educational spectacle was certainly a striking example of the growing impact of visual education in the popularisation of science, most especially palaeontology and geology, which demanded imaginary reconstitution.[15] The new theories of visual education, like those of the Swiss educationalist Johann Heinrich Pestalozzi (1746–1827), which argued that knowledge had to be conveyed directly through the senses, informed the Crystal Palace project. As Waterhouse Hawkins told the Society of Arts in 1854, his master-sculptures were intended to convey knowledge directly through the eye. In his lecture, entitled 'On Visual Education as Applied to Geology', given on 27 May 1854, Hawkins contended that three-dimensional models were an essential component of a new form of education the Crystal Palace Gardens aimed to embody: 'we shall reverse that order of teaching which is described as the names and not the things with which we become acquainted; it will be the things with their names that we shall present to the people'.[16] For M. Digby Wyatt, relying simply on the organs of sight, the visitor needed 'little more than the natural instincts of a child'.[17] This new teaching regime appeared in many reviews of the time. The sudden visualisation of what had been so far unseen was the most striking feature of the park, Routledge insisted, in one of the official guides to the park:

[14] Ralph O'Connor, *The Earth on Show: Fossils and the Poetics of Popular Science, 1802–1856* (Chicago and London: The University of Chicago Press, 2007), p. 279.

[15] O'Connor, *The Earth on Show*, p. 285.

[16] Benjamin Waterhouse Hawkins, *On Visual Education as Applied to Geology, Illustrated by Diagrams and Models of the Geological Restorations at the Crystal Palace* (London: W. Trounce, 1854), qtd. in 'Education Exhibition', *Journal of the Society of Arts*, 78.2 (19 May 1854): 443–9, 444–7, p. 444. This passage is also quoted in Steven McCarthy, *The Crystal Palace Dinosaurs: The Story of the World's First Prehistoric Sculptures* (London: The Crystal Palace Foundation, 1994), p. 89.

[17] M. Digby Wyatt, *Views of the Crystal Palace and Park Sydenham* (Crystal Palace: Day and Son, 1854), p. 10.

6 PREHISTORIC 'BEASTIES' **227**

To attempt to produce, for the instruction of the world, representations of those extinct species which existed in our globe thousands of ages before the birth of man, was a proceeding not less bold than novel on the part of the Directors of the Crystal Palace. Scientific men who had devoted a long life to the accumulation and study of fossil remains, who had put together the skeletons of these gigantic monsters, and seen them in imagination roaming over the pathless forests of our island, had never yet beheld the entire animal reproduced before them; – geologists were for the first time to gaze upon the fruit of their industry, and the results of their science; – to pronounce their verdict upon the truthfulness of the portraits, and to state how far they accorded with their preconceived notions. The opinion of Professor Owen has been given, and it fully justifies the directors for the boldness with which they entrusted to Mr. Waterhouse Hawkins – a gentleman who, in the most remarkable manner, combines in his own person an intimate acquaintance with geology, a profound knowledge of anatomy, and an intense love of art – the construction of these extinct animals.[18]

Hawkins himself promoted his restorations, arguing that the exhibition filled a gap in museology, enabling people to visualise what the fragmentary evidence collected and exhibited in museums had up until then allowed people only to imagine:

We have public museums, it is true, but even our national collection at the British Museum, though containing some of the finest fossils that have been collected throughout the world, from their detached state, there being only two or three skeletons for comparison, offers little more than objects of wonder, literally only dry bones or oddly-shaped stones to the majority who see them. The inevitably fragmentary state of such specimens of course left much to the imagination, even to those who looked at them with some little knowledge of comparative anatomy, and as that amount of knowledge is not found among the average acquirements of the public at large, it was a fallow field, which nothing less than the great enterprise and resources of the Crystal Palace Company could have attempted for the first time to illustrate and realise – the revivifying of the ancient world – to call up from the abyss of time and from the depths of the earth, those vast forms and gigantic beasts which the Almighty Creator designed with fitness to inhabit and precede us in possession of this part of the earth called Great Britain.[19]

---

[18] Routledge, *Routledge's Guide to the Crystal Palace and Park at Sydenham*, pp. 185–6.
[19] Hawkins, 'On Visual Education as Applied to Geology', p. 444.

228   L. TALAIRACH

Thus, the Crystal Palace Gardens' three-dimensional models illustrated a new trend in the popularisation of science through a 'visual language' and 'working objects';[20] the whole site 'designed as a visual treat, a school for the eye'.[21] Furthermore, in many of the articles, the reviewers not only praised the sculptor, whose work had successfully resurrected extinct animals, but also explained the scientific method that had enabled the resurrection itself. This fusion between science and art revealed the magical power of Victorian science, able to unlock the mysteries of the past and lift the veil from nature's secrets. The report of Owen's speech on the night of the dinner in the iguanodon's mould, published in *The Illustrated London News*, emphasised the extent to which the creatures were not so much the product of an artist's genius but of a new type of scientific reasoning, informed by functionalist approaches and interpretations. Art merely 'illustrate[d] [science's] wonderful truths', as the reviewer of the *Illustrated London News* put it.[22] For Routledge,

[m]any of the spectators who for the first time gaze upon these uncouth forms, will be disposed to seek for the authority upon which these antediluvian creatures have been constructed, and to ask if it be possible from a few scattered fossil bones to construct an entire animal. A few years since, the answer to such a question would have involved something of doubt and uncertainty. Now, however, in matters connected with geology, as in many other sciences, brilliant hypotheses, and speculative ingenuity, have given place to well established facts, and clear, well supported deductions from them. The key with which Professor Owen can decipher the records of geological ages, is as certain and as infallible as that with which a Champollion or a Rawlinson can unlock the mysteries of Egyptian or Assyrian inscriptions. The difficulties and seeming contradictions which in its earlier days attached to geology as science have disappeared, and if some of their retiring shadows still darken the horizon, they will, as the day advances, one by one melt away in the increasing light of science.[23]

[20] Martin J. S. Rudwick, 'The Emergence of a Visual Language for Geological Science, 1760–1840', *History of Science*, 14 (1976): 149–95; Lorraine Daston and Peter Galison, 'The Image of Objectivity', *Representations*, 40 (1992): 81–128; Soraya De Chadarevian and Nick Hopwood (eds), *Models: The Third Dimension of Science* (Stanford: Stanford University Press, 2004), p. 2.

[21] James A. Secord, 'Monsters at the Crystal Palace', in Chadarevian and Hopwood (eds), *Models*, pp. 138–69, p. 140.

[22] [Anon.], 'The Crystal Palace at Sydenham', *Illustrated London News* (7 Jan. 1854): 22.

[23] Routledge, *Routledge's Guide to the Crystal Palace and Park at Sydenham*, p. 186.

The palaeontologist was thus both master detective and speculative genius, aligned with earlier decoders of unknown languages, a Champollion for the modern age.[24] Consequently, at a time when scientists were desperately trying to find a language in which to express new scientific realities—such as evolution[25]—the fragmented beasts that palaeontologists were excavating, resurrecting and trying to reconstruct were clearly associated with the search for a language, poised between reality and imagination, the real and the unreal, the natural and the supernatural. Hawkins's Crystal Palace creatures illustrated how scientists and showmen, more generally, used technologies of display to 'play artfully with the boundary between the real and the unreal'.[26] Victorian science did not simply attempt to highlight the wonders of nature; it also sought to push the limits of the real and disrupt the boundaries between the possible and the impossible through its culture of display. As a consequence, as Iwan Morus points out, because science was exhibited, inviting its audiences to experience knowledge through the illusions of modern conjurors handling the latest technology, it contributed to emphasising the Victorian 'culture of wonder'.[27] The process was similar in museums, in which the objects on display aimed to elicit emotional responses in visitors. Fossils, remnants from remote times, whose size evoked wonder, 'shifted the displays from the real toward the fantastic',[28] as Samuel Alberti has shown. By suggesting that science could not be conceived or experienced on a purely rational level, prehistoric displays of extinct creatures thus bound the world of science to the realm of fantasy, fairy tales and, inevitably, to childhood, more especially so when visitors could travel back in time to walk alongside 'curious beasties'.

[24] For more on the links between palaeontologists and detectives, see Lawrence Frank, *Victorian Detective Fiction and the Nature of Evidence: The Scientific Investigations of Poe, Dickens, and Doyle* (London: Palgrave Macmillan, [2003] 2009).

[25] See Gillian Beer, *Darwin's Plots: Evolutionary Narrative in Darwin, George Eliot and Nineteenth-Century Fiction* (Cambridge: Cambridge University Press, [1983] 2000).

[26] Iwan Rhys Morus, '"More the Aspect of Magic than Anything Natural": The Philosophy of Demonstration', in Aileen Fyfe and Bernard Lightman (eds), *Science in the Marketplace: Nineteenth-Century Sites and Experiences* (Chicago and London: The University of Chicago Press, 2007), pp. 336–70, p. 339.

[27] Morus, 'More the Aspect of Magic than Anything Natural', 366.

[28] Samuel J. M. M. Alberti, 'The Museum Affect: Visiting Collections of Anatomy and Natural History', in Fyfe and Lightman (eds), *Science in the Marketplace*, pp. 371–403, p. 389.

The 'Beasties' on Noah's Ark and Hawkins's 'Antediluvian Dragons'

As Martineau reports humorously in the following passage, Hawkins's reconstructed creatures were sometimes believed to be real creatures, stuffed beasts identical to the exotic species on display in the Crystal Palace Natural History Department:

> What is wanted is – more knowledge still. The visitor will observe with pain that the extinct animals have already been roughly handled by strangers – whole rows of their teeth having been pulled out. From the way in which the theft has been more than once seen to be perpetrated, there is reason to believe that the culprits suppose they are carrying away the real teeth of preserved animals. A purloiner here and there, caught in the fact, has appeared so superior to the act of vulgar mischief, that it may be hoped that, if he knew that these teeth were iron, cast for the purpose, after a design requiring science and skill, he would never have done the deed. Meantime, there are the strange creatures – some of them nearly toothless; and, in that fact, very eloquent as to our need of educational exhibitions like this.[29]

The creatures were monstrous and gigantic, yet nonetheless appeared realistic and were placed 'in natural attitudes'. As James Secord explains, 'truth to life' was a key phrase as viewers 'expected realistic poses, balanced proportions and vivid expressions', and the '[a]nimal sculptures had to give the appearance of inner vitality, which created a bond of sympathetic perception', as in zoos.[30] The comparison with zoos is a significant one, because Hawkins, just like Edward Lear, as seen in Chap. 5, had devoted himself to natural history after 1827 and illustrated some specimens of the Earl of Derby's menagerie at Knowsley Park between 1842 and 1847.[31] At the same time, however, the place was 'a dream realized' and the prehistoric reconstructions had 'an unreality about themselves'[32] which reflected how Hawkins's creatures had been inspired by artists, such as John Martin.[33] This may explain why Victorian cartoonists frequently exagger-

[29] Martineau, 'The Crystal Palace', pp. 540–41.

[30] Secord, 'Monsters at the Crystal Palace', p. 147.

[31] Secord, 'Monsters at the Crystal Palace', p. 147.

[32] [John Lindley], 'The Crystal Palace Gardens', *The Athenaeum* (1854): 780.

[33] One of the most famous illustrators of extinct worlds and creatures was certainly John Martin (1789–1854), known for his impressive representations of scenes from prehuman past. His style dramatised and sensationalised death and violence; men were threatened and

ated[34] the gigantic size of Hawkins's extinct creatures, reshaping them as

dwarfed by natural elements, as in *The Deluge* (1826, mezzotint 1828). As shown by 'The Country of the Iguanodon', Martin dramatically transformed the pastoral atmosphere of earlier lithographs, such as Henry de la Beche's *Duria Antiquior* (1830), offering instead a 'nightmarish "Gothick" melodrama'; Martin Rudwick, *Scenes from Deep Time: Early Pictorial Representations of the Prehistoric World* (Chicago and London: University of Chicago Press, [1992] 1995), p. 80. Martin's regard for the sensational frequently undermined the scientific accuracy of his representations. For instance, in 'The Country of the Iguanodon', the animals are 'derived more from the long artistic tradition represented by innumerable paintings of "Saint George and the Dragon"' (Rudwick, *Scenes from Deep Time*, p. 80). The turtles, cycads and ammonite shells in the foreground serve but to enhance the gigantic size of the creatures. Martin's 'The Sea-Dragons as They Lived', which was the frontispiece of Thomas Hawkins's *Book of the Great Sea-Dragons, Ichtyosauri and Plesiosauri, Gedolim Taninim, of Moses. Extinct Monsters of the Ancient Earth* (1840), moved further away from earlier more realistic reconstructions. The pterodactyl's body and wings or the forked tongue of the plesiosaurus are intended more to strike the reader's attention and suggest the venomous and dangerous power of the creatures than convey palaeontological information.

[34] As suggested in the previous note, scientific inaccuracy had from the early nineteenth century marked the popularisation of geology and palaeontology, as in Thomas Milner's popular book, *Gallery of Nature* (1846), which displays creatures from different periods cohabiting together, illustrating, in Rudwick's words, 'the way that all the periods of earth history were still run together, at least imaginatively, in the public eye' (Rudwick, *Scenes from Deep Time*, p. 92). Apart from Henry de la Beche (1796–1855), who was a scientist gifted with artistic skills, the reconstructions which illustrated palaeontological works throughout the nineteenth century were mainly done by professional artists with little scientific training. Some were landscape painters, others book illustrators and very few scientists gave them a prominent role in their works, with the exception of Joshua Trimmer (1795–1857), Gideon Algernon Mantell (1790–1852) and Franz Unger (1800–1870) (Rudwick, *Scenes from Deep Time*, p. 133). Such scientific inaccuracies were nevertheless not merely due to the artist's lack of knowledge in palaeontology or geology; they were sometimes deliberately sensational. Mantell was, for instance, as much a palaeontologist as a populariser of science, and was thus perfectly aware of the selling potential of Martin's romantic or 'Gothick' representations. The illustrations and the increasing visualisation of the threatening extinct creatures were also designed to promote palaeontology as a new scientific discipline. Scientific inaccuracies persisted into the 1850s and 1860s in popular works; Martin's dinosaurs ('The Age of Reptiles', frontispiece of the first edition of George Richardson's *Geology for Beginners* [1842]) reappearing with their forked tongues as in W. F. A. Zimmerman's (a pseudonym of the German writer W. F. Volliner) *The Wonders of the Primitive World* (*Die Wunder der Urwelt*, 1855), and pterodactyls were even drawn as bat-like creatures. For more on scientific inaccuracies in palaeontological illustrations, see Jane P. Davidson, *A History of Paleontology Illustration* (Bloomington and Indianapolis: Indiana University Press, 2008). Within museums themselves, the reconstructions of prehistoric monsters were also often scientifically inaccurate, such as Albert Koch's 'Antediluvian Museum', which presented in the Egyptian Hall in 1841 'an artificially enlarged mastodon' made from the bones of different animals with tusks mounted upside down so as to create dramatic effect; Richard Altick, *The Shows of*

'gigantic neogothic dragons',[35] or why the prehistoric animals were a significant source of inspiration for children's writers. As a matter of fact, the dragons that appeared in the literature and the arts of the second half of the nineteenth century often reflected both recent palaeontological discoveries *and* their unrealistic stylisation in popular science books and events. As Stephen Prickett argues, the Crystal Palace animals radically transformed the old monsters of classical mythology into giant, scaly and long-necked Saurians in the literature and the arts of the period. Size became a keyword, and the human-looking fauns, centaurs or tritons were changed into reptilian creatures.[36] Thus, palaeontological discoveries launched a new type of 'realism', exemplified, for instance, by Martin's illustrations of Milton's *Paradise Lost*.[37] According to Prickett, moreover, Martin's 'The Country of the Iguanodon'[38] and 'The Sea-Dragons as They Lived' influenced nineteenth-century illustrators, such as Gustave Doré, who 'freely adapted the vast and grotesque images of Martin and Hawkins to suit his needs' in his illustrations of Dante, Milton and the Bible.[39] Other examples might be Tenniel's Jabberwocky, in Lewis Carroll's *Through the Looking-Glass, and What Alice Found There* (1871), with 'the leathery wings of a Pterodactyl and the long scaly neck and tail of a sauropod'; Aubrey Beardsley's Questing Beast, in his illustrations for Malory; and H. J. Ford's illustrations for the *Fairy Books* of Andrew Lang, all 'show[ing] the same tradition continuing through various forms of stylization to the end of the century and beyond'.[40]

London (London and Cambridge, Mass.: Belknap Press of Harvard University Press, 1978), p. 289.

[35] Secord, 'Monsters at the Crystal Palace', p. 143. As Ralph O'Connor also explains, some writers 'even attempted to make the word "dragon" into a stable vernacular term for the old saurians, offering the latter as replacements of the fictitious dragons of legend'; Ralph O'Connor, 'Victorian Saurians: The Linguistic Prehistory of the Modern Dinosaur', *Journal of Victorian Culture*, 17.4 (2012): 492–504, p. 499.

[36] Stephen Prickett, *Victorian Fantasy*, 2nd edn (Waco: Baylor University Press, [1979] 2005), pp. 78–9.

[37] Prickett, *Victorian Fantasy*, p. 79.

[38] 'The Country of the Iguanodon', exhibited in Gideon Algernon Mantell's museum, was later turned into a mezzotint print and used as the frontispiece of Mantell's popular *The Wonders of Geology; or, A Familiar Exposition of Geological Phenomena; Being the Substance of a Course of Lectures Delivered at Brighton* (1838).

[39] Prickett, *Victorian Fantasy*, p. 79–80.

[40] Prickett, *Victorian Fantasy*, p. 80.

Furthermore, the motif of Noah's Ark, found in many reviews of the Crystal Palace Gardens, also permeated the children's literature of the decades that followed the opening of the venue. The Ark not only blurred the boundaries between myth and reality; it also undermined those dividing science and religion the better to address middle-class child audiences. For instance, in E. W. Payne's guide to the natural history collections of the British Museum, published by the Religious Tract Society, the motif surfaced regularly to map out the children's journey through the collections of 'beasties', whether living or extinct:

> On ascending the great staircase of the British Museum, the long procession of stuffed animals which first greeted the young visitors' eyes reminded Frank of the Noah's ark of his childish days; but Lucy said it made her think of that verse in the Bible which said that 'God brought all these creatures to Adam to see what he would call them'; and as they stood closer together than the living specimens in the Zoological Gardens, they gave her a clearer idea of the first man's subjects.
> 'Let us go from case to case according to scientific order, please, aunt; for perhaps when the arrangement is complete they will all be so placed', said Frank, who had a very methodical taste.[41]

The comparison calls to mind the idea of Noah as 'ur-collector': an 'extreme case of the collector [who] places his vocation in the service of a higher cause, and who suffers the pathology of completeness at all costs',[42] as John Elsner and Roger Cardinal put it. Although Noah's Ark, one of the most popular nineteenth-century children's toys, is used here both to miniaturise the museum room and safely contain it in a picture adapted to the diminutive world of children, the metaphor also evokes the comprehensiveness of the museum, its curators having, like Noah, 'achieved the complete set'.[43] The guidebook follows the children's exploration of the Natural History Department of the British Museum, taking Lucy and Frank from one room to another in order to teach them all the mysteries of natural history. The children are already budding naturalists, '[o]ne of

[41] E. W. Payne, *More Pleasant Mornings at the British Museum; or, the Handy-Work of Creation. Natural History Department* (London: The Religious Tract Society, 1856), pp. 145–6. All subsequent references are to this edition and will be given parenthetically in the text.

[42] John Elsner and Roger Cardinal, 'Introduction', in John Elsner and Roger Cardinal (eds), *The Cultures of Collecting* (London: Reaktion Books, [1994] 1997), pp. 1–6, p. 1.

[43] Elsner and Cardinal, 'Introduction', *The Cultures of Collecting*, p. 1.

their chief amusements [being] to collect and arrange an assemblage of curiosities which some people contemptuously termed "rubbish", but which they named "a museum"' (p. 1). As their aunt Edith takes them to all the rooms of the museum, she constantly compares the holdings of the British Museum to other European collections, such as when she lectures on reptiles: 'The collection in the British Museum is considered the most complete in Europe, excepting that of Paris. Some specimens are preserved in spirits, but most are stuffed and displayed in glass wall-cases' (p. 101). She describes painstakingly every display case and the synoptic ordering of the specimens according to their (Cuvierian) classification, and points out to the children how the lack of space makes it difficult for curators to faithfully show 'the exact arrangement in scientific order'. Indeed, the guidebook reverberates with the debate about the classification of species, especially in the years preceding the publication of Darwin's *On the Origin of Species*. What is particularly interesting, however, is the way in which concepts relating to museology or museum display potentially (re-)shape the scientific discourse:

> The animals of the British Museum are so numerous, and so various in size, that it is almost impossible to allot space enough for their exact arrangement in scientific order. Indeed, naturalists class them so differently that it would not be easy to meet the views of every one. (p. 111)

The way the aunt brings together the lack of museum space and the variability of classification reflects the significance of museum displays and museum culture more generally in the popularisation of natural history in the nineteenth century. In fact, Payne's children's guide aimed to present the collections of the British Museum's Natural History Department as examples of God's works, especially by using fossils—'the remains of past ages' (p. 36). Among the fossilised animals, Lucy and Frank find a plesiosaurus and an ichthyosaurus, pterodactyls and the perfect skeleton of the *dinornis elephantopus*, with the leg bone of *dinornis giganteus*, 'which must have reached nine feet' (p. 43), as well as casts of the dinotherium, mastodon and megatherium (p. 47). The stress on Richard Owen's work on the specimens, as well as his vast contribution to the field of natural history more generally, provides readers with a comprehensive view of natural history while remaining faithful to the tenets of the Anglican Church. However, as we will see in the next section, the motif of the Ark, which evokes here both creation and children's culture, was often used in children's literature in an attempt to reconcile conflicting contemporary theories about the origins of life on earth—and sometimes even to satirise them.

Noah's Arks, Extinct 'Beasties' and Children's Fiction

Perhaps because Noah, as the first collector, had 'not just casual[ly] [kept] but conscious[ly] rescu[ed]'[44] from extinction the creatures made by God, the arks which appeared in children's fiction in the decades which followed the opening of the Crystal Palace Gardens and up until the turn of the twentieth century often featured extinct creatures, some of them directly borrowed from Benjamin Waterhouse Hawkins's dinosaur park. A case in point is Henry Kingsley's *The Boy in Grey* (1871), written in imitation of both his brother Charles Kingsley's *The Water-Babies, A Fairy Tale for a Land Baby* (1863) and Lewis Carroll's *Alice's Adventures in Wonderland* (1865). As in Charles Kingsley's and Lewis Carroll's fantasies, the narrative is shaped by contemporary definitions of the natural world and recent theories or discoveries in natural history. The fantastic journey enables the child reader, for example, to encounter several New World species which were a source of puzzlement for the naturalists of the time. The narrative opens on the day of Prince Philarete's tenth birthday, and among the gifts the prince receives is a Noah's Ark of a very particular sort:

> Next, a Noah's Ark, in which the elephant was as big as a Newfoundland dog, and would serve for a rantoone; and Noah was as big as your sister's largest doll. All the animals had cloth hides, coloured exactly like nature. If you wound up the brown bear with the watch-key, he would run about for five minutes (that was done by little wheels on his toes), and growl (that was done by bellows in his inside). And the rarer and more recently discovered ones, such as the Wandoodle, the Teezyweezy, and the Rumtifoozle, had only just been made, under the joint superintendence of Mr. Frank Buckland and Mr. Tegetmeier.
>
> There was another great fact about this Noah's Ark of his. Professor Huxley had superintended the models of all the extinct animals, even to that

---

[44] Elsner and Cardinal, 'Introduction', *The Cultures of Collecting*, p. 1.

of the Xylopodotherium, the gigantic armadillo, with a wooden leg, recently discovered in the drift of the Appalachian mountains, near Millidgeville.[45]

The mention of the 'cloth hides' defines the animals as so many stuffed creatures exhibited in museums. Moreover, the stress on verisimilitude ('exactly like nature'), foregrounded by the issue of scale, hints, even if satirically, at the presentation of the Crystal Palace dinosaur park project and its past ecosystems. The Wandoodle, Teezyweezy and Rumtifoozle are humorous echoes of the difficulties encountered by naturalists in their efforts to stabilise the 'curious beasties' as a linguistic category, with comparisons with multiple animals often used to name them. As Ralph O'Connor has shown, dinosaurs and extinct animals 'posed special epistemological problems' and had therefore to be 'constructed from within existing cultural categories' at a time when the 'discourse about nature was particularly multivocal'.[46] The range of signifiers used to define them, by both scientific professionals and the wider public, served, therefore, to frame their identity as much as to reflect taxonomic uncertainty. In this way, they 'unite[d] the features of several animal groups, sometimes from far-flung locations on the Great Chain of Being', sometimes suggesting 'a dizzying range of creatures'.[47] For O'Connor, such descriptions sounded like 'recipes for producing monsters' as the comparisons evoked 'just the kind of creature a mischievous taxidermist might make'.[48]

In the case of Prince Philarete's Ark, mischief comes from the association of an opponent and a supporter of Darwinism, which rewrites Charles Kingsley's combination of Richard Owen and T. H. Huxley in his Professor Ptthmllnsprts. The fresh additions to the Ark (the Wandoodle, the Teezyweezy and the Rumtifoozle) result from the work of Francis Trevelyan Buckland (1826–80), son of the geologist and palaeontologist William Buckland (whose menagerie was particularly famous because he

---

[45] Henry Kingsley, *The Boy in Grey and other Stories and Sketches* (New York: Longmans, Green & Co., 1899), p. 2. The first issue of Kingsley's *The Boy in Grey* appeared in *Good Words for the Young*, on 1 March 1869 (pp. 239–63).

[46] O'Connor, 'Victorian Saurians: The Linguistic Prehistory of the Modern Dinosaur', p. 493.

[47] O'Connor, 'Victorian Saurians: The Linguistic Prehistory of the Modern Dinosaur', p. 497.

[48] O'Connor, 'Victorian Saurians: The Linguistic Prehistory of the Modern Dinosaur', p. 497.

ate many of the exotic species he owned),[49] and William Bernhardt Tegetmeier (1816–1912), another English naturalist who advocated evolutionary theory. In addition, Henry Kingsley swaps Richard Owen (who supervised Hawkins's work for the Crystal Palace Gardens, as mentioned above) for T. H. Huxley, mixing once again friends and foes of Darwin. Furthermore, the reference to Huxley's 'Xylopodetherium, the gigantic armadillo' probably hints at his work on *Hoplophoridæ*,[50] which described remains of animals identical to Glyptodons—extinct relatives of the armadillo from the Pleistocene discovered in South America, of which Darwin had found fossils—then believed to belong to the Megatherium, or giant ground sloth. Thus, the child's miniature zoo, with its exotic, extinct and unfamiliar creatures, resonates with contemporary debates, research and museal projects.[51] The illustration, furthermore, adds to the satirical nod to the anatomists of the time. Huxley's Xylopodotherium, with its spines from the hips to the tip of the tail, strongly resembles the model of Hylæosaurus displayed in the Crystal Palace Gardens, especially because it is seen from the back, just like many representations of the Sydenham models (Fig. 6.1).

As this example reveals, in both Victorian popular science works designed for children and children's fiction, dinosaurs were often modelled on Benjamin Waterhouse Hawkins's extinct creatures, such as his model of Hylæosaurus, whose spines were clearly visible along its back. The same might be said of the *Punch* cartoons representing the Crystal Palace dinosaur park, such as 'A Visit to the Antediluvian Reptiles at Sydenham—Master Tom Strongly Objects to Having His Mind Improved',

---

[49] See George C. Bompas, *Life of Frank Buckland* (London: Smith & Elder, 1885) and Richard Girling, *The Man Who Ate the Zoo: Frank Buckland, Forgotten Hero of Natural History* (London: Vintage, 2017).

[50] Huxley's lecture on the genus Glyptodon was given on 28 January 1864 at the Royal College of Surgeons. Huxley had presented to the Royal Society a specimen discovered in 1860 in the province of Buenos Ayres and acquired by the Royal College of Surgeons on 14 November 1862.

[51] Several passages from the novel typify Henry Kingsley's attempt at reconciling evolutionary theory with religious belief: 'It was only a little bird, one of the smallest birds in existence; it was only a little grey bird, without any "tropical" colour at all; but yet the most astoundingly beautiful bird in creation. Conceive a tiny wren, which has a grey tail more delicate than that of the Bird of Paradise. How did it get that tail? I, believing in a good God, cannot tell you. Competition and selection I do believe in, but I have seen things which tell me that there is a great *Will*, and the emu wren is one'; Kingsley, *The Boy in Grey and other Stories and Sketches*, p. 59.

mentioned earlier, or 'The Effect of a Hearty Dinner After Visiting the Antediluvian Department at the Crystal Palace' (1855). In both cases, the Hylæosaurus is also displayed from the back in order to emphasise its spines and reptilian shape. Likewise, John Cargill Brough's *The Fairy Tales of Science: A Book for Youth* (1859) opens with a chapter on dinosaurs ('The Age of Monsters'), whose descriptions and illustrations owe much to Hawkins's sculptures.[52] With the exception of the Cetiosaurus, the 'monsters'[53] described are those displayed in the Crystal Palace Gardens (the Plesiosaurus, Hylæosaurus, Megalosaurus, Iguanodon and Pterodactyl). The Hylæosaurus is one of the most fearful creatures featured, especially because 'with the help of the spines along his back he contrives to inflict some severe wounds upon the huge body of his opponent'.[54] The spines reinforce the dinosaur's monstrosity, making the animal readily comparable to the dragons of fairy tales:

> How shall we describe this monster of the old world, which is so unlike any modern inhabitant of the woods? Its body, which is at least twenty feet long, is upheld by legs of proportional size, and a massive tail, which drags upon the ground and forms a fifth pillar of support. Its head is hideously ugly, its immense jaws and flat forehead recalling the features of those grim monsters which figure in our story-books. Its dragon-like appearance is still further increased by the ridge of large triangular bones or spines which extends along its back. We should not be at all surprised were we to see streams of fire issuing from the mouth of this creature, and we look towards the palm-forest half expecting a St George to ride forth on his milk-white charger.[55]

Another significant example is Brough's later description of the Iguanodon, which he compares to a cow 'quietly grazing' or a rhinoceros bathing, thus recalling Richard Owen's revisions of Gideon Mantell's reconstruction of the creature, changing the lizard-like creature into a

---

[52] The illustration of 'The Age of Monsters' shows the 'monster' from the side so that its spines may be seen, exactly like in the *Punch* cartoons.

[53] As Gowan Dawson explains, '*Monster* … was the customary terminology for almost all prehistoric creatures'; Gowan Dawson, *Show Me the Bone: Reconstructing Prehistoric Monsters in Nineteenth-Century Britain and America* (Chicago and London: University of Chicago Press, 2016), p. 162.

[54] John Cargill Brough, *The Fairy Tales of Science: A Book for Youth* (London: Griffith and Farran, 1859), p. 9.

[55] Brough, *The Fairy Tales of Science*, pp. 7–8.

6 PREHISTORIC 'BEASTIES'   239

Fig. 6.1 Henry Kingsley, 'The Boy in Grey', *Good Words for the Young* (1869). (Courtesy of Armstrong Browning Library. Baylor University, Waco, Texas)

240   L. TALAIRACH

rhinocerine animal.[56] Similarly, Brough's Megalosaur, far from sprawling reptile-like, stands on its four legs, mammal-like:

> Yonder is one of these extraordinary monsters. He has just emerged from the forest, and is marching towards the lake slowly and majestically, a regular moving mountain! His legs are like trunks of trees, and his body, which rivals that of the elephant in bulk, is covered with scales. In length and height he equals the great lizard we have already described, but his whole appearance is far less awe-inspiring. There is a good-humoured expression in his face, and his teeth are not nearly so formidable as those of his predacious neighbour, being blunt and short, and evidently fitted for the mastication of vegetable food.
>
> Look! he is quietly grazing on those luxuriant ferns which lie in his path. ... Having stript the fallen stem of its sword-like leaves, he plunges in the lake, and flounders in the water as though the bath were his greatest source of enjoyment.[57]

Likewise, in Tom Hood's *From Nowhere to the North Pole. A Noah's Ark-Æcological Narrative* (1875), Tom's ark is 'one of the largest Noah's Arks that ever bothered a bishop or disconcerted a Darwin', although the narrator playfully notes that:

> [T]he camelopards were at least a head taller than Noah, though certainly the elephants only came up to his waist, and the foxes came up to the elephant's shoulder; while the raven was quite as large as the lion, who, mane and all, was no taller than the cock. But then the ducks were not more than half as big as the deer and the camels were quite a hump taller than the leopards, who had beautiful spotted tails like currant roley-poleys.[58]

[56] Because Owen's research sought to challenge the Lamarckian theory of descent, his restorations aimed to suggest that dinosaurs were highly evolved reptiles, close to mammals in the evolutionary scale. As Adrian Desmond underlines, Owen even 'equipped them with mammal-like four-chambered hearts and near-perfect circulatory systems', thus 'cunningly using Lamarck's cardiovascular criteria to place dinosaurs on the top rung'; Adrian Desmond, *Archetypes and Ancestors: Paleontology in Victorian London, 1850–1875* (Chicago and London: The University of Chicago Press, 1982), p. 119. Huxley later dismantled Owen's rhinocerine dinosaur theory, by 'throwing the reptile into an evolutionary light and making it the ancestor of birds'; Desmond, *Archetypes and Ancestors*, p. 124.

[57] Brough, *The Fairy Tales of Science*, p. 11.

[58] Tom Hood, *From Nowhere to the North Pole. A Noah's Ark-Æcological Narrative* (London: Chatto and Windus, 1875), p. 6.

Despite the fact that there is no direct reference here to the Crystal Palace Gardens, the issue of scale, related to three-dimensional models of animals likely to 'disconcert[–] a Darwin', echoes nonetheless the scientific debates which the dinosaur park crystallised, especially because Hood later uses models of animals to convey 'the nonsensicality of evolution'[59] and promote natural theology, as mentioned in Chap. 5.

Connections between the Crystal Palace dinosaur park and representations of extinct creatures in children's literature were found time and again in the fiction of the second half of the nineteenth century and the beginning of the twentieth century. Edith Nesbit's fantasies, published at the turn of the Edwardian era, are good examples of the impact of such three-dimensional models of prehistoric 'beasties' on juvenile literature over a period of fifty years. Her children's books repeatedly play upon the motif of Noah's Ark; they also feature numerous collections and exhibitions of animals, be they exotic, extinct or fantastic. But her hints at Hawkins's open-air museum of extinct creatures reveal more strikingly perhaps the role of museums which, as Virginia Zimmerman observes, 'offer individuals the opportunity to place themselves in context and thus provide a material encounter that shapes identity'.[60] It will be argued that Nesbit's extinct 'beasties' enable her to propose 'coming-of-age narratives'[61] which match her exploration of imperial identity, as she constructs her strange creatures as the victims of both colonialism and mass consumerism, proposing thereby a journey through the looking-glass.

Edith Nesbit's Dinosaurus and Megawhatsitsname

As the granddaughter of Anthony Nesbit, whose *Essay on Education* (1840) posited that 'the prehistoric animals had become extinct because Noah had left them out of the Ark',[62] Nesbit was much inspired by Victorian museums and exhibits. Both writer and poet, she remains famous today above all for her children's books, notably *The Story of the Treasure Seekers* (1899), *The Wouldbegoods* (1901), the Psammead series (*Five Children and It* [1902], *The Phoenix and the Carpet* [1904], *The Story of*

---

[59] Jessica Straley, *Evolution and Imagination in Victorian Children's Literature* (Cambridge: Cambridge University Press, 2016), p. 91.

[60] Virginia Zimmerman, 'The Curating Child: Runaways and Museums in Children's Fiction', *The Lion and the Unicorn*, 39.1 (2015): 42–62.

[61] Zimmerman, 'The Curating Child', p. 42.

[62] Prickett, *Victorian Fantasy*, p. 75.

the Amulet [1906]), *The Railway Children* (1906), *The Enchanted Castle* (1907) and *The Magic City* (1910). Animals of sorts—particularly of the 'curious' variety—are found time and again in her children's novels. They are discovered in the natural world, stuffed and used as toys by children, exhibited in museums and shows, or purchased in shops. The museum venues she mentions or alludes to, just like the curious creatures that inhabit her works, frame a discourse that oscillates between fantasy and science, the real and the imaginary, education and entertainment.[63] Quite logically, the Crystal Palace Gardens, with their dinosaurs and giant beasts from the Pleistocene, are central to Nesbit's children's fiction.

Indeed, dinosaurs and prehistoric creatures inform her books for young audiences from the outset. The dragons that haunt her fairy tales, as in *The Book of Dragons* (1901)[64] or 'The Last of the Dragons', are presented as scientific specimens; their necks and tails often more reminiscent of sauropods than mythical creatures. In 'The Deliverers of their Country', dragons are studied by scientists under a microscope. Monstrous varieties of all sizes invade Britain and are contrasted with St George's dragon: some have four limbs, long caudal appendages and traces of bat-like wings; others look like 'tiny lizard[s], about half an inch long … with scales and wings' (p. 42).[65] In 'Kind Little Edmund', a *Draco centipedis* looks 'like a long, rattling worm – or perhaps more like a monstrous centipede' (p. 145), whilst 'Billy the King'[66] features a two-headed dragon, with a

---

[63] Many of the artists who illustrated children's fiction in the second half of the nineteenth century and early twentieth century were also involved in scientific publications. Alice Bolingbroke Woodward (1862–1951), for instance, one of the first scientific illustrators, illustrated both the Iguanodon at the end of the nineteenth century for *The Illustrated London News* (reproduced in popular science works, such as H. N. Hutchinson's *Extinct Monsters: A Popular Account of Some of the Larger Forms of Ancient Animal Life* [1897]) and classical children's books by Juliana Horatia Ewing, Mrs Molesworth, Lewis Carroll, Charles Kingsley and J. M. Barrie.

[64] *The Book of Dragons* was serialised in *The Strand* in 1899 and published in volume form in 1901. It comprised 'The Book of Beasts', 'Uncle James, or The Purple Stranger', 'The Deliverers of Their Country', 'The Ice Dragon, or Do as You Are Told', 'The Island of the Nine Whirlpools', 'The Dragon Tamers', 'The Fiery Dragon, or the Heart of Stone and the Heart of Gold' and 'Kind Little Edmund, or the Caves and the Cockatrice'. 'The Last of the Dragons' was published posthumously and added later to the collection.

[65] Edith Nesbit, *The Book of Dragons* [1901] (Mineola, New York: Dover Publications, 2004), p. 45. All subsequent references are to this edition and will be given parenthetically in the text.

[66] E. Nesbit, 'Billy the King', in *Fairy Stories*, ed. by Naomi Lewis (London and Tonbridge: Ernest Benn Limited, 1977), pp. 1–22.

6 PREHISTORIC 'BEASTIES' 243

pig's head and a lizard's head. Moreover, her tales play time and again with the boundary between science and myth, as in 'The Fiery Dragon', or merge modern science with technology, as in 'The Dragon Tamers' where a dragon's plate needs to be fixed.

By playing on the connexion between science and myth, Nesbit followed in the footsteps of mid-Victorian science popularisers whose prose hovered between romance and reality, the better to catch their readers' attention. As in many popular science works of the second half of the nineteenth century, she interwove scientific objects, specimens, models and fairy-tale motifs (or mythical creatures, such as the Manticora and the Hippogriff, in 'The Book of Beasts'). The fairy tales of *Beauty and the Beast* and *Sleeping Beauty* permeate *The Enchanted Castle*, for instance, sealing the union of antediluvian monsters with the beasts and time-frame of fairy tales. By combining education and entertainment in the fairy-tale mode,[67] Nesbit's fantasies are thoroughly in keeping with Victorian pedagogical methods. Her stress on the power of the imagination and the

---

[67] 'You show the child many things, all strange, all entrancing ... You tell it that the stars, which look like pin-holes in the floor of heaven, are really great lonely worlds, millions of miles away; that the earth, which the child can see for itself to be flat, is really round; that nuts fall from the trees because of the force of gravitation, and not, as reason would suggest, merely because there is nothing to hold them up. And the child believes; it believes all the seeming miracles. Then you tell it of other things no more miraculous and no less; of fairies, and dragons, and enchantments, of spells and magic, of flying carpets and invisible swords. The child believes in these wonders likewise. Why not? If very big men live in Patagonia, why should not very little men live in flower-bells? If electricity can move unseen through the air, why not carpets? The child's memory becomes a storehouse of beautiful and wonderful things which are or have been in the visible universe, or in that greater universe, the mind of man. Life will teach the child, soon enough, to distinguish between the two. But there are those who are not as you and I. These say that all the enchanting fairy romances are lies, that nothing is real that cannot be measured or weighed, seen or heard or handled. Such make their idols of stocks and stones, and are blind and deaf to the things of the spirit. These hard-fingered materialists crush the beautiful butterfly wings of imagination, insisting that pork and pews and public-houses are more real than poetry; that a looking-glass is more real than love, a viper than valour. These Gradgrinds give to the children the stones which they call facts, and deny to the little ones the daily bread of dreams'; Edith Nesbit, *Wings and the Child; or the Building of Magic Cities* (London: Hodder and Stoughton, 1913), pp. 24–6. A very similar remark appears in *The Enchanted Castle*: 'When you are young so many things are difficult to believe, and yet the dullest people will tell you that they are true – such things, for instance, as that the earth goes round the sun, and that it is not flat but round. But the things that seem really likely, like fairy-tales and magic, are, so say the grown-ups, not true at all'; Edith Nesbit, *The Enchanted Castle* [1907] (London: Puffin Classics, 1994), p. 27. All subsequent references are to this edition and will be given parenthetically in the text.

significance of belief explains why the Crystal Palace appeared in many of her works. Education in Nesbit's conception was shaped like a Palace,[68] and the Crystal Palace, with its dinosaur park, epitomised a now vanished ability to combine beauty and knowledge:

> Think of the imagination, the feeling for romance that went to the furnishing of the old Crystal Palace. There was a lake in the grounds of Penge Park ... How did these despised mid-Victorians deal with it? They set up, amid the rocks and reeds and trees of the island in that lake, life-sized images of the wonders of a dead world. On a great stone crouched a Pterodactyl, his vast wings spread for flight. A mammoth sloth embraced a tree, and I give you my word that when you came on him from behind, you, in your six years, could hardly believe that he was not real, that he would not presently leave the tree and turn his attention to your bloused and belted self. ... There was an Ichtyosaurus too, and another chap whose name I forget, but he had a scalloped crest all down his back to the end of his tail. And the Dinosaurus ... he had a round hole in his antediluvian stomach: and, with a brother ... to give you a leg-up, you could explore the roomy interior of the Dinosaur with feelings hardly surpassed by those of bandits in a cave. It is almost impossible to overestimate the dinosaurus as an educational influence.[69]

Even though Nesbit was ambivalent concerning the nature of the imagination,[70] most of her children's books likewise combine fantasy with reality, often by foregrounding the imaginative potential of science, from geology and palaeontology to mathematics and physics, or turning curious prehistoric 'beasties' and technological advances into wonders.[71] Unsurprisingly, the Crystal Palace is referred to in many of her works, many of which pivot around time and/or space travel. This is the case, for

[68] 'In the Palace of Education which the great minds have designed and are designing, many stones will be needed – and so I bring the little stone I have hewn out and tried to shape, in the hope that it may fit into a corner of that great edifice'; Nesbit, *Wings and the Child*, p. 16.

[69] Nesbit, *Wings and the Child*, p. 48.

[70] As we will see later, many of her books deal with the dangers of books and of the imagination, such as 'The Book of Beasts', first published in *The Strand* in March 1899.

[71] Science is also subject to irony. In *Five Children and It*, for instance, the Psammead's explanation for fossils (wished-for objects turned into stone) may be read as a potential ironical reversal of contemporary 'far-fetched "explanations" for fossils, such as that of the fundamentalist Philip Henry Gosse'; Julia Briggs, *Edith Nesbit: A Woman of Passion* (Stroud: Tempus, [1987] 2007), p. 233.

example, with the beginning of 'The Ice Dragon, or Do as You Are Told', in which the protagonists go to the North Pole, in 'Whereyouwantogoto',[72] and in her longer fantasies: *The Magic City, The Phoenix and the Carpet, The Story of the Amulet* and *The Enchanted Castle*.

The Enchanted Castle features a park with a lake and stone dinosaurs which come to life at night, directly alluding to Hawkins's models. Hawkins's dinosaurs are here reimagined, appearing as the 'dinosaurus' with 'a round hole in his antediluvian stomach',[73] while the Crystal Palace becomes an 'enchanted castle' with a lake surrounded by the statues of the Immortals (Roman gods and goddesses). Nesbit's revisiting of the Crystal Palace Gardens in *The Enchanted Castle* is not simply a historical reference or an allusion to a place she liked as a child; the fantasy also brings into play the ideas that underlay the whole dinosaur park project. The narrative contends, for instance, that reality is a construct and is highly subjective ('It will be real to you … there is no other realness' [p. 228]), and the fantasy, sometimes playing on suspense as in detective narratives, is reminiscent of the type of epistemological and narratological issues inspired by such nineteenth-century scientific disciplines as geology, archaeology, palaeontology and evolutionary biology, all of which attempted to reconstruct the past from fragmentary or inadequate evidence.

The Enchanted Castle opens with the children's discovery of a labyrinthine castle where a sleeping princess lies, and whom they can only reach by following a red thread. The princess at the heart of the maze, like the fairy tale embedded in the frame narrative, must be reconstructed from fragments—the princess's body parts reproducing the reconstitution of extinct creatures:

> He walked forward, winding the red thread round his fingers as he went. And it *was* a clue, and it led them right up in the middle of the maze. And in the middle of the maze they came upon a wonder.
>
> The red clue led them up two stone steps to a round grass plot. There was a sun-dial in the middle, and all round against the yew hedge a low, wide marble seat. The red clue ran straight across the grass and by the sun-dial, and ended in a small brown hand with jewelled rings on every finger. The hand was, naturally, attached to an arm, and that had many bracelets on it,

---

[72] Edith Nesbit, 'Whereyouwantogoto', in *Nine Unlikely Tales* (London: T. Fisher Unwin, 1901), pp. 49–84.

[73] As seen above, Nesbit later mentioned the 'dinosaurus' in her non-fiction pedagogical essay *Wings and the Child* (1913).

sparkling with red and blue and green stones. The arm wore a sleeve of pink and gold brocaded silk, faded a little here and there but still extremely imposing, and the sleeve was part of a dress, which was worn by a lady who lay on the stone seat asleep in the sun. (p. 20)

The discovery of the 'wonder' thus lies at the end of a causal chain of clues, foreshadowing how later on in the tale the children play detectives, referring on several occasions to Sherlock Holmes. Moreover, the sleeping princess, supposedly protected by dragons, spurs the narrative's interest in time: just as the spell freezes the princess's life for a hundred years, the children see extinct prehistoric creatures, statue-like in daytime, come to life at night. As a result, Nesbit's fairy-tale realm looks very much like a geological site, especially as the children discover the castle by walking down into a cave, as if delving into the subterranean depths of the earth. Furthermore, the fantasy starts when a magical ring makes one of the characters, Mabel, invisible. Dinosaurs modelled on Hawkins's extinct creatures then come to life, Nesbit making explicit the link between her narrative and the Crystal Palace park:

Something enormously long and darkly grey came crawling towards him, slowly, heavily. The moon came out just in time to show its shape. It was one of those great lizards that you see at the Crystal Palace, made in stone, of the same awful size which they were millions of years ago when they were masters of the world, before Man was. ... As it writhed past him he reached out a hand and touched the side of its gigantic tail. It was of stone. It had not 'come alive' as he had fancied, but was alive in its stone. (p. 85)

The story reaches a moment of climactic intensity when Kathleen, standing inside the dinosaur's stomach wishes she were a statue. Instantly turned into one, she wakes up at night as the dinosaur is moving. The wish has therefore enabled her to come to life at the same time as the dinosaur. This connection between the extinct 'beastie' and the child character is a motif that informs many of Nesbit's fantasies, as will be seen presently. Here, the symbolic parturition, as the dinosaur walks faster and faster, swinging her from side to side, before slipping into the lake among the lily-pads (another allusion to the *Victoria Regia* floating in ponds at Sydenham),[74] signals the girl's rebirth and evolution. The dinosaur thus

[74] See Martineau's review: 'For the next class – the school-children, the artisans and trades-people, Natural History is still the most attractive. After sitting down, on their arrival, to

metaphorises a passage towards another stage—a new reality directly inspired by palaeontological research. The children must learn to see what is not there, whether it is the princess telling them that believing in magic will make them see what they wish to see, or when they later participate in the immortals' feast and learn to pick fruits they cannot see.[75] Consequently, when Nesbit closes her narrative with Lord Yalding promising to tell the children 'all about the anteddy-something animals – it means before Noah's Ark' (p. 253), she combines her vision of the children's growth and evolution with a didactic lesson related to the Crystal Palace creatures:

> The antediluvian animals are set in a beech-wood on a slope at least half a mile across the park from the castle. The grandfather of the present Lord Yalding had them set there in the middle of the last century, in the great days of the late Prince Consort, the Exhibition of 1851, Sir Joseph Paxton, and the Crystal Palace. Their stone flanks, their wide, ungainly wings, their lozenge crocodile-like backs show grey through the trees a long way off. (p. 255)

As the children visualise 'the great beasts ... gigantic lizards with wings – dragons they lived as in men's memories' (p. 286), the ending suggests that they have gained in both maturity and knowledge—knowledge about the history and evolution of the earth, and of humankind.

Although Kathleen's connection with the 'dinosaurus' must be read, above all, in metaphorical terms, since it creates parallels between the evolution of the earth and of the human child, numerous references or allusions to Hawkins's extinct creatures and children's education pepper Nesbit's other fantasies, which often use prehistoric 'beasties' to tackle the issue of humans' relationships with animals and with natural resources more generally. *The Magic City*, published at the close of the Edwardian period, provides a good illustration of the way in which Nesbit repeatedly compared Crystal Palace Gardens and prehistoric beasts on the one hand

---

their lumps of cake or their sandwiches, they go first to the savages and the beasts. But they have a prospective eye for the fountains, and for the future Victoria Regia, which is to float at either end of the ponds in the nave. They can fancy the climbing plants covering the red shafts of the pillars, and festooning the girders, and tempering the noonday light'; Martineau, 'The Crystal Palace', p. 540.

[75] It is interesting to note that H. R. Millar's illustrations present Nesbit's Dinosaurus as a diplodocus, a representation which may easily be connected with the display of Dippy on 12 May 1905 in the hall of reptiles at the London Natural History Museum. Diplodocuses were believed to be lake-dwellers because of their bulk, making the creatures fit perfectly Nesbit's tale.

and mythical and legendary creatures on the other. The fantasy plays with scale, both spatial and temporal, as the hero, Philip Haldane, who lives 'in a little red-roofed house in a little red-roofed town',[76] falls asleep and wakes up in the city he has built with Lucy's books and toys, finding that he is now much smaller than the animals of the girl's Noah's Ark. Lucy is the daughter of his sister's new husband, and her nursery 'is full of toys of the most fascinating kind': 'A rocking-horse as big as a pony, the finest dolls' house you ever saw, boxes of tea-things, boxes of bricks – both the wooden and the terra-cotta sorts – puzzle maps, dominoes, chessmen, draughts, every kind of toy or game that you have ever had or ever wished to have' (p. 19). Lucy's nursery illustrates the growth of manufactured toys in the nineteenth century, demonstrating how 'the pleasure of commercial consumption [was] a key element of children's play, which itself was becoming more central to the lives of children and their parents'.[77] Yet, the entry to Lucy's nursery is first forbidden to the little boy, who, at the beginning of the fantasy, longs for 'a pony of [his] very own' (p. 16). The toys thus feed his desire to consume, driving the boy's relationship with animals as material possessions. When the nurse leaves the house for a few days, Philip is allowed to take some of Lucy's toys, builds a city with them as well as with some books from the library. Philip's construction is eventually 'as good as a peep-show', as the parlour-maid puts it, recalling the 'picture post-cards [her] brother in India sends [her]. All them pillars and domes and things – and the animals too' (p. 21). As part of Lucy's collection of toys, the Noah's Ark playset serves to anchor the fantasy in the world of childhood. But as the parlour-maid underlines, it also suggests that Philip has travelled to the zoo, as if to see the originals of his reproductions. Furthermore, Philip's journey is also an (antediluvian) voyage back in time, as a Hippogriff—a creature whose name evokes contemporary palaeontological research, since prehistoric animals were frequently named after or compared to mythical creatures[78]—was this time not too large to go into the Ark:

---

[76] Edith Nesbit, *The Magic City* (1910) (s.l.: BiblioBazaar, 2007), p. 13. All subsequent references are to this edition and will be given in the text.

[77] Teresa Michals, 'Experiments before Breakfast: Toys, Education and Middle-Class Childhood', in Dennis Denisoff (ed.), *The Nineteenth-Century Child and Consumer Culture* (Aldershot: Ashgate, 2008), pp. 29–42, p. 32.

[78] Archæopteryx, much involved in contemporary scientific debates concerning the link between reptiles and birds, was for instance also called 'Griphosaurus'; Gowan Dawson, *Show Me the Bone: Reconstructing Prehistoric Monsters in Nineteenth-Century Britain and America*

... there were not only horses here, but every sort of animal that has ever been ridden on. Elephants, camels, donkeys, mules, bulls, goats, zebras, tortoises, ostriches, bisons and pigs. And in the last stall of all, which was not of common wood but of beaten silver, stood the very Hippogriff himself, with his long, white mane and his long, white tail, and his gentle, beautiful eyes. His long, white wings were folded neatly on his satin-smooth back, and how he and the stall got here was more than Philip could guess. All the others were Noah's Ark animals, alive, of course, but still Noah's Arky beyond possibility of mistake. But the Hippogriff was not Noah's Ark at all. (pp. 101–2)

Philip's encounter with mythical and extinct creatures, such as the Great Sloth, one of the 'beasties' modelled in the Crystal Palace Gardens, reinforces the parallels between Nesbit's fantasy and Hawkins's dinosaur park. Moreover, in H. R. Millar's illustration, the tail of the Great Sloth is drawn as a tripod, hinting at Richard Owen's contention that the creature (in that case more precisely a Mylodon) used its tail as a third hind leg in this way.[79]

However, the animal toys have not all been realistically manufactured, but are made up of both real and imaginary creatures, with the pack of 'beasties' evoking Edward Lear's or Tom Hood's nonsensical menageries.

(Chicago and London: University of Chicago Press, 2016), p. 319. For more on the use of metaphors and analogy in palaeontology, see Laurence Talairach-Vielmas, 'Shaping the Beast: The Nineteenth-Century Poetics of Palaeontology', in Maria Freddi, Barbara Korte, Joseph Schmied (eds), *The Rhetoric of Science, EJES (European Journal of English Studies)*, 17.3 (2013): 269–82.

[79] The implication was that the animal, thus raised up on his hind legs and tail, would have been able to push over or wrench out trees with its forelegs. Owen wrote two essays on the Megatherium in which he studied the megatheroid specimens of 1841 and 1845: *Description of the Skeleton of an Extinct Gigantic Sloth, Mylodon robustus, Owen: with Observations on the Osteology, Natural Affinities, and Probable Habits of the Megatherioid Quadrupeds in General* (London: R. and J. E. Taylor, 1842) and *Memoir on the Megatherium, or the Giant Ground-Sloth of America (Megatherium americanum, Cuvier)* (London: Williams and Norgate, 1861). The idea that the Mylodon could have accidentally smashed its head with falling trees was humorously seized on by William Buckland—until a Mylodon skeleton arrived in 1841, fractured in two places on the skull, one of the fractures being healed and the other partially so. The healed fractures seemed to confirm Owen's hypothesis, and restorations of Mylodons subsequently featured the animal standing on its hind legs. Nicolaas A. Rupke, *Richard Owen: Victorian Naturalist* (New Haven and London: Yale University Press, 1994), p. 130. My interpretation of Millar's illustration disagrees with Prickett's, who argues that Millar's illustrations 'make it clear that this creature is some kind of sauropod, not unlike the "dinosaurus" of Yalding Castle', Prickett, *Victorian Fantasy*, p. 88.

250   L. TALAIRACH

Interestingly, those which do not look like real animals and are named according to what colours the manufacturers find or fancy—the vertoblancs, graibeestes, chockmunks and pinkuggers—are the animals that are being hunted for food:

> When they got out into the courtyard of the castle, they found it full of a crowd of animals, any of which you may find in the Zoo, or in your old Noah's ark if it was a sufficiently expensive one to begin with, and if you have not broken or lost too many of the inhabitants. Each animal has its rider and the party rode out on to the beach.
>
> 'What *is* it they hunt?' Philip asked the parrot, who had perched on his shoulder.
>
> 'All the little animals in the Noah's ark that haven't any names', the parrot told him. 'All those are considered fair game. Hullo! Blugraiwee!' it shouted, as a little grey beast with blue spots started from the shelter of a rock and made for the cover of a patch of giant seaweed. Then all sorts of little animals got up and scurried off into places of security.
>
> 'There goes a vertoblanc', said the parrot, pointing to a bright green animal of uncertain shape, whose breast and paws were white, 'and there's a graibeeste'.
>
> The graibeeste was about as big as a fox, and had rabbit's ears and the unusual distinction of a tail coming out of his back just half-way between one end of him and the other. But there are graibeestes of all sorts and shapes.
>
> You know when people are making the animals for Noah's arks they make the big ones first, elephants and lions and tigers and so on, and paint them as nearly as they can the right colour. Then they get weary of copying nature and begin to paint the animals pink and green and chocolate colour, which in nature is not the case. These are the chockmunks, and vertoblancs and the pinkuggers.
>
> And presently the makers get sick of the whole business and make the animals any sort of shape and paint them all one grey – these are the graibeestes. And at the very end a guilty feeling of having been slackers comes over the makers of Noah's arks, and they paint blue spots on the last and littlest of the graibeestes to ease their consciences. This is the bluegraiwee. (pp. 118–20)

Nesbit's use of colour seems, to a certain extent, to mirror the way in which animals appeared to have been valued, as Andrew Flack explains, 'according to somewhat crude criteria that reflected attitudes towards beauty and violence in the natural world'.[80] This led the more 'savage' beasts, such as jaguars, to be considered as pests and therefore unworthy of preservation. Nesbit's beasts are here hunted because they do not resemble familiar animals (i.e. those found at the zoo) and as a result are less aesthetically pleasing. Paradoxically, as game, 'skinned and cut up in the courtyard' (p. 121), the beasts are also, inevitably, more 'animal'-like and wilder, and play a key part in the ecosystem. Thus, the blugraiwee, which feeds on periwinkles, becomes food for the (human) hunters. Although the issue of animal pain is raised by Lucy ('I don't a bit mind killing things, but I do hate hurting them' [p. 106]) and Mr Noah encourages the children to be 'humane' and 'study the natural taste of the animals in [their] charge' (p. 131), the question of who eats whom, as in Carroll's *Alice's Adventures in Wonderland*, suggests that Noah's museum of animals is also conceived in terms of predation.

The idea that the colouring of animals makes them less wild, and hence invisible (to the hunter), is confirmed later on in the story and may be related, it could be argued, to Nesbit's stance upon animal shows and the commodification of animals. As Philip finds himself among giant beasts, he is told by Mr Noah that he needs to perform seven Great Deeds to become the Deliverer, some of which involve killing animals. While imbued with (Christian) mythology, the tale nonetheless plays with 'curious beasties' (that have escaped from the same book) the better to collapse boundaries between extinct, real and imaginary creatures. This is illustrated by the similarities between Mr Noah, who keeps animals on his ark,[81] and his son, who owns 'a shop that said outside "Universal Provider. Expeditions fitted out at a moment's notice. Punctuality and dispatch"' (p. 102). Both father and son play a part in the tasks Philip must perform to become the Deliverer. The young boy is provided with animals bought from Mr Noah's son's shop, such as 'one best-quality talking parrot' (p. 102), and invited to choose a creature to ride amongst Mr Noah's animals. Noah's son's commerce hints at colonial expeditions and the commodification of animals purchased by collectors and museums. The camel kept by Mr Noah is compared to those of the London Zoological

---

[80] Andrew Flack, *The Wild Within: Histories of a Landmark British Zoo* (Charlottesville and London: University of Virginia Press, 2018), p. 45.

[81] It is also noticeable that the Great Sloth is 'enormous, as big as a young elephant', walks 'on its hind legs like a gorilla' and 'yawn[s] like a hippopotamus' (p. 171), confirming these echoes between extinct and living creatures.

Gardens, albeit less melancholy in character ('the camel seemed less a prey than usual to that proud melancholy which you must have noticed in your visits to the Zoo as his most striking quality' [p. 103]). Hence the Noahs, father and son, are dealers in 'curious beasties'. Furthermore, Philip's slaying of the dragon and the lions in the desert, respectively the first and fourth tasks that Philip must perform, connote imperial conquest, intermingling mythical (or extinct) 'beasties' and contemporary animals.

As suggested above, the life and death of the animals throughout the fantasy are bound to their body colouring. Helped by Lucy, the princess Philip has rescued, the children decide to kill the lions by having dogs lick the paint off the lions' legs so that they feel nothing when the axe breaks them. The children then restrain the animals with ropes whilst the dogs perform their duty and turn the creatures' legs from 'real writhing resisting lion-leg' into wood. When the lions have become weak from loss of paint, Philip chops off their legs with 'the explorer's axe' (p. 109). As the legs vanish from the children's sight, therefore, so does the animals' pain: the lions are killed, and Philip's fourth task is successful. The lions, just like the vertoblancs, graibeestes, chockmunks and pinkuggers, may therefore be hunted and killed, it seems, not because they no longer look like animals, but because their colour (or lack thereof) allows the children to be less concerned with animal welfare. The lions only exist through the codes of representation the children know (here, colour); once these vanish the creatures may feel the bite of the explorer's axe. Seen through the prism of the imperial experience, the animals' colourlessness becomes, as a result, a sign of wildness, revealing the colourless lions as game. The properly coloured 'beasties', on the contrary, appear 'real' to the children only because they still look like the children's toys, that is to say representations of (living) animals, which makes them, paradoxically, invisible.

The issue of the disappearance of animals is interesting because, as Flack argues, 'performance and spectatorship have ... been at the core of our interactions with animals in modernity'.[82] And yet, according to John Berger, the disappearance of animals has informed our relationship with them since the nineteenth century. This idea is arguably connected with Nesbit's use of extinct 'beasties' inspired by Hawkins's models and of the reproductions of zoo animals on Mr Noah's Ark. For Berger, it was in the nineteenth century that animals disappeared: first aesthetically, as romantic painting showed images 'of animals *receding* into a wildness that existed

---

[82] Flack, *The Wild Within*, p. 90.

only in the imagination'.[83] Secondly, because they were 'transformed into spectacle', turned into 'pictures which carry with them numerous indications of [the animals'] *invisibility*'.[84] Lastly, and perhaps most importantly, because the public zoo emerged 'at the beginning of the period which was to see the disappearance of animals from daily life', and is therefore 'a monument to the impossibility of such encounters'.[85] As he argues, from 'an endorsement of modern colonial power', illustrating how the 'capturing of animals was a symbolic representation of the conquest of all distance and exotic lands',[86] the zoo thus became in the twentieth century 'the living monument of [animals'] own disappearance'.[87] This, I contend, is what many of Nesbit's fantasies highlight, through representations of animal displays that mix extinct and contemporary beasts and map out precisely such impossible encounters.

The story of the animals' escape from the book of beasts in *The Magic City* has often been read as a significant example of Nesbit's discourse on the danger of books and of the imagination. Some of the 'beasties' conjured up by Philip as he is sleeping symbolise dream mechanisms and random associations, as in Hood's *From Nowhere to the North Pole*, in which the creatures seem to be bits of animals randomly taken from a zoo or the natural history collection at the British Museum and haphazardly glued together. Philip's adventure is eventually explained as a dream surging up like a 'flood' (p. 57); the creatures from the depths of the earth metaphorising the character's psychic depths in order to symbolise the child's growth and evolution. However, it is also interesting to highlight that throughout the fantasy the creatures, whether prehistoric, mythical or wild and exotic, are found in places that can, or should, only be imagined, in particular through reading books. Animals figure on pages as they would stand on arks. As the talking parrot explains, they are classified according to resemblances or differences; the words, sentences and narratives capturing the creatures as cages would do. Danger emerges when they fall out of the book, thus leaving the realm of representation, and become 'real':

[83] John Berger, 'Why Look at Animals', *About Looking* (New York: Vintage International [1980] 1991), pp. 3–28, p. 17.

[84] Berger, 'Why Look at Animals', p. 16.

[85] Berger, 'Why Look at Animals', p. 21.

[86] Berger, 'Why Look at Animals', p. 21.

[87] Berger, 'Why Look at Animals', p. 26.

'Many years ago, in repairing one of the buildings, the masons removed the supports of one of the books which are part of the architecture. The book fell. It fell open, and out came the Hippogriff. Then they saw something struggling under the next page and lifted it, and out came a megatherium. So they shut the book and built it into the wall again'.

'But how did the megawhatsitsname and the Hippogriff come to be the proper size?'

'Ah! that's one of the eleven mysteries. Some sages suppose that the country gave itself a sort of shake and everything settled down into the size it ought to be. I think myself that it's the air. The moment you breathe this enchanted air you become the right size. *You* did, you know'.

'But why did they shut the book?'

'It was a book of beasts. Who knows what might have come out next? A tiger perhaps. And ravening for its prey as likely as not'. …

'And where did you come from, Polly, dear?'

'I', said the parrot modestly, 'came out of the same book as the Hippogriff. We were on the same page. My wings entitled me to associate with him, of course, but I have sometimes thought they just put me in as a contrast. My smallness, his greatness; my red and green, his white'. (pp. 113–14)

By bringing together various extinct 'beasties' from different geological periods, the Crystal Palace park functioned in ways very similar to zoos, which brought together animals from different countries and continents. Nesbit's use of a child's Noah's ark, in which extinct, mythical, fanciful and contemporary animals come together, some from the Crystal Palace Gardens, others from the zoo, indicates that her creatures' journey from the reproduction to the original could serve, like the Crystal Palace creatures, a didactic agenda rooted in conservation and ecology; thus raising people's awareness about the reality of extinction—of prehistoric creatures, just as much as that of the captives of the Zoological Gardens. But Noah's animals, as playthings, are also constructed as commodities, devised for children's consumption. Noah's collection of animals, as a representation of the complete collection, defines animals as possessions brought together to satisfy a child's desires. Such a consumerist ethos, in particular related to animals, permeates Nesbit's Psammead series, which interweaves commercial transactions and imperial power. In the following section, we will see that the trilogy owes a great deal to Hawkins's prehistoric beasties. As a Sand-fairy is brought back to life, it resembles Hawkins's three-dimensional models of extinct creatures, aimed at resurrecting the prehistoric past. Like them, the Psammead enables the children to travel in space and time, as they visit exotic places and encounter 'savages'. Like the dinosaur park's visitors, the children gain knowledge, discovering above all how excessive consumption may lead to want, as well as, perhaps, to extinction.

'THREE POUNDS AND A QUARTER OF SOLID PSAMMEAD'[88]

In many ways a visit to the Zoo is one of the most delightful things in the world. The majority of the animals are light-minded creatures, who appear well content with their lot. And for most of them, life is rendered as agreeable as possible under the circumstances. But one rarely visits it without bringing away some picture to haunt one, of an animal pacing its cage with despair and misery in its eye. Two sights to stir one's pity with hunched-up shoulders on a perch in a small white-washed cave, and the Polar bear, who occupies a small asphalt enclosure with an imitation pond. The eagle is the picture of soul-consuming despair – the Polar bear of noble scorn and implacable hatred of man.[89]

As Alice Dew-Smith's words underline, by the turn of the twentieth century, late-Victorian children's writers placed more and more emphasis on the idea that the zoo constructs 'animal as other, caged for display; people regarding the spectacle as confirmation of empowerment',[90] to quote Randy Malamud's words. Although, as we have seen, nineteenth-century animal shows materialised the British empire and invited juvenile audiences to participate in the colonial experience, late-Victorian children's fiction often deconstructed the narratives of knowledge, power and violence that informed collections of exotic fauna. Malamud's study of twentieth-century 'zoo stories' argues that cultural representations are 'capable of altering what brought them into being' and 'promoting new ways of perceiving the world – ways that enhance our appreciation of a holistic paradigm for coexisting with animals'.[91] The idea that zoos

[88] Edith Nesbit, *The Story of the Amulet* (s.l.: Amazon, n.d.), p. 16. All subsequent references are to this edition and will be given parenthetically in the text.

[89] Alice Dew-Smith, *Tom Tug and Others. Sketches in a Domestic Menagerie* (London: Seeley and Co. Limited, 1898), p. 70.

[90] Randy Malamud, *Reading Zoos: Representations of Animals and Captivity* (New York: New York University Press, 1998), p. 271.

[91] Malamud, *Reading Zoos*, pp. 12, 35.

'represent a cultural danger, a deadening of our sensibilities'[92] filters through Nesbit's use of her iconic 'beastie'—the Psammead—a prehistoric Sand-fairy resurrected by five children and which is central to Nesbit's ecological fantasy.

The Psammead trilogy includes *Five Children and It*, *The Phoenix and the Carpet* and *The Story of the Amulet*. The Psammead is a 'curious beastie' *par excellence*, discovered by the five children—Cyril, Robert, Anthea, Jane and Lamb—living in a gravel pit while they are on holiday in the countryside. Whilst the children dig a hole (believing, like Lewis Carroll's Alice, that they will reach Australia and see exotic creatures, from kangaroos and opossums to blue-gums and Emu Brand birds), they find a creature which has been 'buried in the sand for thousands of years' (*Amulet*, p. 3). Whilst prehistoric and the last of its species, the Psammead remains a living animal which, as the children say in the last volume of the trilogy, 'ought to be in the Zoo' (*Amulet*, p. 9). In the space of the three narratives, the Psammead is captured and exchanged several times. Because it has become prisoner of a society driven by consumption, as we will see, it first begs the children to buy it so that he can recover its freedom before teaching them that consumption can only lead to more frustration. The series closes with the creature's last wish to vanish for ever, thus saving the Psammead from the grip of capitalism and putting an end to animal consumption.

Because the Psammead is the last of its kind, the creature, unlike zoo animals, does not function as an 'embodied gateway[-] to knowledge … to understand the fundamental nature of [the] species',[93] metonymically representing all the other Psammeads in the world. It functions in reverse, denoting, on the contrary, its rarity, and symbolising the threat of extinction. The Psammead thus becomes representative of Nesbit's counter-zoo, much inspired by the other 'curious beasties' which typified nineteenth-century museum culture and which found their way into the children's literature of the period. Nesbit's Psammead is also reminiscent of other nineteenth-century fantasies informed by natural history practice, such as Kingsley's *Water-Babies*, which, as we have seen, deals with a creature which will not be caught up, objectified and exhibited in a glass full of spirits, and becomes a foil to the bottled specimen displayed on museum shelves. Although, as argued earlier in this book, Kingsley's water-baby

[92] Malamud, *Reading Zoos*, p. 5.
[93] Flack, *The Wild Within*, p. 94.

aims to bring to light how Victorian science depended on scientists' belief in impossible things (to explain and popularise evolutionary theory), the 'fairy tale' also alludes to the threat of extinction as part of the evolution of creatures—or the lack of it. The fantasy argues, for instance, that species which fail to evolve will inevitably become extinct, as demonstrated by the example of the last Gairfowl, which refuses to share the attributes of lower—more recent—creatures. Lady Gairfowl, whose wings have become 'two little feathery arms' with which she fans herself, is doomed to extinction because of her aristocratic arrogance; wings serving, in her opinion, the 'vulgar creatures' which seek to 'rais[e] themselves above their proper station'.[94] The story of Lady Gairfowl relates, in fact, the extinction of the great auk, the giant seabird which became extinct in the middle of the nineteenth century due to humans' overhunting, as sailors took their eggs, partly for food, but also to provide specimens for collectors and museums.

In the last decades of the nineteenth century, as Victorians witnessed several indigenous species driven to extinction by avid naturalists, the issue of extinction permeated more and more books for children. This was due, in part, to new models of animal biographies, as seen in Chap. 4 which increasingly helped raise young audiences' awareness of wild or exotic 'beasties' through offering insights into the subject with a view to their scientific study. Alongside animal biographies, fantasies could also advocate the cause of 'curious beasties', as illustrated by Ethel C. Pedley's *Dot and the Kangaroo* (1899). The fantasy, aimed at denouncing animal cruelty, started with a dedication by the author warning against the threat of extinction faced by Australian animals as a result of human activity: 'To the children of Australia, in the hope of enlisting their sympathies for the many beautiful, amiable, and frolicsome creatures of their land; whose extinction, through ruthless destruction, is being surely accomplished'.[95]

Pedley's fantasy echoes strongly Nesbit's depiction of her children's encounter with the Psammead. The heroine, lost in the bush, meets a kangaroo which introduces her to a platypus, which in turn explains to the little girl how naturalists have gained knowledge about its species. As in Kingsley's *Water-Babies*, Pedley points out the interdependence of (natural history) knowledge, power and violence: 'Their idea of learning all

[94] Charles Kingsley, *The Water-Babies, A Fairy Tale for a Land Baby* [1863] (London: Penguin, 1995), p. 249.

[95] Ethel C. Pedley, *Dot and the Kangaroo* [1899] (Sydney: Angus and Robertson, 1906), n.p. All subsequent references are to this edition and will be given parenthetically in the text.

258   L. TALAIRACH

about a creature was to dig up its home, and frighten it out of its wits, and kill it' (p. 26). However, when Dot encounters the platypus, the description of the animal is deflated by its comparison to 'an empty fur bag ... fished out of the water', which offends the creature's vanity:

> When the little girl reached the pool, she was still more surprised, on a nearer view of the Platypus, that the Kangaroo should think so much of it. At her feet she beheld a creature like a shapeless bit of wet matted fur. She thought it looked like an empty fur bag that had been fished out of the water. Projecting from the head, that seemed much nearer to the ground than the back, was a broad duck's bill, of a dirty grey colour; and peeping out underneath were two fore feet that were like a duck's also. Altogether it was such a funny object that she was inclined to laugh, only the Kangaroo looked so serious, that she tried to look serious too, as if there was nothing strange in the appearance of the Platypus.
> 'I am the *Ornithorhynchus Paradoxus*!' said the Platypus pompously.
> 'I am Dot', said the little girl. (pp. 24–5)

Commodified as a 'fur bag' or 'funny object', the 'curious beastie', which introduces itself through its scientific name, then asks Dot—whose own name binds the little girl to the activity of writing—if she is going to write a book about him. Humans tell stories, or 'fibs' as the odd creature calls them, before proceeding to recall the numerous scientists who have misunderstood what platypuses are and have even argued that such creatures are examples of fake taxidermy:

> 'Then I'll try to believe you', said the Platypus, clumsily waddling towards some grass, amongst which it settled itself comfortably. 'But it's very difficult to believe you Humans, for you tell such dreadful fibs', it continued, as it squirted some dirty water out of the bag that surrounded its bill, and swallowed some water beetles, small snails and mud that it had stored there. 'See, for instance, the way you have all quarrelled and lied about me! First one great Human, the biggest fool of all, said I wasn't a live creature at all, but a joke another Human had played upon him. Then they squabbled together one saying I was a Beaver; another, that I was a Duck; another, that I was a Mole, or a Rat. Then they argued whether I was a bird, or an animal, or if we laid eggs, or not; and everyone wrote a book, full of lies, all out of his head. That's the way Humans amuse themselves. They write books about things they don't understand, and keep the game going by each new book saying the others are all wrong. It's a silly game, and very insulting to the creatures they write about. ... Us! whose ancestors knew the world mil-

lions of years before the ignorant Humans came on the earth at all!' The
Platypus spluttered out more dirty water, in its indignation. (pp. 25–6)

The animal's speech points out how species classification and the reality of
animal species are fictions—narratives of knowledge, power and violence
embodied by museum specimens, as seen in the previous chapter. Through
the fantasy, therefore, Pedley's platypus lays bare the 'imperial fiction'
Thomas Richards sees in the archive—'the fantasy of knowledge collected
and united in the service of state and Empire'[96]—just as the numerous
books written on the 'curious beastie' mirror the number of specimens
killed, dissected, stuffed or displayed. Moreover, the scene rewrites humorously the relationship between humans and animals since the human child
is, in fact, a simple 'dot', whilst the animal tells its own (natural) (his)story.
As a subversive instructor to the child, challenging humans' superiority
over animals, Pedley's platypus is reminiscent of Nesbit's fairy. The creature is introduced into the narrative because 'there was nothing like that
in [Dot's] Noah's ark' (p. 19). The reason, as the kangaroo explains, is
due to the animal's being made up 'of so many kinds' (p. 19) that it is
refused in the ark and forced to swim for itself—an explanation often given
for the exclusion of dinosaurs and other extinct creatures. This is confirmed by the animal's reference to prehistory, as the 'beastie' posits the
shortness of humans' time-span compared to platypuses':

> 'A million years is a very long time', said Dot; unable at the moment to
> think of anything better to say. But this remark angered the Platypus more,
> for it seemed to suspect Dot of doubting what it said.
>
> It clambered up into a more erect position, and its little brown eyes
> became quite fiery.
>
> 'I didn't say a million; I said millions! I can prove by a bone in my body
> that my ancestors were the Amphitherium, the Amphilestes, the
> Phascolotherium, and the Stereognathus!' almost shrieked the little crea-
> ture. ... 'I trace my ancestry back to the oolite age. Where does man
> come in?'
>
> 'I don't know', said Dot.
>
> 'Of course you don't!', replied the Platypus, contemptuously, 'Humans
> are so ignorant! That's because they are so new. When they have existed a
> few more million years, they will be more like us of old families; they will

---

[96] Thomas Richards, *The Imperial Archive: Knowledge and the Fantasy of Empire* (London
and New York: Verso, 1993), p. 6.

respect quiet, exclusive living, like that of the *Ornithorhynchus Paradoxus*, and will not be so inquisitive, pushing, and dangerous as now. The age will come when they will understand, and will cease to write books, and there will be peace for everyone'. (p. 26)

Pedley's instructor, which equates here writing (about animal species) with violence, is very similar to Nesbit's Sand-fairy, which also teaches the five children about the cruelty of collecting specimens.

As in Pedley's fantasy, *Five Children and It* draws upon natural history. The trilogy opens with the discovery of a creature likely to be collected by the children as an unknown specimen, with the children presented as amateur naturalists and collectors, going on excursions in search of shells, ferns and other collectibles. They observe the creature ('Let's look'[97]), using their senses to see, feel (its fur) or hear (its voice), in order to understand what kind of animal the Psammead may be and assign it to a species. The Psammead recalls the wonder associated with exotic specimens brought back to England, and the children's holiday to the countryside is redolent of the fictional journeys offered to young readers by popularisers of natural history. The 'curious beastie' seems to be an imaginary compound of five different species (symbolising the five children), recalling Australian 'unusual composites'.[98]

> It was worth looking at. Its eyes were on long horns like a snail's eyes, and it could move them in and out like telescopes; it had ears like a bat's ears, and its tubby body was shaped like a spider's and covered with thick soft fur; its legs and arms were furry too, and it had hands and feet like a monkey's. (*Five*, p. 19)

The description of the Psammead hints at comparative anatomy and functionalist deductions. The association of several species, from land and flying mammals to gastropods and arachnids, mixing vertebrates and invertebrates, recalls both analyses of Australasian mammals and of fossil finds, often seen as *collages* of different species too, as already argued.

---

[97] Edith Nesbit, *Five Children and It* [1902] (London: Penguin, 1995), p. 17. All subsequent references are to this edition and will be given parenthetically in the text.

[98] David Rudd, 'Where It Was, There Shall Five Children Be: Staging Desire in *Five Children and It*', in Raymond E. Jones (ed.), *E. Nesbit's Psammead Trilogy: A Children's Classic at 100* (Lanham, Toronto, Oxford: Children's Literature Association and the Scarecrow Press, 2006), pp. 135–49, p. 140.

However, as the story maps out the evolution of the earth,[99] it also points to the possibility of extinction: the Psammead, part-rat, part-snake, 'brown and furry and fat' (*Five*, p. 18), is an endangered (in fact, already condemned) species and is the last of its kind.

If the Psammead trilogy is thus concerned with the issue of extinction, it also denounces that of the commodification of animals, linking animal extinction and commodification through its stance upon animal displays and, therefore, nineteenth-century museum culture. The last volume opens with the children wandering through the streets of London, where can be seen shops 'filled with cages, and all sorts of beautiful birds in them', and 'cats, but the cats were in cages' (*Amulet*, p. 6), whilst 'the dog-shop … [was] not a happy thing to look at either, because all the dogs were chained or caged' (*Amulet*, p. 7). The Psammead is found in a shop that 'only sold creatures that did not much mind where they were – such as goldfish and white mice, and sea-anemones and other aquarium beasts, and lizards and toads, and hedgehogs and tortoises, and tame rabbits and guinea-pigs' (*Amulet*, p. 7). The creature appears 'much thinner than when [they] had last seen it … dusty and dirty … its fur … untidy and ragged' (*Amulet*, p. 8). Kept in dreadful conditions, the now captive Psammead—'the only one ever seen in London'—is above all constructed as a commodity linked to nineteenth-century wild animal trafficking and the rise of exotic pets: the children purchase the animal for 'it'll be only like having a pet dog' (*Amulet*, p. 11).[100] Nevertheless, they then put it 'into a flat bass-bag that had come from Farringdon Market with two pounds of filleted plaice in it' (*Amulet*, p. 16), thereby aligning the extinct 'beastie' with food: '[the flat bass-bag] now contained about three pounds and a quarter of solid Psammead' (*Amulet*, p. 16).

---

[99] Nesbit's interest in geological processes and the history of the earth and its inhabitants filters through many other of her stories, such as 'Uncle James, or The Purple Stranger'. In this short story, in which the children go to the Zoological Gardens, the narrator explains the spinning of the earth. However, a loose piece of earth hits a piece of hard rock and goes spinning the wrong way, leading the animals that live on this island to grow to the wrong size: guinea pigs become the size of elephants and elephants that of dogs. If partaking of the fantasy world of the Kingdom of Rotundia, Nesbit's world yet draws upon insular dwarfism or island gigantism.

[100] The children recurrently treat the Psammead like a pet. Anthea strokes it (*Amulet*, p. 55) and the little black girl the children meet puts it in her lap, taking it for a cat or an organ-monkey (*Amulet*, p. 90).

The creature's shift from prehistoric beast dug out from the sand to living animal displayed at the zoo, and then from pet to food, is interesting, for it follows the trajectory of the evolution of humans' relationship with animals, as Berger explains. Before the nineteenth century in Western Europe, 'animals constituted the first circle of what surrounded man'; 'such centrality was of course economic and productive ... men depended upon animals for food, work, transport, clothing'.[101] In the first stages of the Industrial Revolution, however, animals were more and more used as machines, whilst 'in the so-called post-industrial societies, they were treated as raw materials ... required for food [and] processed like manufactured commodities'.[102] Published at the turn of the twentieth century, Nesbit's fantasy, it could be argued, illuminates the shift Berger describes from machines to commodities (pets, as useless animals, being also considered as commodities), because the 'curious beastie' (mechanically) grants wishes to the children which all turn out to be useless. The lesson ultimately learnt by the five children is thus that of the elusiveness of desire, and the 'curious beastie'—as both a scientific specimen and an impossible (whether magical or extinct) creature—embodies that very desire.[103]

Jean Baudrillard's definition of the collection and systems of collecting is useful to understand the Psammead's passage from an object that 'can be utilized' to one that 'can be possessed'.[104] For Baudrillard, utilisation and possession are the two mutually exclusive functions of any given object. Both imply the subject's control of the object—a 'practical control within the real world' in the first case; 'an enterprise of abstract mastery whereby the subject seeks to assert himself as an autonomous totality outside the world' in the second.[105] In the second case, moreover, the possession of the object 'is always both satisfying and frustrating', since the object, whose meaning is dependent upon the subject, systematically points to 'an extension beyond itself and upsets its solitary status'[106]— hence the collection. In fact, for Baudrillard, the object is 'the perfect pet',

[101] Berger, 'Why Look at Animals', pp. 3–4.

[102] Berger, 'Why Look at Animals', p. 13.

[103] For an analysis of the Psammead as a representation of the children's Id, see Rudd, 'Where It Was, There Shall Five Children Be: Staging Desire in *Five Children and It*', pp. 135–49.

[104] Jean Baudrillard, 'The System of Collecting', translated by Roger Cardinal, in Elsner and Cardinal, *The Cultures of Collecting*, pp. 7–24, p 8.

[105] Baudrillard, 'The System of Collecting', p. 8.

[106] Baudrillard, 'The System of Collecting', p. 8.

6 PREHISTORIC 'BEASTIES'   263

'the ideal mirror [which] reflects images not of what is real, but only of what is desirable'.[107] In Nesbit's fantasy, the Psammead conflates the two functions: by granting wishes, the creature is useful to the children; as a 'curious beastie', it spurs the children's wish to collect the animal. Yet in so doing the creature mirrors not what is real—the Psammead, as an animal—but the children's desire. The prehistoric fairy's wishes thus shape the narrative into a fantasy of consumption which hinges upon the 'dynamic of desire and disenchantment'[108] that, according to Andrew H. Miller, lies at the heart of commodity culture. This is even more true as the Psammead is the last of its species, dooming from the start the children's wish to possess the animal by preventing the collection (the extension beyond itself; the wish for completion) and so, at the same time, their capacity to exercise control over the world.

The wishes around which the narrative revolves find their source in 'The Three Wishes', a fairy tale directly mentioned by one of the children in *Five Children and It* ('I daresay you have often thought what you would do if you had three wishes given to you, and have despised the old man and his wife in the black-pudding story' (*Five*, p. 24)), which was one of the tales collected and published by Joseph Jacobs in *More English Fairy Tales* (1893).[109] But the Sand-fairy's wishes also connect the fantasy to Victorian consumer society. As the Psammead explains, in prehistoric times, the creature was a key link in the chain of consumption, providing pterodactyls, megatheria or plesiosaurs (prehistoric versions of more contemporary animals) to hungry children who consumed animals thoughtlessly, without thinking about the future:

'Where do you get your Megatheriums from now?'
'What?' said the children all at once. It is very difficult always to remember that 'what' is not polite, especially in moments of surprise or agitation.
'Are Pterodactyls plentiful now?' the Sand-fairy went on.
The children were unable to reply.
'What do you have for breakfast?' the Fairy said impatiently, 'and who gives it you?'

---

[107] Baudrillard, 'The System of Collecting', p. 11.

[108] Andrew H. Miller, *Novels behind Glass: Commodity Culture and Victorian Narrative* (Cambridge: Cambridge University Press, 1995), p. 5.

[109] Teya Rosenberg, 'Generic Manipulation and Mutation: E. Nesbit's Psammead Series as Early Magical Realism', in E. Jones (ed.), *E. Nesbit's Psammead Trilogy*, pp. 63–88, p. 72.

264   L. TALAIRACH

'Eggs and bacon, and bread-and-milk, and porridge and things. Mother gives it us. What are Mega-what's-its-names and Ptero-what-do-you-call-- thems? And does anyone have them for breakfast?'

'Why, almost everyone had Pterodactyl for breakfast in my time! Pterodactyls were something like crocodiles and something like birds – I believe they were very good grilled. You see it was like this: of course there were heaps of sand-fairies then, and in the morning early you went out and hunted for them, and when you'd found one it gave you your wish. People used to send their little boys down to the seashore early in the morning before breakfast to get the day's wishes, and very often the eldest boy in the family would be told to wish for a Megatherium, ready jointed for cooking. It was as big as an elephant, you see, so there was a good deal of meat on it. And if they wanted fish, the Ichthyosaurus was asked for – he was twenty to forty feet long, so there was plenty of him. And for poultry there was the Plesiosaurus; there were nice pickings on that too. Then the other children could wish for other things. But when people had dinner-parties it was nearly always Megatheriums; and Ichthyosaurus, because his fins were a great delicacy and his tail made soup'. (*Five*, pp. 21–2)

In the first volume of the trilogy, the five children are allowed one wish per day, each wish turning to stone at sunset. Unfortunately, each wish systematically leads the children to more frustration: their wish to be beautiful causes them not to be recognised; their wish to be rich provides them with money which is no longer in circulation and cannot be used; their wish for wings leads them to the top of a church bell tower from which they are unable to get down; their wish to be bigger turns Robert into a giant, who is then exhibited at a fair and so on. Each wish thus leads to increased dissatisfaction until the children agree never to ask for any more wishes.

It is significant that in the process the children should also learn about their proximity to animals. Because the children's encounter with the monkey-like Psammead[110] whose ancestors have all become extinct evokes humanity's own ancestry, the creature establishes links between the children and animals which are developed throughout the fantasy and its sequels.[111] After her first wish, Anthea, who wishes them all to be 'beautiful as the day' (*Five*, p. 24), dreams that she is walking in the Zoological

---

[110] The Psammead's simian features are strengthened in the last volume as it is described as a 'mangy monkey' (*Amulet*, p. 9).

[111] Edith Nesbit, *The Story of the Amulet* (London: T. Fisher Unwin, 1906), p. 24. *The Story of the Amulet* even makes explicit that because the Psammead is a prehistoric creature,

Gardens and hears animals growling. The growling is, in fact, her sister's breathing, but the dream suggests that the girls' physical transformation—their 'beautification', as Alice would put it in Wonderland—has symbolically revealed their own bestiality, perhaps because the girls were 'donkeys enough to ask for [them] all to be beautiful as the day' (*Five*, p. 32). The children, whose nicknames—Lamb, Panther, Pussy, Squirrel—associate them with animals, are constantly equated with 'savages', freaks or animals. Robert, who cannot control himself, becomes 'a perfect savage' (*Five*, p. 145), and is ultimately turned into a freak—a giant—exhibited at a fair. Likewise, Cyril and Anthea believe they are 'beast[s]' (*Five*, pp. 166, 167) and have to face 'untutored savages' (*Five*, p. 191)—Indians—who want to scalp and roast them. Cyril becomes chief Squirrel of the Moning Congo tribe, while Anthea is the 'Black Panther', chief of the Mazawattee tribe and Jane Wild Cat, leader of the Phiteezi tribe. The same fate is reserved for Lamb, who appears 'hungry as a lion', scratches 'like a cat', bellows 'like a bull' (*Five*, p. 27) and roars.

The association of the children with animals, with prehistoric creatures (the Psammead) and with 'savage or barbarous peoples, [creatures who] lacked the civilized virtues, behaving like children (the Victorian "little savages")',[112] allows Nesbit to denounce Britain's imperialism and exploitation of natural resources. Eitan Bar-Yosef has shown how Nesbit's trilogy tackles the 'imperial crisis' and hinges upon a 'reworking of the reverse colonization narrative'.[113] As he argues, Nesbit's subversion of the conventions of imperial gothic tales (notably through 'ridiculing the clichés of imperial popular culture') 'offer[s] the children of Empire an alternative political – and literary – ethos'.[114] Mavis Reimer, drawing upon Bar-Yosef's analysis, contends further that the five children, having been turned into 'exotic and strange creatures, with bodies out of their control', 'manage their imperial adventures in part by taking the role of the objectified other'.[115] But as they experience otherness, the children also understand

---

it may not be as mentally developed as the children: 'For a creature that had in its time associated with Megatheriums and Pterodactyls, its quickness was really wonderful' (*Amulet*, p. 68).

[112] Carole G. Silver, *Strange and Secret Peoples: Fairies and Victorian Consciousness* (Oxford: Oxford University Press, 1999), p. 150.

[113] Eitan Bar-Yosef, 'E. Nesbit and the Fantasy of Reverse Colonization: How Many Miles to Modern Babylon?', *English Literature in Transition*, 46.1 (2003): 5–28, p. 6.

[114] Bar-Yosef, 'E. Nesbit and the Fantasy of Reverse Colonization', pp. 11, 6.

[115] Mavis Reimer, 'The Beginning of the End: Writing Empire in E. Nesbit's Psammead Books', in Jones (ed.), *E. Nesbit's Psammead Trilogy*, pp. 39–62, p. 44.

how, as Randy Malamud's puts it, 'anthropocentric fantasies of natural supremacy' 'distort[–] our biological role as one part of an interrelated ecosystem'.[116] Indeed, from the beginning of the fantasy, the children want to see the Psammead the better to possess it, like miniature nineteenth-century natural history collectors. Both magical and natural, hence shaped as a natural wonder, the Psammead is instantly commodified: it is 'worth looking at'. This objectification climaxes when the creature is called a 'thing', whilst the pronoun 'it' is used throughout to qualify the creature. In addition, the first thing Cyril wants to do is to tame it ('If it is a snake, I'll tame it, and it will follow me everywhere' [*Five*, p.17]), and Jane proposes to take it home. Because the children fear that the creature may 'get away' (*Five*, p. 17), they thus hurry to capture it. Ironically, the Psammead, as an extinct creature magically resurrected, symbolises humans' unreasonable consumption of natural resources. As the animal revisits a fairy-tale tradition of creatures granting wishes, it reworks in so doing the 'desire to *possess* animals, whether as trophies, or as adoptees',[117] as we saw in Chap. 4. Whilst in *The Phœnix and the Carpet*, the children must learn to let go of the phœnix and its egg, as well as of the carpet, by the third volume of the trilogy, the ultimate wish is that the Psammead return to the past. The fantasies close therefore with the children's dispossession of their 'curious beasties', putting an end to Nesbit's 'fantasies of consumption',[118] to borrow Andrew Miller's words.

We saw in the first part of this section how the Crystal Palace Gardens were used to map out the growth of Nesbit's young characters, pointing out in so doing the role that museums could play in children's evolution. According to Ruth Hoberman, 'the role of the museum as an incarnation of the "nation" shaping the lives of its characters' was widely explored in Edwardian fiction. It could offer 'a kind of conversation about the relationship of the self to the past, to materiality, and to the nation' or function, on the contrary, as 'an oppressive presence against and within which characters define their desires'.[119] As she explains, at the turn of the century, the museum became a much 'contested and conflicted site'; standing, on the one hand, for 'the ideal of a unified system of knowledge' and,

---

[116] Malamud, *Reading Zoos*, p. 271.

[117] Diana Donald, *Women against Cruelty: Protection of Animals in Nineteenth-Century Britain* (Manchester: Manchester University Press, 2020), p. 149.

[118] Andrew H. Miller, *Novels behind Glass*, p. 1.

[119] Ruth Hoberman, *Museum Trouble. Edwardian Fiction and the Emergence of Modernism* (Charlottesville and London: University of Virginia Press, 2011), p. 25.

on the other, 'the limits of scientific materialism'.[120] Hoberman's study traces the evolution of museums in the second half of the nineteenth century, underlining the way in which the growth of museum collections echoed 'the development of mass production, advertising, and department stores, all of which work[ing] together to make the commodity ... the "master fiction around which society organized and condensed its cultural life and political ideology"'.[121] As a consequence, she explains, the tension 'between the ideal of aesthetic transcendence' and 'the realities of commodification'[122] was often negotiated in Edwardian fiction.

This is undoubtedly the case in the Psammead series. The last part of the trilogy illuminates Nesbit's stance on imperialism and its collections of riches by drawing upon 'the perfect specimen' (*Amulet*, p. 135)—the complete amulet of which the children only own half, purchased in a shop, and which they eventually give to the 'learned gentleman'[123] to be displayed in the British Museum. For Bar-Yosef, the British Museum, as a 'great "Imperial Archive"', is 'both a metaphor for and a metonymy for British imperialism' which 'becomes a central reference-point in [the children's] lives'.[124] Moreover, as Barbara Black argues, the amulet found by the children in a curiosity shop 'is a synecdoche for the museum itself':[125] it 'reminds us of the exotic nature of museums', and 'as the archway for the children's time travel, [it] emphasizes the inclusiveness of museums'.[126] Moreover, according to Hoberman, in *The Story of the Amulet*, the museum functions as a 'refuge from commodification'.[127] By ultimately placing the complete amulet in the British Museum, the children realise 'the whole project of the museum ... as an effort to resacralize displayed

[120] Hoberman, *Museum Trouble*, p. 41.

[121] Hoberman, *Museum Trouble*, pp. 10–11. Hoberman cites Thomas Richards, *The Commodity Culture of Victorian England: Advertising and Spectacle, 1851–1914* (Stanford: Stanford University Press, 1990), p. 53.

[122] Hoberman, *Museum Trouble*, p. 41.

[123] The 'learned gentleman' is based on the figure of Ernest Wallis Budge (1857–1934), Keeper of the Egyptian and Assyrian antiquities at the British Museum, who helped Nesbit in her research. *The Story of the Amulet* is dedicated to him.

[124] Bar-Yosef, 'E. Nesbit and the Fantasy of Reverse Colonization', p. 12.

[125] Barbara J. Black, *On Exhibit: Victorians and their Museums* (Charlottesville and London: University Press of Virginia, 2000), p. 159. Bar-Yosef also sees the amulet as the 'Lilliputian version that is the learned gentleman's study, or the British Museum itself'; Bar-Yosef, 'E. Nesbit and the Fantasy of Reverse Colonization', p. 21.

[126] Black, *On Exhibit*, pp. 159–60.

[127] Hoberman, *Museum Trouble*, p. 25.

objects to compensate for mass commodification'.[128] Thus, *The Story of the Amulet* 'dramatizes precisely [the] contrast between commodification and the museum' by constructing the museum 'as a space from which desire has been expelled, a space that provides a safe haven from the collector's desires, the dealer's greed, and the marketplace uncertainties'.[129]

The children's fantastic adventures throughout the trilogy are, as a matter of fact, spurred by their desires, as already argued, especially their appetite for curious and exotic items. In *The Phœnix and the Carpet*, the children's voyage through space and time[130] results from the purchase of a carpet for the children's mother's collection of 'Indian things' for her 'bazaar'.[131] By drawing upon the marketing and consumption of foreign commodities, Nesbit presents colonialism as 'a vehicle of capitalism'.[132] The carpet functions in ways similar to the Psammead in *Five Children and It*, as it grants the children three wishes a day and transports them anywhere they wish. According to Teya Rosenberg, the children's attitude to the carpet 'is imperialist, for they make use of the carpet's qualities with little regard to its cultural significance or the possible limits of its resources', whilst the Phoenix, on the other hand, functions as 'the anti-imperialist voice, tr[ying] a number of times to draw attention to the problems inherent in that attitude'.[133] Indeed, the possibility of journeying through space and time offered by the carpet gives the children a chance to experience the British empire, as the children choose to explore the North Pole, the Equator and the diamond mines of Golconda (*Phœnix*, p. 94), and to

[128] Hoberman, *Museum Trouble*, p. 41.

[129] Hoberman, *Museum Trouble*, p. 42.

[130] In *The Phoenix and the Carpet*, the phoenix, or Psammead of the Desert, takes the lead. Both the Psammead and the phoenix are, to some extent, resurrected creatures. Moreover, the stress on the phoenix's eggs, together with the space travel the phoenix enables the children to experience, calls to mind the world of palaeontology which informs *Five Children and It*, highlighting the way in which the emerging scientific discipline impacted contemporary debates on the conceptions of time and space and the issue of extinction.

[131] Edith Nesbit, *The Phœnix and the Carpet* [1904] (Ware: Wordsworth Editions, 1995), p. 78. All subsequent references are to this edition and will be given in the text.

[132] For Bar-Yosef, Nesbit follows the English economist John Atkinson Hobson's view of 'imperialism as a direct result of underconsumption at home and capitalism's consequent search for ever-expanding markets abroad', constructing therefore imperialism as 'a retrograde social development, a backsliding towards barbarism'. By de-colonising or reverse colonisation, her 'socialist vision would be realized'; Bar-Yosef, 'E. Nesbit and the Fantasy of Reverse Colonization', p. 18.

[133] Rosenberg, 'Generic Manipulation and Mutation', p. 85.

bring back specimens to England, thereby plundering the colonies' natural resources. Desire informs their quest for objects typifying the coloniser's paraphernalia ('We ought to have an explorer's axe... I shall ask father to give me one for Christmas' [*Phœnix*, p. 62]), whilst the children discover exotic places, with 'tangled creepers with bright, strange shaped flowers', 'curtains of creepers with scented blossoms hanging from the trees, and brilliant birds' (*Phœnix*, p. 62), and encounter 'savages' who might be 'cannibals' and who wear 'hardly any clothes' (*Phœnix*, p. 63). Their exploration of the island and their discoveries, hinting at natural history through the descriptions and collections of unknown species,[134] are systematically bound up with a desire to own 'foreign curiosities' (*Phœnix*, p. 180).

For instance, when they explore the island, their gaze instantly turns the natural specimens into purchasable commodities, as when they see 'rich tropic shells of the kind you would not buy in the Kentish Town Road under at least fifteen pence a pair' (*Phœnix*, p. 64), or when the carpet brings 199 'beautiful objects' which are, in fact, 'Persian cats' (*Phœnix*, p. 142). The fantasy thus shows how adventure fiction 'rehearsed the colonial experience and offered a conduit to capitalism's economic and individual desires' by 'codif[ying] and mapp[ing] – in other words taxonomically classif[ying] – the empire's exotic objects, lands and peoples so that they could be more readily consumed by a wide spectrum of the public, in particular, the middle class', as Ymitri Mathison points out.[135] Moreover, the commodification of exotic specimens is linked with their display, and defined above all as (consumerist) spectacle, as when Robert believes that the 199 Persian cats are 'a zoological garden of some sort' (*Phœnix*, p. 142). As the narrative is punctuated by references to the Crystal Palace (*Phœnix*, p. 32), as well as to Paris museums and picture galleries (*Phœnix*, p. 179), it soon becomes obvious that the children's knowledge about the unknown world they encounter is framed by narratives of power which define the objects with which they come into contact.

From the beginning of the narrative, the narrator invites readers to check their knowledge of mummies by visiting the British Museum, Kew

---

[134] One of the characters (the clergyman) wanders 'about the island collecting botanical specimens' (*Phœnix*, p. 177).

[135] Ymitri Mathison, 'Maps, Pirates and Treasure: The Commodification of Imperialism in Nineteenth-Century Boys' Adventure Fiction', in Dennis Denisoff (ed.), *The Nineteenth-Century Child and Consumer Culture* (Aldershot: Ashgate, 2008), pp. 173–85, p. 174.

Gardens, the Tower of London, Madame Tussaud's Exhibition, the Crystal Palace, the Zoological Gardens—all venues which punctuate *The Story of the Amulet* and represent Britain's imperialist culture. Museums are compared and contrasted as when Anthea wonders if the pharaoh's house in Egypt is 'like the Egyptian Court in the Crystal Palace' (*Amulet*, p. 101). The Phœnix is likewise recognised because the children have 'seen a picture' (*Phœnix*, p. 24) of it. Similarly, the children's mother believes the Phœnix to be 'an orange-coloured cockatoo' (*Phœnix*, p. 208), which testifies to her familiarity with exotic species then exhibited in menageries. Moreover, animals are systematically associated with representation/performance, as when the children's play 'wild beasts in a cage under the table' (*Phœnix*, p. 213) or the 'Noah's Ark game' (*Phœnix*, p. 33), a game which consists in choosing an animal and becoming that animal, at least through a piece of poetry that turns them into a bear, eel, hedgehog, snake, crocodile or rabbit, although 'some of the animals, like the zebra and the tiger, haven't got any poetry, because they are so difficult to rhyme to' (*Phœnix*, p. 33). *The Story of the Amulet* is also shot through with links between the commodification of animals and their display or representation: animals and animal parts are sold; 'hippopotamus flesh' and 'ostrich-feathers' (*Amulet*, p. 41) are used by the natives the children encounter, and the doorway of one of the huts has 'a curtain of skins' hanging over it (*Amulet*, p. 42). In addition, the children recognise objects or animals which they have seen at the British Museum (the amulet, a necklace [*Amulet*, p. 69], a shilling compass [*Amulet*, p. 125]) or at the Zoo; they ride creatures that are wilder than those they can ride at the London Zoo (*Amulet*, p. 82); and meet people who make collections (of wives, such as the King [*Amulet*, p. 59], or of wild beasts, like the Pharaoh, who wants to keep the Psammead, believing it to be '[a] very curious monkey' [*Amulet*, p. 104]).

The central part played by the British Museum in *The Story of the Amulet* illuminates how the 'magic' of the story revolves around objects which can be collected, purchased and displayed in glass cases. By enabling travel through space and time, museum objects illuminate the fantasy's 'narrative arc'.[136] The stories embodied by each object punctuate and drive the children's quest, from the past to the present, from ancient civilisations to London—the 'modern Babylon' (*Amulet*, p. 14). In so doing, as Virginia Zimmerman has shown, the museum setting 'highlights the

[136]Zimmerman, 'The Curating Child', p. 42.

relationship between objects and identity, whether cultural or personal'. In the last volume of the Psammead series, the amulet thus 'serves at once to lead the children to assorted cultures and times and to reunite their family'.[137] Moreover, functioning symbolically, and containing the essence of foreignness, exoticism or past worlds, Nesbit's objects become therefore portals, allowing entry into fairyland—in fact, imperial Britain—reduced, as a result, to the realm of the marvellous and that of fiction. Seen through the lens of the Victorian museum culture, the five children's journey into colonisation thus ultimately lays bare the imperial myth, denying in so doing the latter's role in the formation of the children's identity.

However, although the amulet eventually finds a safe place in the museum, it is significant that Nesbit's 'curious beasties' challenge Hoberman's construction of the museum in the fantasy as a 'refuge from commodification'.[138] Unlike the inanimate amulet, both the Phoenix—'the anti-imperialist voice'[139]—and the Psammead vanish at the end of the fantasies, subverting thereby the imperial capture/closure. Indeed, even if both the amulet and the Psammead disappear from the shop, the Psammead nonetheless refuses to be 'made a show of' (p. 137), wishing himself rather safe in the past. The prehistoric fairy thus refuses to board the Ark; it gives a twist to the 'completist's' dream and prevents the 'control over existence itself'[140] that the complete amulet, like Noah's collection of 'beasties', represents, as icons of the perfect museum. It is therefore no coincidence that the Psammead's dream comes true when the 'learned gentleman', politely, yet unconsciously, repeats the Sand-fairy's wish. The 'learned gentleman' is 'the repository of knowledge about other cultures and, hence, the preserver of British imperial ideology',[141] as Raymond E. Jones observes. While the children 'relinquish their right to be imperial travellers'[142] and give him the amulet to keep it safe in the British Museum, it may be thought that imperial ideology is preserved and will remain on display in the galleries of the museum for ever—that, as Bar-Yosef puts it,

[137] Zimmerman, 'The Curating Child', p. 42.

[138] Hoberman, *Museum Trouble*, p. 25.

[139] Rosenberg, 'Generic Manipulation and Mutation', p. 85.

[140] Elsner and Cardinal, 'Introduction', *The Cultures of Collecting*, p. 3.

[141] Raymond E. Jones, 'Introduction', Jones (ed.), *E. Nesbit's Psammead Trilogy*, pp. xii–xxv, p. xix.

[142] Reimer, 'The Beginning of the End: Writing Empire in E. Nesbit's Psammead Book', p. 58.

the 'Imperial Archive is stronger than magic'.[143] Yet, the prehistoric 'beastie' is, simultaneously, whisked away for ever, refusing to be kept in the 'Imperial Archive'. This dual resolution, which reveals the limits of Nesbit's 'attack on imperial politics', as Bar-Yosef has argued, nevertheless unambiguously frees the animal from the collection. It is tempting to believe that the lesson taught to the five children and their readers by the creature's last wish is that extinct 'beasties' on display, like the tigers, lions, elephants and platypuses exhibited at the zoological gardens, could offer merely impossible encounters. If so, the sense of self Nesbit's children may have gained in the course of their adventures through Victorian museum culture is then, perhaps, above all, ecological.

[143] Bar-Yosef, 'E. Nesbit and the Fantasy of Reverse Colonization', p. 21.

CHAPTER 7

Epilogue

Poor Tradescant, if he could see his ark now, would no doubt weep over the disappearance of his 'Dragon's Egge', and head of a Griffin, though it might be some consolation to find that the head and leg of his Dodo had been saved from destruction. But his spirit would at last be gratified and surprised when, as he passed through the hall and galleries of the new museum, under the courtly and pleasant guidance of the present keeper, he saw what a noble palace, full of rich treasures, his old hobby, like a second Aladdin's lamp, had conjured up.[1]

We have seen throughout *Animals, Museum Culture and Children's Literature in Nineteenth-Century Britain: Curious Beasties* that over the course of the Georgian, Victorian and Edwardian periods, children's literature was inextricably bound up with the material culture of the age. This body of literature reflected constantly the ways in which worlds were mapped, collections made and public venues opened up for visitors eager to see curious animals, peoples and objects. It also brought together 'curious beasties' and museum spaces, offering young readers 'portals to other worlds' in ways similar to zoos, 'evok[ing] both wild things *and* their wild

---

[1] Rev. H. H. Wood, 'Marvels, Ancient and Modern', *Aunt Judy's Christmas Volume for Young People* (London: Bell and Daldy, 1867), pp. 94–8, p. 96.

© The Author(s), under exclusive license to Springer Nature Switzerland AG 2021
L. Talairach, *Animals, Museum Culture and Children's Literature in Nineteenth-Century Britain*, Palgrave Studies in Animals and Literature, https://doi.org/10.1007/978-3-030-72527-3_7

273

worlds'.[2] Like in a museum, children's writers displayed new, old, extinct or impossible creatures for children to see, devising stories that made the invisible more and more visible. Just like palaeontologists resurrected creatures of the prehistoric past, children's books conjured up the British empire and its booming capitalist economy through adventures where characters could both feel and see the urge to possess and subjugate these 'curious beasties'. From the origins of children's literature in the middle of the eighteenth century to the end of the Edwardian period, children were therefore invited to participate, albeit vicariously, in the colonial experience. At times, they were also given the opportunity to question the ideologies of imperialism.

The 'curious beasties' we have encountered throughout this study were not just, therefore, or not simply, vehicles for instruction. As Amy Ratelle has argued, children's literature often 'present[s] the boundary between humans and animals as, at least, permeable and in a state of flux'.[3] Indeed, this survey of children's literature from the end of the Georgian era to the Edwardian period has demonstrated that zoological and palaeontological displays presenting amazing 'beasties' to children reveal in striking fashion the tensions informing contemporary discourses on the relationship between humans and animals. Displays of curious literary 'beasties' could shift easily from 'well-regulated space[s]' reflecting the coherence of imperial Britain to Wonderlands—fantastic cabinets of curiosities foregrounding their 'resistance to order'.[4]

Moreover, the 'curious beasties' we have examined in this book were representations of creatures which borrowed from or alluded to real animals exhibited in myriad forms throughout the nineteenth century—whether living, stuffed, skeletonised, dried or bottled in preserving fluid. These literary representations tricked readers in much the same way that taxidermied animals deceived audiences in a museum. Pretending to be alive yet dead, authentic yet products of mimetic realism, just like artificially mounted or reconstructed creatures, literary 'curious beasties' were used by Georgian educationalists as symbols of royal power to mould children into 'sensible' citizens; they also functioned as signs of imperial power

[2] Andrew Flack, *The Wild Within: Histories of a Landmark British Zoo* (Charlottesville and London: University of Virginia Press, 2018), p. 69.

[3] Amy Ratelle, *Animality and Children's Literature and Film* (New York: Palgrave Macmillan, 2015), p. 4.

[4] Ruth Hoberman, *Museum Trouble. Edwardian Fiction and the Emergence of Modernism* (Charlottesville and London: University of Virginia Press, 2011), p. 135.

in travel writing, adventure fictions and popular science articles alike. Tied as they were to realistic representation, their existence as a whole was a chimera upon which to build pedagogical strategies to raise children. Furthermore, seen through the prism of museum culture, these 'curious beasties' lay bare the way in which materiality and narrative are constantly woven together into unfamiliar creatures, made up of twisted threads, as in nonsense texts, which playfully reveal the skein of colonial ideology once disentangled. They crystallised the imperial myth, functioning as stories-within-stories, morbid mementoes of distant lands and times that captured the story of their own capture; signifiers able to prompt 'experiential readings', to quote Rachel Poliquin,[5] and that helped to question the anthropocentric belief in humans' ascendancy over nature.

Thus, it has been suggested throughout this study of 'curious beasties' in the long nineteenth century that Georgian, Victorian and Edwardian pedagogues, by systematically shaping wild, exotic or unknown creatures as objects of scientific investigation and knowledge, foregrounded the human-animal relationship. Although the strange creatures remained objectified, commodified and seen through an anthropocentric lens, the increased insight into their behaviour proposed by these writers nevertheless revealed changes in the constructions of such unfamiliar animals—sometimes rewriting the relationship between wildness and lack of concern for animal preservation, especially in the last decades of the nineteenth century. The idea that species were superabundant and 'available for almost endless acquisition'[6] was a striking presence in many children's texts foregrounding wild and exotic animals. It was an idea that helped provide the rationale for the mass destruction of wildlife in order to supply 'a consistently comprehensive stock of animals'.[7] At the cusp of the twentieth century, however, whilst 'museums and government surveys dispatched dozens or scores of expeditions a year to collect specimens and map the biogeography of species',[8] more and more campaigns for nature conservation emerged, gradually shaping a 'system of animal valuation' that positioned 'the natural environments of the world as simultaneously

---

[5] Rachel Poliquin, 'The Matter and Meaning of Museum Taxidermy', *Museum and Society*, 6.2 (2008): 123–34, p. 130.

[6] Flack, *The Wild Within*, p. 32.

[7] Flack, *The Wild Within*, p. 32.

[8] Robert E. Kohler, *Naturalists, Collectors, and Biodiversity, 1850–1950* (Princeton and Oxford: Princeton University Press, 2006), p. xi. Kohler's study focuses on American collections and collectors.

worth cherishing and also – still – extrinsically valuable as resources in the contexts of science and commerce'.[9] As Harriet Ritvo has shown, moreover, the development of wildlife sanctuaries and national parks in the late nineteenth century resulted from 'shifts in scientific understanding, which redefined high-end predators as a necessary element of many natural ecosystems'.[10] In addition, in natural history museums, as habitat dioramas appeared in the 1890s, stuffed creatures were 'more and more displayed in less static and more dramatic poses', and 'realistic exhibits of animal groups [presented] in strikingly naturalistic settings'.[11]

Over the same period, children's literature remained a significant barometer of the evolution of display practices. While not always an obvious agent of change in itself, this body of literature nonetheless reflected growing concerns about animal presentation and exploitation. Although John Tradescant's 'Dragon's Egge' and Griffin head disappeared from his Ark, children's writers dreamt of zoos that acquired a new dynamic quality—places where 'curious beasties' were free to roam the world and embark upon a voyage back to their native lands, as Alice Dew-Smith imagined. Let us therefore board her 'second Noah's Ark' to conclude this study:

> As a child, one of my favourite dreams was that of a second Noah's Ark. Into it all the animals at the Zoo and elsewhere in captivity should march two by two, and, journeying round the world, it should deposit them all on the shores of their native land. It was a pleasant dream...[12]

---

[9] Flack, *The Wild Within*, p. 45.

[10] Harriet Ritvo, 'Calling the Wild', in Joan B. Landes, Paula Young Lee and Paul Yougquist (eds), *Gorgeous Beasts: Animal Bodies in Historical Perspective* (University Park: Penn State University Press, 2012), pp. 105–16, p. 107.

[11] Kohler, *Naturalists, Collectors, and Biodiversity*, p. 49.

[12] Alice Dew-Smith, *Tom Tug and Others. Sketches in a Domestic Menagerie* (London: Seeley and Co. Limited, 1898), p. 70.

SELECT BIBLIOGRAPHY

Aesop. *Aesop's Fables. With instructive morals and reflections, abstracted from all party considerations, adapted to all capacities; and design'd to promote religion, morality, and universal benevolence* (London: J. F. and C. Rivington, T. Longman, B. Law, W. Nicol, G. G. J. and J. Robinson, T. Cadell, R. Balwin, S. Hayes, W. Goldsmith, W. Lowndes, and Power and Co., ?1775).

Aesop. *Bewick's Select Fables, In Three Parts* (Newcastle: Thomas Saint, 1784).

Aesop. *Old Friends in a New Dress; or, Select Fables of Aesop, in verse* (London: Darton & Harvey, 1809).

Aikin, John, and Anna Laetitia Barbauld. *Evenings at Home; or, the Juvenile Budget Opened. Consisting of a Variety of Miscellaneous Pieces, for the Instruction and Amusement of Young Persons* (London: J. Johnson, 1792).

Alberti, Samuel J. M. M. 'The Museum Affect: Visiting Collections of Anatomy and Natural History', in Aileen Fyfe and Bernard Lightman (eds), *Science in the Marketplace: Nineteenth-Century Sites and Experiences* (Chicago and London: The University of Chicago Press, 2007), pp. 371–403.

Allen, David Elliston. *The Naturalist in Britain: A Social History* (Princeton: Princeton University Press, [1976] 1994).

Allman, George James. 'Critical Notes on the New Zealand *Hydroida*', *Proceedings of the New Zealand Institute*, 8 (1875): 298–302.

Allman, George James. 'Description of Australian, Cape and other Hydroida, mostly new, from the collection of Miss H. Gatty', *Journal of the Linnean Society*, 19 (1885): 132–61.

© The Author(s), under exclusive license to Springer Nature Switzerland AG 2021
L. Talairach, *Animals, Museum Culture and Children's Literature in Nineteenth-Century Britain*, Palgrave Studies in Animals and Literature, https://doi.org/10.1007/978-3-030-72527-3

278 SELECT BIBLIOGRAPHY

Allman, George James. *Letters to Miss Gatty on coelenterates, from G. J. Allman and G. Busk.* Zoology Manuscripts MSS GAT. Natural History Museum, London. 1875–1896.

A. L. O. E. [Charlotte Maria Tucker]. *The Rambles of a Rat* [1857] (London, Edinburgh and New York: T. Nelson and Sons, 1864).

Altick, Richard. *The Shows of London* (London and Cambridge, Mass.: Belknap Press of Harvard University Press, 1978).

Anderson, Vicky. *The Dime Novel in Children's Literature* (Jefferson and London: McFarland and Company, 2005).

[Anon.]. *The ABC of Animals* (New York: McLoughlin Bros., c. 1880).

[Anon.]. *The ABC of Animals* (New York: McLoughlin Bros., c. 1899).

[Anon.]. *The Animal Picture Book for Kind Little People* (London: Ward, Lock and Tyler, ?1873).

[Anon.]. 'The Aquatic Vivarium at the Zoological Gardens, Regent's Park', *The Illustrated London News* (28 May 1853): 420.

[Anon.]. 'Aunt Mary's Pets', *Aunt Judy's Christmas Volume for Young People* (London: Bell and Daldy, 1870), pp. 299–305.

[Anon.]. *Birds, Beasts, Fish and Insects to Teach Little Folks to Read* (London: W. Darton, 1823).

[Anon.]. *The Cat's Concert* (London: C. Chapple, 1808).

[Anon.]. 'The Crystal Palace at Sydenham', *Illustrated London News*, 23 (31 Dec. 1853): 599.

[Anon.]. 'Education Exhibition', *Journal of the Society of Arts*, 78.2 (19 May 1854): 443–9.

[Anon.]. *The Fancy Fair; or Grand Gala at the Zoological Gardens* (London: John Harris, 1832).

[Anon.]. 'Fifty Years Ago. Some of an Old Man's Recollections of London in His Childhood', *Aunt Judy's Christmas Volume for Young People* (London: Bell and Daldy, 1874), pp. 748–52.

[Anon.]. 'Foundling Birdie', *Aunt Judy's Yearly Volume for Young People* (London: Bell and Daldy, 1869), pp. 15–18.

[Anon.]. 'Greenhouse forbidden', *Aunt Judy's Christmas Volume for 1875* (London: Bell and Daldy, 1875), pp. 586–91.

[Anon.]. *An Historical Account of the Curiosities of London and Westminster, in Three Parts* (London: Newbery and Carnan, 1769).

[Anon.]. *The History of Little Goody Two-Shoes; otherwise called, Mrs. Margery Two-Shoes: with the Means by which she acquired her Learning and Wisdom, and in Consequence thereof her Estate* [1765] (London: T. Carnan and F. Newbery, 1772).

[Anon.]. 'Involuntary Contributions', *Aunt Judy's Magazine for Young People. The Christmas Volume for 1866* (London: Bell and Daldy, 1866), pp. 12–17.

SELECT BIBLIOGRAPHY 279

[Anon.]. 'The Little Bird Who Told Stories', *Aunt Judy's Magazine for Young People. The Christmas Volume for 1866* (London: Bell and Daldy, 1866), pp. 332–5.

[Anon.]. *The Menageries. Quadrupeds, Described and Drawn from Living Subjects*, Vol. 1, The Library of Entertaining Knowledge (London: Charles Knight, 1829).

[Anon.]. 'More Last Words', *Punch*, 15 (1848): 243.

[Anon.]. *My Zoo Animals* (London, New York: Raphael Tuck and Sons, c. 1890).

[Anon.]. *New Battledore of Natural History* (London: G. Martin, c. 1810).

[Anon.]. 'Our Eye-Witness and a Salamander', *All the Year Round* (19 May 1860): 140–44.

[Anon.]. 'Parlour Aquaria', *Family Friend*, 2 (1856): 192–7.

[Anon.]. 'Rather a Long Walk', *Aunt Judy's Yearly Volume for Young People* (London: Bell and Daldy, 1869), pp. 351–8.

[Anon.]. 'Scene in an Elephant Kraal. Extract from a Letter of the Bishop of Colombo', *Aunt Judy's Yearly Volume for Young People* (London: Bell and Daldy, 1869), pp. 367–8.

[Anon.]. 'Sea-Gardens', *Household Words*, 14 (1856): 244.

[Anon.]. *A Short and Easie Method to give Children an Idea or True Notion of Celestial and Terrestrial Beings* (London: n.p., 1710).

[Anon.]. *The Zoological Gardens* (London: Frederick Warne and Co., c. 1875).

[Anon.]. *The Zoological Gardens* (London: Frederick Warne and Co., c. 1880).

[Anon.]. 'Zoological Society of London Secretary's Report', *The Zoologist*, 12 (28 February 1854): 4277.

Antinucci, Raffaella. '"Sensational Nonsense": Edward Lear and the (Im)purity of Nonsense Writing', *English Literature*, 2.3 (Dec. 2015): 291–311.

Audubon, John James. *The Birds of America* [1827–38] (London: The Natural History Museum, 2011).

Austen, Jane. *Sense and Sensibility* [1811] (Oxford: Oxford University Press, [1980] 2008).

[B., W]. *The Elephant's ball and Grand Fete Champetre* (London: J. Harris, 1808).

Bagnold, E. S. H. 'Friends and Acquaintances', *Aunt Judy's Christmas Volume for Young People* (London: Bell and Daldy, 1874), pp. 501–508; 559–66; 631–36; 654–63; 728–38.

Ball, Mieke. 'Telling Objects: A Narrative Perspective on Collecting', in John Elsner and Roger Cardinal (eds), *The Cultures of Collecting* (Cambridge, Mass.: Harvard University Press, 1994), pp. 97–115.

Barbauld, Anna Laetitia. *Lessons for Children* [1778–79] (London: J. F. Dove, 1830).

Barber, Lyn. *The Heyday of Natural History: 1820–1870* (Garden City, New York: Doubleday and Company, Inc., 1980).

Barker, Mary Anne. 'About Monkeys', *Good Words for the Young* (1 Feb. 1870): 196–201.

Barker, Mary Anne. 'Adventures', *Good Words for the Young* (1 Feb. 1871): 191–8.

280 SELECT BIBLIOGRAPHY

Barker, Mary Anne. 'Aunt Annie's Stories about Horses', *Good Words for the Young* (1 Sept. 1870): 600–605.

Barker, Mary Anne. 'Aunt Annie's Story about Jamaica. Part II', *Good Words for the Young* (1 July 1870): 518–23.

Barker, Mary Anne. *Boys* (London: G. Routledge & Sons, 1874).

Barker, Mary Anne. 'A Chapter of Accidents', *Good Words for the Young* (1 June 1871): 438–45.

Barker, Mary Anne. 'Four Months in Camp', *Good Words for the Young* (1 Oct. 1870): 645–57.

Barker, Mary Anne. 'The Marble Cross', *Good Words for the Young* (1 Aug. 1871): 530–36.

Barker, Mary Anne. 'More Adventures', *Good Words for the Young* (1 March 1871): 267–76.

Barker, Mary Anne. *Sibyl's Book* (London: Macmillan & Co., [1872] 1874).

Barker, Mary Anne. *Station Amusements in New Zealand* [1873] (Christchurch: Whitcombe & Tombs, 1953).

Barker, Mary Anne. *Station Life in New Zealand* (s.l.: Macmillan, 1870).

Barker, Mary Anne. *Stories About:–* [1870] (Leipzig: Bernhard Tauchnitz, 1876).

Barnes, Barry. *Interests and the Growth of Knowledge* (London: Routledge and Kegan Paul, 1977).

Bar-Yosef, Eitan. 'E. Nesbit and the Fantasy of Reverse Colonization: How Many Miles to Modern Babylon?', *English Literature in Transition*, 46.1 (2003): 5–28.

Baudrillard, Jean. 'The System of Collecting', translated by Roger Cardinal, in John Elsner and Roger Cardinal (eds), *The Cultures of Collecting* (London: Reaktion Books, [1994] 1997), pp. 7–24.

Beer, Gillian. *Darwin's Plots: Evolutionary Narrative in Darwin, George Eliot and Nineteenth-Century Fiction* (Cambridge: Cambridge University Press, [1983] 2000).

Bellanca, Mary Ellen. 'Science, Animal Sympathy and Anna Barbauld's "The Mouse's Petition"', *Eighteenth-Century Studies*, 37.1 (Fall 2003): 47–67.

Bennett, Edward Turner. *The Gardens and Menagerie of the Zoological Society Delineated* (London: Thomas Tegg, 1830).

Bennett, Edward Turner. *The Tower Menagerie, comprising the natural history of the animals contained in that establishment, with anecdotes of their characters and history* (London: Robert Jennings, 1829).

Berger, John. *About Looking* (New York: Vintage International [1980] 1991).

Bevis, Matthew. 'Edward Lear's Lines of Flight', *Journal of the British Academy*, 1 (2013): 31–69.

Bewell, Alan. *Natures in Translation: Romanticism and Colonial Natural History* (Baltimore: Johns Hopkins University Press, 2017).

Bewick, Thomas. *A History of British Birds, Vol. 1 containing the history and description of land birds* [1797] (Newcastle: Edw. Walker, 1826).

Bewick, Thomas. *A History of British Birds, Vol. 2, containing the History and Description of Water Birds* [1804] (Newcastle: Edw. Walker, 1826).

Bewick, Thomas, and Ralph Beilby. *A General History of Quadrupeds* (Newcastle upon Tyne: s.n., 1820).

Bickerstaffe, Mona B. 'A Bit of Moss', *Aunt Judy's Christmas Volume for Young People* (London: Bell and Daldy, 1867), pp. 339–48.

Bishop, James. *Henry and Emma's Visit to the Zoological Gardens in the Regent's Park. Interspersed with a Familiar Description of the Manners and Habits of the Animals Contained Therein. Intended as a Pleasing Companion to Juvenile Visitors of this Delightful Place of Recreation and Fashionable Resort* (London: Dean and Munday; A. K. Newman and Co., [1829] 1830).

Black, Barbara J. *On Exhibit: Victorians and their Museums* (Charlottesville and London: University Press of Virginia, 2000).

Blackwell, M. (ed.). *The Secret Life of Things: Animals, Objects, and It-Narratives in Eighteenth-Century England* (Lewisburg: Bucknell University Press, 2007).

Bompas, George C. *Life of Frank Buckland* (London: Smith and Elder, 1885).

[Boreman, Thomas]. *Curiosities in the Tower of London* (London: T. Boreman, 1741).

Boreman, Thomas. *A Description of a Great Variety of Animals and Vegetables... Extracted from the most considerable writers of natural history* (London: F. Gyles, 1736).

Boreman, Thomas. *A Description of Three Hundred Animals; viz. Beasts, Birds, Fishes, Serpents and Insects* [1730] (London: R. Ware, 1753).

Briggs, Asa. *Victorian Things* (Stroud: Sutton Publishing [1988] 2003).

Briggs, Julia. '"Delightful Task!" Women, Children, and Reading in the Mid-Eighteenth Century', in Donelle Ruwe (ed.), *Culturing the Child, 1690–1914. Essays in Memory of Mitzi Myers* (Lanham, Toronto, Oxford: Scarecrow Press, 2005), pp. 67–82.

Briggs, Julia. *Edith Nesbit: A Woman of Passion* (Stroud: Tempus, [1987] 2007).

Brightwen, Elizabeth. *Inmates of my House and Garden* (New York and London: Macmillan and Co., 1895).

Brightwen, Elizabeth. *More about Wild Nature* (London: T. Fisher Unwin, [1892] 1893).

Brightwen, Elizabeth. *Wild Nature Won by Kindness*, 5th edn (London: T. Fisher Unwin, [1890] 1909).

Brough, John Cargill. *The Fairy Tales of Science: A Book for Youth* (London: Griffith and Farran, 1859).

Brown, Daniel. *The Poetry of Victorian Scientists: Style, Science and Nonsense* (Cambridge: Cambridge University Press, 2013).

Buffon, George-Louis Leclerc, comte de. *Buffon's Natural History, Containing a Full and Accurate Description of the Animated Beings in Nature* (Halifax: Milner and Sowerby, 1800).

282 SELECT BIBLIOGRAPHY

Burroughs, John. 'Real and Sham Natural History', in Ralph Lutts (ed.), *The Wild Animal Story* (Philadelphia: Temple University Press, [1998] 2001), pp. 129–43.

Busk, George. *Catalogue of Marine Polyzoa in the Collection of the British Museum* (London: Trustees of the British Museum, 1854).

Butt, John, and Kathleen Tillotson. *Dickens at Work* (London: Methuen, 1957).

Camden, Charles [Richard Rowe]. 'The Boys of Axleford – I – Fibbing Bill', *Good Words for the Young* (1 Jan. 1869): 145–8.

Camden, Charles [Richard Rowe]. 'The Boys of Axleford—IV—Shy Dick', *Good Words for the Young* (1 April 1869): 292–4.

Camden, Charles [Richard Rowe]. 'When I was Young – V – Patty Thomas and her Children', *Good Words for the Young* (1 April 1871): 324–28.

Camden, Charles [Richard Rowe]. *The Travelling Menagerie* (London: Henry S. King and Co., 1873).

Carey, M. R. 'Rachel's visit to Devonshire', *Aunt Judy's Christmas Volume for Young People* (London: Bell and Daldy, 1874).

Carrington, George. 'An Abstract View of a Sea Voyage', *Aunt Judy's Yearly Volume for Young People* (London: Bell and Daldy, 1869), pp. 312–17.

Carrington, George. 'Adventures among the Blacks in Australia', *Aunt Judy's Yearly Volume for Young People* (London: Bell and Daldy, 1869), pp. 345–52.

Carrington, George. 'The City of the Sultans', *Aunt Judy's Christmas Volume for Young People* (London: Bell and Daldy, 1870), pp. 275–80.

Carrington, George. 'Commonplace Journey Notes', *Aunt Judy's Christmas Volume for 1873* (London: Bell and Daldy, 1873), pp. 105–10; 172–7; 301–6; 359–64; 403–8; 487–93.

Carrington, George. 'A Day in the Australian Bush', *Aunt Judy's Yearly Volume for Young People* (London: Bell and Daldy, 1869), pp. 224–7.

Carrington, George. 'A Family-Man for Six Days', *Aunt Judy's Christmas Volume for Young People* (London: Bell and Daldy, 1868), pp. 114–21.

Carrington, George. 'Gold Digging', *Aunt Judy's Christmas Volume for Young People* (London: Bell and Daldy, 1871), pp. 147–53.

Carrington, George. 'With the Children in Australia', *Aunt Judy's Yearly Volume for Young People* (London: Bell and Daldy, 1869), pp. 173–8.

Carroll, Lewis. *Alice's Adventures in Wonderland* [1865], in *The Annotated Alice*, ed. Martin Gardner (London: Penguin, 2001).

Carroll, Lewis. *The Annotated Snark* [*The Hunting of the Snark: an agony, in eight fits*] [1876], ed. Martin Gardner (New York: Simon & Schuster, 1962).

Carroll, Lewis. *Sylvie and Bruno* [1889] (New York: Dover Publications, 1988).

Carroll, Lewis. *Sylvie and Bruno Concluded* (London: Macmillan & Co., 1893).

Carroll, Lewis. *Through the Looking-Glass, and What Alice Found There* [1871], in *The Annotated Alice*, ed. Martin Gardner (London: Penguin, 2001).

SELECT BIBLIOGRAPHY   283

Chadarevian, Soraya De, and Nick Hopwood (eds). *Models: The Third Dimension of Science* (Stanford: Stanford University Press, 2004).

Cockle, Mary. *The Fishes Grand Gala* (London: J. Harris, 1808).

Cohen, Jeffrey Jerome. 'Inventing with Animals in the Middle Ages', in Barbara A. Hanawalt and Lisa J. Kiser (eds), *Engaging with Nature: Essays on the Natural World in Medieval and Early Modern Europe* (Notre Dame, IN: University of Notre Dame Press, 2008), pp. 39–62.

Cohen, Morton N. (ed.). *The Selected Letters of Lewis Carroll* (Basingstoke: Macmillan [1982] 1989).

Cohen, Morton N., and Edward Wakeling (eds). *Lewis Carroll and his Illustrators. Collaborations & Correspondence, 1865–1898* (Ithaca and New York: Cornell University Press, 2003).

Colley, Ann C. 'Edward Lear's Anti-Colonial Bestiary', *Victorian Poetry*, 30.2 (Summer 1992): 109–20.

Colley, Ann C. 'Edward Lear and Victorian Animal Portraiture', in Raffaella Antinucci and Anna Enrichetta Soccio (eds), *Edward Lear in the Third Millennium: Explorations into his Art and Writing*, *Rivista di Studi Vittoriani*, 2.1 (2012–13): 11–24.

Colley, Ann C. *Wild Animal Skins in Victorian Britain: Zoos, Collections, Portraits and Maps* (Farnham: Ashgate, 2014).

Colombat, Jacqueline. 'Mission Impossible: Animal Autobiography', *Cahiers Victoriens et Edouardiens*, 39 (April 1994): 37–49.

Cornish, Charles John. *Life at the Zoo. Notes and Traditions of the Regent's Park Gardens...With illustrations, etc.* (London: Seeley & Co., 1895).

Cosslett, Tess. *Talking Animals in British Children's Fiction, 1786–1914* (Aldershot: Ashgate, 2006).

Cowie, Helen. *Exhibiting Animals in Nineteenth-Century Britain: Empathy, Education, Entertainment* (Basingstoke: Palgrave Macmillan, 2014).

Crockford, C. 'About Philip', *Good Words for the Young* (1 April 1870): 328–31.

Cutt, Nancy. *Mrs Sherwood and her Books for Children* (London: Oxford University Press, 1974).

Danahay, Martin A. 'Nature Red in Hoof and Paw: Domestic Animals and Violence in Victorian Art', in Deborah Denenholz Morse and Martin A. Danahay (eds), *Victorian Animal Dreams: Representations of Animals in Victorian Literature and Culture* (Aldershot: Ashgate, 2007), pp. 97–119.

Darton, F. J. Harvey. *Children's Books in England: Five Centuries of Social Life*, 2nd edn (Cambridge: Cambridge University Press, [1932] 1970).

[Darton, William], *Trifles for Children* (London: Darton and Harvey, [1796] 1804).

Darwin, Charles. *On the Origin of Species by Means of Natural Selection, or the Preservation of favoured races in the struggle for life* [1859] (Oxford: Oxford University Press, 1998).

284 SELECT BIBLIOGRAPHY

Daston, Lorraine and Peter Galison. 'The Image of Objectivity', *Representations*, 40 (1992): 81–128.

Davidson, Jane P. *A History of Paleontology Illustration* (Bloomington and Indianapolis: Indiana University Press, 2008).

Dawson, Gowan. *Show Me the Bone: Reconstructing Prehistoric Monsters in Nineteenth-Century Britain and America* (Chicago and London: University of Chicago Press, 2016).

Dean, Dennis R. *Gideon Mantell and the Discovery of Dinosaurs* (Cambridge: Cambridge University Press, 1999).

Denisoff, Dennis (ed.). *The Nineteenth-Century Child and Consumer Culture* (Aldershot: Ashgate, 2008).

Desmond, Adrian. *Archetypes and Ancestors: Paleontology in Victorian London, 1850–1875* (Chicago and London: The University of Chicago Press, 1982).

Dew-Smith, Alice. *Tom Tug and Others. Sketches in a Domestic Menagerie* (London: Seeley and Co. Limited, 1898).

Dickens, Charles. 'Fairyland in fifty-four', *Household Words*, 193 (3 December 1853): 313–17.

Dickens, Charles. *Hard Times* [1854], ed. David Craig (Harmondsworth: Penguin, 1982).

Dickens, Charles. 'Scotland Yard – Sketches by "Boz", n° II (New Series)', *Morning Chronicle* (4 Oct. 1836): n.p.

Dickens, Charles. 'Sketches on London', *Morning Chronicle* (7 Feb. 1835): n.p.

Dobson, Jessie. 'John Hunter's Animals', *Journal of the History of Medicine and Allied Sciences*, 17.4 (Oct. 1962): 479–86.

Donald, Diana. '"Beastly Sights": The Treatment of Animals as a Moral Theme in Representations of London, c.1820–1850', *Art History*, 22.4 (Nov. 1999): 514–44.

Donald, Diana. *Picturing Animals in Britain, 1750–1850* (New Haven and New York: Yale University Press, 2007).

Donald, Diana. *Women against Cruelty: Protection of Animals in Nineteenth-Century Britain* (Manchester: Manchester University Press, 2020).

Donato, Eugenio. 'The Museum's Furnace: Notes Towards a Contextual Reading of *Bouvard and Pécuchet*', in Josué Harari (ed.), *Textual Strategies: Perspectives in Post-Structuralist Criticism* (Ithaca, New York: Cornell University Press, 1979), pp. 213–38.

[Dorset, Catherine]. *The Lion's Masquerade* (London: J. Harris and B. Tabart, 1807).

[Dorset, Catherine]. *The Lioness's Ball; being a Companion to the Lion's Masquerade* (London: C. Chapple, B. Tabart, J. Harris, Darton and Harvey, c. 1808).

[Dorset, Catherine]. *The Peacock 'At Home'* (London: Newbery, 1807).

Drayson, A. W. 'At Home with the Python', *Good Words for the Young* (1 Feb. 1871): 211–13.

SELECT BIBLIOGRAPHY   285

Drayson, A. W. 'A Night in an African Tree', *Good Words for the Young* (1 Dec. 1871): 174–6.

Drayson, A. W. 'On the Trail of the Wild Elephant in Africa', *Good Words for the Young* (1 Sept. 1871): 644–8.

Drayson, A. W. 'A Ramble in the New Forest', *Good Words for the Young* (1 Aug. 1871): 559–62.

Drayson, A. W. 'Unusual Fishing', *Good Words for the Young* (1 July 1871): 469–72.

Drayson, A. W. 'Venomous Serpents and their habits', *Good Words for the Young* (1 June 1871): 422–6.

Dugaw, Dianne. 'Folklore and John Gay's Satire', *Studies in English Literature 1500–1900*, 31.3 (Summer 1991): 515–33.

Dyer, Gertrude P. *Elsie's Adventures in Insect-Land* (London: Ward and Co., 1882).

[Eastlake, Elizabeth]. 'The Crystal Palace', *Quarterly Review*, 192 (March 1855): 303–55.

Eden, Horatia K. F. *Juliana Horatia Ewing and her Books* (London: Society for Promoting Christian Knowledge, 1896).

Elsner, John, and Roger Cardinal (eds). *The Cultures of Collecting* (London: Reaktion Books, [1994] 1997).

Endersby, Jim. 'Classifying Sciences: Systematics and Status in mid-Victorian Natural History', in Martin Daunton (ed.), *The Organisation of Knowledge in Victorian Britain* (Oxford: Oxford University Press, 2005), pp. 61–85.

Enfield, The Viscountess. 'A Little Natural History', *Good Words for the Young* (1 Nov. 1870): 6–9.

Ewing, Alexander. 'The Prince of Sleona', *Aunt Judy's Magazine for Young People. The Christmas Volume for 1866* (London: Bell and Daldy, 1866), pp. 28–36; 94–9; 151–66; 216–25; 277–90; 344–54.

Ewing, Juliana Horatia. 'Among the Merrows. A Sketch of a Great Aquarium', *Aunt Judy's Christmas Volume for 1873* (London: George Bell and Sons, 1873), pp. 44–57.

Ewing, Juliana Horatia. 'In Memoriam, Margaret Gatty, Daughter of the Rev. Alexander John Scott', *Aunt Judy's Christmas Volume for Young People* (London: Bell and Daldy, 1874), pp. 4–7.

Ewing, Juliana Horatia. 'May-Day, Old Style and New Style', *Aunt Judy's Christmas Volume for Young People* (London: Bell and Daldy, 1874), pp. 424–39.

Ewing, Juliana Horatia. 'Mrs Overtheway's Remembrances', *Aunt Judy's Magazine for Young People. The Christmas Volume for 1866* (London: Bell and Daldy, 1866), pp. 15–28; 82–90; 167–78.

Ewing, Juliana Horatia. 'Mrs Overtheway's Remembrances', *Aunt Judy's May-Day Volume for Young People* (London: Bell and Daldy, 1867), pp. 65–78; 134–43; 193–202.

286 SELECT BIBLIOGRAPHY

Ewing, Juliana Horatia. 'Our Garden', *Aunt Judy's Christmas Volume for Young People* (London: Bell and Daldy, 1874), pp. 308–10.

Ewing, Juliana Horatia. 'Three Little Nest-Birds', *Aunt Judy's Christmas Volume for Young People* (London: Bell and Daldy, 1874), pp. 726–8.

[Ewing, Juliana Horatia, and/or H. K. F. Gatty]. 'What to Pick up on the Sea-Shore', *Aunt Judy's Christmas Volume for Young People* (London: Bell and Daldy, 1874), pp. 492–500.

Feuerstein, Anna. '*Alice in Wonderland*'s Animal Pedagogy: Governmentality and Alternative Subjectivity in Mid-Victorian Liberal Education', *Victorian Review*, 44.2 (Fall 2018): 233–50.

Findlen, Paula. *Possessing Nature: Museums, Collecting and Scientific Culture in Early Modern Italy* (Berkeley, Los Angeles, London: University of California Press, 1996).

Flack, Andrew. *The Wild Within: Histories of a Landmark British Zoo* (Charlottesville and London: University of Virginia Press, 2018).

Frank, Lawrence. *Victorian Detective Fiction and the Nature of Evidence: The Scientific Investigations of Poe, Dickens, and Doyle* (London: Palgrave Macmillan, [2003] 2009).

Fraser, Louis. *Catalogue of the Knowsley Collections* (Knowsley: s.n., 1850).

Fyfe, Aileen. 'Reading Children's Books in Late Eighteenth-Century Dissenting Families', *The Historical Journal*, 43.2 (2000): 453–73.

Fyfe, Aileen. *Science and Salvation: Evangelical Popular Science Publishing in Victorian Britain* (Chicago and London: The University of Chicago Press, 2004).

Fyfe, Aileen. 'Tracts, Classics and Brands: Science for Children in the Nineteenth Century', in Julia Briggs, Dennis Butts and M. O. Grenby (eds), *Popular Children's Literature in Britain* (Aldershot: Ashgate, 2008), pp. 209–28.

Fyfe, Aileen, and Bernard Lightman (eds). *Science in the Marketplace: Nineteenth-Century Sites and Experiences* (Chicago and London: The University of Chicago Press, 2007).

Gates, Barbara T. *Kindred Nature: Victorian and Edwardian Women Embrace the Living World* (Chicago and London: University of Chicago Press, 1998).

Gatty, Alfred, and Juliana Horatia Ewing. 'Cousin Peregrine's Wonder Stories. Waves of the Great South Seas. Founded on fact', *Aunt Judy's Christmas Volume for Young People* (London: Bell and Daldy, 1874), pp. 415–31.

Gatty, Alfred. 'The Prince of Sleona', *Aunt Judy's Magazine for Young People. The Christmas Volume for 1866* (London: Bell and Daldy, 1866), pp. 28–36; 94–99; 151–66; 216–25; 277–90; 344–54.

Gatty, Horatia Katherine Frances. 'A New Collection of Old Oyster-Shells', *Aunt Judy's Christmas Volume for 1876* (London: George Bell and Sons, 1876), pp. 157–65.

Gatty, Margaret. *British Sea-Weeds* (London: Bell and Sons, 1863).

Gatty, Margaret. 'Coral', *Aunt Judy's Magazine for Young People. The Christmas Volume for 1866* (London: Bell and Daldy, 1866), pp. 38–42; 90–93.

Gatty, Margaret. *The Fairy Godmothers* (London: s.n., 1851).

Gatty, Margaret. 'Introduction', *Aunt Judy's Magazine for Young People. The Christmas Volume for 1866* (London: Bell and Daldy, 1866), pp. 1–2.

Gatty, Margaret. 'Microscoping Objects', *Aunt Judy's Christmas Volume for 1870* (London: Bell and Daldy, 1870), pp. 311–16.

Gatty, Margaret. 'Nights at the Round Table', *Aunt Judy's Magazine for Young People. The Christmas Volume for 1866* (London: Bell and Daldy, 1866), pp. 42–55; 99–111.

Gatty, Margaret. *Parables from Nature* [1855–71] (Chapel Hill: Yesterday's Classics, 2006).

[Gatty, Margaret]. [Review of *Alice's Adventures in Wonderland*], *Aunt Judy's Magazine for Young People. The Christmas Volume for 1866* (London: Bell and Daldy, 1866).

Gay, John. *Fables by the Late John Gay, in one volume complete* (London: J. F. Rivington and B. White, 1743).

Gifford, Isabella. *The Marine Botanist: An Introduction to the Study of Algology, containing descriptions of the commonest British sea-weeds*, 2nd edn (London: Darton & Co., [1840] 1848).

Girling, Richard. *The Man Who Ate the Zoo: Frank Buckland, Forgotten Hero of Natural History* (London: Vintage, 2017).

Gosse, Philip Henry. *Actinologia Britannica: A History of the British Sea-Anemones and Corals* (London: Van Voorst, 1860).

Gosse, Philip Henry. *The Aquarium: An Unveiling of the Wonders of the Deep Sea* (London, Bath: s.n., 1854).

Gosse, Philip Henry. *Evenings at the Microscope* (s.l.: Society for Promoting Christian Knowledge, 1859).

Gosse, Philip Henry. *A Manual of British Zoology for the British Isles* (London: s.n., 1855–6).

Gosse, Philip Henry. *A Naturalist's Rambles on the Devonshire Coast* (s.l.: s.n., 1853).

Gosse, Philip Henry. *The Ocean* [1845] (London, s.n., 1854).

Gosse, Philip Henry. *Tenby: A Sea-Side Holiday* (London: Van Voorst, 1856).

Gould, Charles. *Mythical Monsters* (London: W. Allen, 1886).

Gould, John. *Birds of Europe* (London: s.n., 1837).

Gray, John Edwards. *Gleanings for the Menagerie and Aviary at Knowsley Hall* (Knowsley: s.n., 1846).

Green, Roger Lancelyn. *The Diaries of Lewis Carroll*, 2 vols (London: Cassell, 1953).

Gresswell, Albert and George. *The Wonderland of Evolution* (London: Field & Tuer, 1884).

288 SELECT BIBLIOGRAPHY

Greville, Robert Kaye. *Algae Britannicae; or, descriptions of the marine and other inarticulated plants of the British Islands, belonging to the order Algae: with plates, illustrative of the genera* (Edinburgh: s.n., 1830).

Griset, Ernest. *The Alphabet of Animals* (London: Frederick Warne and Co., c. 1880).

Gunter, R. T. *Early Science in Cambridge* (Oxford: Oxford University Press, 1937).

Gwynfryn [Jones, Dorothea]. 'From Bex to St Bernard', *Aunt Judy's May-Day Volume for Young People* (London: Bell and Daldy, 1867), pp. 38–49.

H., W. B. 'An Adventure with a Boa Constrictor', *Aunt Judy's May-Day Volume for Young People* (London: Bell and Daldy, 1868), pp. 117–19.

Hahn, Daniel. *The Tower Menagerie: The Amazing True Story of the Royal Collection of Wild Beasts* (London: Simon and Schuster, 2003).

Hancher, Michael. *The Tenniel Illustrations to the 'Alice' Books* (Columbus: Ohio University Press, 1985).

Harvey, William H. *A Manual of British Algae* (London: s.n., 1841).

Harvey, William H. *The Sea-Side Book; being an introduction to the natural history of the British coasts* (London: s.n., 1849).

Hawkins, Benjamin Waterhouse. *On Visual Education as Applied to Geology, Illustrated by Diagrams and Models of the Geological Restorations at the Crystal Palace* (London: W. Trounce, 1854).

Henchman, Anna. 'Fragments out of Place: Homology and the Logic of Nonsense', in James Williams and Matthew Bevis (eds), *Edward Lear and the Play of Poetry* (Oxford: Oxford University Press, 2016), pp. 183–201.

Henning, Michelle. 'Anthropomorphic Taxidermy and the Death of Nature: The Curious Art of Hermann Ploucquet, Walter Potter, and Charles Waterton', *Victorian Literature and Culture*, 35 (2007): 663–78.

Hill, Kate. *Women and Museums 1850–1914: Modernity and the Gendering of Knowledge* (Manchester: Manchester University Press, 2016).

Hilton, Mary. *Women and the Shaping of the Nation's Young: Education and Public Doctrine in Britain 1750–1850* (Aldershot: Ashgate, 2007).

Hoberman, Ruth. *Museum Trouble. Edwardian Fiction and the Emergence of Modernism* (Charlottesville and London: University of Virginia Press, 2011).

Hood, Thomas. 'Address to Mr. Cross, of Exeter 'Change, on the Death of the Elephant', *The New Monthly Magazine and Literary Journal*, 16 (London: Henry Colburn, 1826), pp. 343–4.

Hood, Tom. *From Nowhere to the North Pole. A Noah's Ark-Æcological Narrative* (London: Chatto and Windus, 1875).

Hooper-Greenhill, Eilean. *Museums and the Shaping of Knowledge* (London and New York: Routledge, 1992).

Hopley, Catherine C. 'The Deirodon; or, Neck-Toothed Snake', *Aunt Judy's Christmas Volume for Young People* (London: Bell and Daldy, 1874), pp. 626–30.

SELECT BIBLIOGRAPHY   289

Hopley, Catherine C. 'How Snakes Feed', *Aunt Judy's Christmas Volume for Young People* (London: Bell and Daldy, 1874), pp. 561–6.

Hopley, Catherine C. 'How a Snake Walks', *Aunt Judy's Christmas Volume for Young People* (London: Bell and Daldy, 1874), pp. 658–62.

Hopley, Catherine C. 'Snake Destroyers', *Aunt Judy's Christmas Volume for Young People* (London: Bell and Daldy, 1874), pp. 164–70.

Horton, E. 'Flamborough and Flamborough Head', *Aunt Judy's Yearly Volume for Young People* (London: Bell and Daldy, 1869), pp. 103–110, pp. 104–5.

Howe, Edward [Richard Rowe]. 'The Boy in the Bush', part II–'Up a Sunny Creek', *Good Words for the Young* (1 Jan. 1871): 136–40.

Howe, Edward [Richard Rowe]. 'Bush Neighbours', *Good Words for the Young* (1 Dec. 1869): 93–9.

Howe, Edward [Richard Rowe]. 'The Iguana's Eyes', *Good Words for the Young* (1 Nov. 1870): 25–7.

Hunt, Peter (ed.) *et al. Children's Literature: An Illustrated History* (Oxford: Oxford University Press, 1995).

Hutchinson, H. N. *Creatures of Other Days: Popular Studies in Palaeontology* (London: Chapman and Hall, 1896).

Hutchinson, H. N. *Extinct Monsters: A Popular Account of Some of the Larger Forms of Ancient Animal Life* (London: Chapman and Hall, 1897).

Ito, Takashi. *London Zoo and the Victorians, 1828–1859* (Woodbridge: The Boydell Press, 2014).

Jackson, Mary V. *Engines of Instruction, Mischief and Magic: Children's Literature in England from its Beginnings to 1839* (Lincoln: University of Nebraska Press, 1989).

Jamrach, Charles. 'My Struggle with a Tiger', *Boy's Own Paper*, 1.3 (1 Feb. 1879): 1–2.

Johns, B. J. 'Among the Butterflies', *Good Words for the Young* (1 Jan. 1870): 132–40.

Johnston, George H. *A History of British Zoophytes* [1838], 2nd edn (s.l.: s.n., 1847).

Jones, Dorothea. *Friends in Fur and Feathers* (London: s.n., 1869).

Jones, Jo Elwyn and J. F. Gladstone. *The Alice Companion: A Guide to Lewis Carroll's* Alice *Books* (New York: New York University Press, 1998).

[Keary, Annie]. 'The Cousins and their Friends', *Aunt Judy's Magazine for Young People. The Christmas Volume for 1866* (London: Bell and Daldy, 1866), pp. 3–15; 65–79; 129–44; 204–13; 269–76; 357–66.

[Keary, Annie]. 'The Cousins and their Friends', *Aunt Judy's May-Day Volume for Young People* (London: Bell and Daldy, 1867), pp. 27–37; 79–86; 144–55; 202–10; 278–89; 335–43.

Ketabgian, Tamara. *The Lives of Machines: The Industrial Imaginary in Victorian Literature and Culture* (Ann Arbor: The University of Michigan Press, 2011).

290 SELECT BIBLIOGRAPHY

Kendall, Edward Augustus. *Keeper's Travels in Search of His Master*, 2nd edn (London: Newbery, [1798] 1799).

Kenyon-Jones, Christine. *Kindred Brutes: Animals in Romantic Period Writing* (Aldershot: Ashgate, 2001).

[Kilner, Dorothy]. *The Life and Perambulation of a Mouse*, 2 vol. (London: John Marshall, 1784).

Kingsley, Charles. *Glaucus, or the Wonders of the Shore* (Cambridge, London: s.n., 1855).

Kingsley, Charles. *Madam How and Lady Why; or, first lessons in earth lore for children* [1869] (London: Macmillan, 1879).

Kingsley, Charles. *The Water-Babies, a Fairy Tale for a Land Baby* [1863] (London: Penguin, 1995).

Kingsley, Henry. *The Boy in Grey and other Stories and Sketches* (New York: Longmans, Green and Co., 1899).

Knoepflmacher, U. C. *Ventures into Childland: Victorians, Fairy Tales, and Femininity* (Chicago and London: University of Chicago Press, 1998).

Koenigsberger, Kurt. *The Novel and the Menagerie: Totality, Englishness, and Empire* (Columbus: Ohio State University Press, 2007).

Kohler, Robert E. *Naturalists, Collectors, and Biodiversity, 1850–1950* (Princeton and Oxford: Princeton University Press, 2006).

Kutzer, Daphne M. *Empire and Imperialism in Classic British Children's Books* (New York and London: Garland Publishing, Inc., 2000).

Lamb, Mary. 'The Beasts in the Tower', in Charles and Mary Lamb, *Poetry for Children* [1809], ed. William MacDonald (London: J. M. Dent & Co., 1903), pp. 132–4.

Landes, Joan B., Paula Young Lee and Paul Yougquist (eds). *Gorgeous Beasts: Animal Bodies in Historical Perspective* (University Park: Penn State University Press, 2012).

Landsborough, David. *A Popular History of British Sea-Weeds, comprising their structure, fructification, specific characters, arrangement and general distribution, with notices of some of the Fresh-Water Algae* (London: s.n., 1849).

Landsborough, David. *A Popular History of British Zoophytes and Corallines* (s.l.: s.n., 1852).

Lear, Edward. *The Complete Nonsense and Other Verse*, ed. Vivien Noakes (London: Penguin, 2002).

Lecercle, Jean-Jacques. *Philosophy of Nonsense: The Intuitions of Victorian Nonsense Literature* (London and New York: Routledge, [1994] 2002).

Lee, Mrs R. *Little Neddie's Menagerie* (London: Griffith and Farran, c. 1884).

Leighton, Mary Elizabeth and Lisa Surridge. 'The Empire Bites Back: The Racialized Crocodile of the Nineteenth Century', in Deborah Denenholz Morse and Martin A. Danahay (eds), *Victorian Animal Dreams: Representations of Animals in Victorian Literature and Culture* (Aldershot: Ashgate, 2007), pp. 249–70.

Lewes, G. H. *Sea-Side Studies at Ilfracombe, Tenby, the Scilly Isles, and Jersey* (Edinburgh: s.n., 1858).

Lightman, Bernard. *Victorian Popularizers of Science: Designing Nature for New Audiences* (Chicago and London: The University of Chicago Press, 2007).

Lightman, Bernard. '"The Voices of Nature": Popularizing Victorian Science', in Bernard Lightman (ed.), *Victorian Science in Context* (Chicago and London: The University of Chicago Press, 1997), pp. 187–211.

[Lindley, John]. 'The Crystal Palace Gardens', *The Athenaeum* (1854): 780.

Locke, John. *The Educational Writings of John Locke*, ed. James L. Axtell (Cambridge: Cambridge University Press, 1968).

Losano, Antonia. 'Performing Animals/Performing Humanity', in Laurence W. Mazzeno and Ronald D. Morrison (eds), *Animals in Victorian Literature and Culture: Contexts for Criticism* (London: Palgrave Macmillan, 2017), pp. 129–46.

Loudon, Jane. *The Young Naturalist; or, the Travels of Agnes Merton and her mamma*, 3rd edn (London: Routledge, Warne, and Routledge, [1840] 1860).

Lovell-Smith, Rose. 'The Animals of Wonderland: Tenniel as Carroll's Reader', *Criticism*, 45.4 (Fall 2003): 383–415.

Lovell-Smith, Rose. 'Eggs and Serpents: Natural History Reference in Lewis Carroll's Scene of Alice and the Pigeon', *Children's Literature*, 35 (2007): 27–53.

Lutts, Ralph (ed.). *The Wild Animal Story* (Philadelphia: Temple University Press, [1998] 2001).

Lydekker, R. *The History of the Collections Contained in the Natural History Departments of the British Museum*, vol. 2 (London: British Museum, 1906).

[Lydekker, R.]. *Guide to the Specimens of the Horse Family (Equidae) Exhibited in the Department of Zoology, British Museum (Natural History)* (London: British Museum, 1907).

M., A. D. *The Butterfly's Birthday, St. Valentine's Day, and Madame Whale's Ball: Poems to Instruct and Amuse the Rising Generation* (London: J. Harris, 1808).

MacKenzie, John M. 'Chivalry, Social Darwinism and Ritualised Killing: The Hunting Ethos in Central Africa up to 1914', in David Anderson and Richard Grove (eds), *Conservation in Africa: Peoples, Policies and Practice* (Cambridge: Cambridge University Press, 1987), pp. 41–61.

MacKenzie, John M. *The Empire of Nature: Hunting, Conservation and British Imperialism* (Manchester: Manchester University Press, 1988).

Macleod, Norman. 'Our Holiday in the West Highlands', *Good Words for the Young* (1 Nov. 1870): 20–24.

Macleod, Norman. 'Talks with the Boys about India', *Good Words for the Young* (1 Dec. 1870): 116–19.

292 SELECT BIBLIOGRAPHY

McCarthy, Steven. *The Crystal Palace Dinosaurs: The Story of the World's First Prehistoric Sculptures* (London: The Crystal Palace Foundation, 1994).

McCarthy, William. *Anna Letitia Barbauld: Voice of the Enlightenment* (Baltimore: The Johns Hopkins University Press, 2008).

Malamud, Randy. *Reading Zoos: Representations of Animals and Captivity* (New York: New York University Press, 1998).

[Manning, Anne]. 'Home with a Hooping-Cough; or; How they made the best of it', *Aunt Judy's Magazine for Young People. The Christmas Volume for 1866* (London: Bell and Daldy, 1866), pp. 193–204, 257–69, 321–31.

Mangum, Teresa. 'Animal Angst: Victorians Memorialize their Pets', in Deborah Denenholz Morse and Martin A. Danahay (eds), *Victorian Animal Dreams: Representations of Animals in Victorian Literature and Culture* (Aldershot: Ashgate, 2007), pp. 15–34.

Mantell, Gideon. *The Wonders of Geology; or, A Familiar Exposition of Geological Phenomena; Being the Substance of a Course of Lectures Delivered at Brighton* (1838).

[Marcet, Jane]. *Conversations on Chemistry, etc.* (London: Longman, Hurst, Rees & Orme, [1805] 1809).

Martin, Sarah Catherine. *The Comic Adventures of Old Mother Hubbard and Her Dog* (London: John Harris, 1805).

Martineau, Harriet. 'The Crystal Palace', *Westminster Review*, 6 (1854): 534–50.

Mathison, Ymitri. 'Maps, Pirates and Treasure: The Commodification of Imperialism in Nineteenth-Century Boys' Adventure Fiction', in Dennis Denisoff (ed.), *The Nineteenth-Century Child and Consumer Culture* (Aldershot: Ashgate, 2008), pp. 173–85.

Maxwell, Christabel. *Mrs Gatty and Mrs Ewing* (London: Constable & Co., 1949).

Mayer, Jed. '"Come Buy, Come Buy": Christina Rossetti and the Victorian Animal Market', in Laurence W. Mazzeno and Ronald D. Morrison (eds), *Animals in Victorian Literature and Culture. Contexts for Criticism* (London: Palgrave Macmillan, 2017), pp. 213–31.

Mazzeno, Laurence W. and Ronald D. Morrison (eds). *Animals in Victorian Literature and Culture: Contexts for Criticism* (London: Palgrave Macmillan, 2017).

Measom, G. *The Crystal Palace Alphabet: A Guide for Good Children* (London: Dean and Son, 1855).

Meyer, Susan E. *A Treasury of the Great Children's Books Illustrators* (New York: Harry N. Abrams, 1997).

Michals, Teresa. 'Experiments before Breakfast: Toys, Education and Middle-Class Childhood', in Dennis Denisoff (ed.), *The Nineteenth-Century Child and Consumer Culture* (Aldershot: Ashgate, 2008), pp. 29–42.

Miller, Andrew H. *Novels behind Glass: Commodity Culture and Victorian Narrative* (Cambridge: Cambridge University Press, 1995).

SELECT BIBLIOGRAPHY   293

Miller, John. *Empire and the Animal Body: Violence, Identity and Ecology in Victorian Adventure Fiction* (London: Anthem Press, 2014).

Monkhouse, William Cosmo. *The Life and Works of Sir John Tenniel* (London: Art Union Monthly Journal, Easter Art Annual, 1901).

Moore-Park, Carton. *An Alphabet of Animals* (London: Blackie and Son, 1899).

Morrison, Ronald D. 'Dickens, *Household Words*, and the Smithfield Controversy at the Time of the Great Exhibition', in Laurence W. Mazzeno and Ronald D. Morrison (eds), *Animals in Victorian Literature and Culture: Contexts for Criticism* (London: Palgrave Macmillan, 2017), pp. 41–63.

Morse, Deborah Denenholz and Martin A. Danahay (eds). *Victorian Animal Dreams: Representations of Animals in Victorian Literature and Culture* (Aldershot: Ashgate, 2007).

Morus, Iwan Rhys. '"More the Aspect of Magic than Anything Natural": The Philosophy of Demonstration', in Aileen Fyfe and Bernard Lightman (eds), *Science in the Marketplace: Nineteenth-Century Sites and Experiences* (Chicago and London: The University of Chicago Press, 2007), pp. 336–70.

Murphy, Ruth. 'Darwin and 1860s Children's Literature: Belief, Myth, or Detritus', *Journal of Literature and Science*, 5.2 (2012): 5–21.

Nagai, Kaori. 'Counting Animals: Nonhuman Voices in Lear and Carroll', in Jane Spencer, Derek Ryan and Karen L. Edwards (eds), *Reading Literary Animals: Medieval to Modern Perspectives on the Non-Human in Literature and Culture* (New York: Routledge, 2019), pp. 124–39.

Noakes, Richard. 'The *Boy's Own Paper* and Late-Victorian Juvenile Magazines', in Geoffrey Cantor, Gowan Dawson, Graeme Gooday, *et al.* (eds), *Science in the Nineteenth-Century Periodical: Reading the Magazine of Nature* (Cambridge: Cambridge University Press, 2004), pp. 151–71.

Nesbit, Edith. *The Book of Dragons* [1901] (Mineola, New York: Dover Publications, 2004).

Nesbit, Edith. *The Enchanted Castle* [1907] (London: Puffin Classics, 1994).

Nesbit, Edith. *Fairy Stories*, ed. Naomi Lewis (London and Tonbridge: Ernest Benn Limited, 1977).

Nesbit, Edith. *Five Children and It* [1902] (London: Penguin, 1995).

Nesbit, Edith. *The Magic City* [1910] (n.p.: BiblioBazaar, 2007).

Nesbit, Edith. *Nine Unlikely Tales* (London: T. Fisher Unwin, 1901).

Nesbit, Edith. *The Phœnix and the Carpet* [1904] (Ware: Wordsworth Editions, 1995).

Nesbit, Edith. *The Story of the Amulet* (London: T. Fisher Unwin, 1906).

Nesbit, Edith. *Wings and the Child; or the Building of Magic Cities* (London: Hodder and Stoughton, 1913).

[Newbery, F.]. *The Natural History of Birds by T. Telltruth. Embellished with Curious Cuts* (London: F. Newbery, 1778).

294 SELECT BIBLIOGRAPHY

[Newbery, John]. *A Little Pretty Pocket Book, intended for the instruction and amusement of little Master Tommy and pretty Miss Polly. With two letters from Jack the Giant-Killer ... To which is added a little Song-Book, etc.* (London: John Newbery, 1744).

[Newbery, John]. *The Philosophy of Tops and Balls; or, The Newtonian System of Philosophy, adapted to the capacities of young gentlemen and ladies, and familiarized and made entertaining by objects with which they are intimately acquainted: being the substance of six lectures read to the Lilliputian Society, by* TOM TELESCOPE, *A. M. and collected and methodized for the benefit of the youth of these kingdoms, by their old friend,* MR. NEWBERY, *in St. Paul's Church Yard* (London: John Newbery, 1761).

O'Connor, Ralph. *The Earth on Show: Fossils and the Politics of Popular Science, 1802–1856* (Chicago: Chicago University Press, 2007).

O'Connor, Ralph. 'Victorian Saurians: The Linguistic Prehistory of the Modern Dinosaur', *Journal of Victorian Culture*, 17.4 (2012): 492–504.

O'Gorman, Francis. '"More interesting than all the books, save one": Charles Kingsley's Construction of Natural History', in Juliet John and Alice Jenkins (eds), *Rethinking Victorian Culture* (Houndmills, Basingstoke: Macmillan, 2000), pp. 146–61.

O'Reilly, Mrs Robert. 'London Daisies', *Aunt Judy's Christmas Volume for Young People* (London: Bell and Daldy, 1874), pp. 350–58.

Opper, Frederick Burr. *Museum of Wonders, and what young folks saw there explained in many pictures* (London, New York: George Routledge and Sons, 1884).

Owen, Richard. *Description of the Skeleton of an Extinct Gigantic Sloth, Mylodon robustus, Owen: with Observations on the Osteology, Natural Affinities, and Probable Habits of the Megatherioid Quadrupeds in General* (London: R. and J. E. Taylor, 1842).

Owen, Richard. *Geology and Inhabitants of the Ancient World* (London: Crystal Palace Library and Bradbury and Evans, 1854).

Owen, Richard. *Memoir on the Megatherium, or the Giant Ground-Sloth of America (Megatherium americanum, Cuvier)* (London: Williams and Norgate, 1861).

Pan [Shute, Anna Clara]. 'The Bell-Bird. Poem', *Aunt Judy's Christmas Volume for Young People* (London: Bell and Daldy, 1870), pp. 337–8.

Pastan, Elizabeth Carson. 'Fables, Bestiaries, and the Bayeux Embroidery: Man's Best Friend Meets the "Animal Turn"', in Laura D. Gelfand (ed.), *Our Dogs, Our Selves. Dogs in Medieval and Early Modern Art, Literature, and Society* (Leiden and Boston: Brill, 2016), pp. 97–126.

Payne, E. W. *More Pleasant Mornings at the British Museum; or, the Handy-Work of Creation. Natural History Department.* (London: The Religious Tract Society, c. 1858).

SELECT BIBLIOGRAPHY 295

Pearce, Susan. *Museums, Objects, and Collections: A Cultural Story* (Washington, DC: Smithsonian Institution, 1992).

Pearson, Susan J. '"Infantile Specimens": Showing Babies in Nineteenth-Century America', *Journal of Social History*, 42.2 (Winter 2008): 341–70.

Pedley, Ethel C. *Dot and the Kangaroo* [1899] (Sydney: Angus and Robertson, 1906).

Percival, Thomas. *A Father's Instruction; Consisting of Moral Tales, Fables, and Reflections Designed to Promote the Love of Virtue, a Taste for Knowledge, and an Early Acquaintance with the Works of Nature* [1776], 8th edn, 2 vol. (London: Warrington, 1793).

Perkins, David. *Romanticism and Animal Rights* (Cambridge: Cambridge University Press, 2003).

Pickering, Samuel F. Jr., *John Locke and Children's Books in the Eighteenth Century* (Knoxville: The University of Tennessee Press, 1981).

Pielak, Chase. *Memorializing Animals during the Romantic Period* (Farnham: Ashgate, 2015).

Plumb, Christopher. 'Exotic Animals in Eighteenth-Century Britain', PhD. diss. (University of Manchester, 2010).

Plumb, Christopher. '"Strange and Wonderful": Encountering the Elephant in Britain, 1675–1830', *Journal of Eighteenth-Century Studies*, 33.4 (2010): 525–43.

Plumb, J. H. 'The New World of Children in Eighteenth-Century England', *Past and Present*, 67.1 (May 1975): 64–95.

Poliquin, Rachel. *The Breathless Zoo: Taxidermy and the Cultures of Longing* (Pennsylvania: The Pennsylvania State University Press, 2012).

Poliquin, Rachel. 'The Matter and Meaning of Museum Taxidermy', *Museum and Society*, 6.2 (2008): 123–34

Pollard, William. *The Stanleys at Knowsley: A History of that Noble Family* (London: Frederick Warne, 1869).

Pomian, Krzysztof. *Collectionneurs, amateurs et curieux. Paris, Venise: XVIᵉ– XVIIIᵉ siècles* (Paris: Gallimard, 1987).

Pratt, Anne. *Chapters on Common Things of the Sea-Side* (London: Society for Promoting Christian Knowledge, [1850] 1853).

Prickett, Stephen. *Victorian Fantasy*, 2nd edn (Waco: Baylor University Press, [1979] 2005).

Ratelle, Amy. *Animality and Children's Literature and Film* (New York: Palgrave Macmillan, 2015).

Reichertz, Ronald. *The Making of the Alice Books: Lewis Carroll's Uses of Earlier Children's Literature* (Montreal and Kingston, London, Ithaca: McGill-Queen's University Press, 1997).

Reimer, Mavis. 'The Beginning of the End: Writing Empire in E. Nesbit's Psammead Books', in Raymond E. Jones (ed.), *E. Nesbit's Psammead Trilogy:*

296 SELECT BIBLIOGRAPHY

A Children's Classic at 100 (Lanham, Toronto, Oxford: Children's Literature Association and the Scarecrow Press, 2006), pp. 39–62.

Richards, Thomas. *The Commodity Culture of Victorian England: Advertising and Spectacle, 1851–1914* (Stanford: Stanford University Press, 1990).

Richards, Thomas. *The Imperial Archive: Knowledge and the Fantasy of Empire* (London and New York: Verso, 1993).

Richardson, Alan. *Literature, Education and Romanticism. Reading as Social Practice, 1780–1832* (Cambridge: Cambridge University Press, 2010).

Ridley, Glynis. *Clara's Grand Tour: Travels with a Rhinoceros in Eighteenth-Century Europe* (New York: Grove Press, 2004).

Ridley, Glynis. 'Introduction: Representing Animals', *Journal for Eighteenth-Century Studies*, 34.4 (2010): 431–6.

Ritvo, Harriet. *The Animal Estate: The English and Other Creatures in the Victorian Age* (Cambridge, Mass.: Harvard University Press., 1987).

Ritvo, Harriet. 'Calling the Wild', in Joan B. Landes, Paula Young Lee and Paul Yougquist (eds), *Gorgeous Beasts: Animal Bodies in Historical Perspective* (University Park: Penn State University Press, 2012), pp. 105–16.

Ritvo, Harriet. *The Platypus and the Mermaid and Other Figments of the Classifying Imagination* (Cambridge, Mass. and London: Harvard University Press, 1997).

Roscoe, Sidney. *John Newbery and his Successors, 1740–1814: A Bibliography* (Wormsley: Five Owls Press, 1973).

Roscoe, William. *The Butterfly's Ball, and the Grasshopper's Feast* (London: Newbery, [1807] 1808).

Rosenberg, Teya. 'Generic Manipulation and Mutation: E. Nesbit's Psammead Series as Early Magical Realism', in Raymond E. Jones (ed.), *E. Nesbit's Psammead Trilogy: A Children's Classic at 100* (Lanham, Toronto, Oxford: Children's Literature Association and the Scarecrow Press, 2006), pp. 63–88.

Rothfels, Nigel. *Savages and Beasts: The Birth of the Modern Zoo* (Baltimore: John Hopkins University Press, 2002).

Routledge, G. *Routledge's Guide to the Crystal Palace and Park at Sydenham: with descriptions of the principal works of science and art, and of the terraces, fountains, geological formations and restoration of extinct animals therein exhibited* (London: G. Routledge, 1854).

Rowe, Richard. 'Running Away to See', *Good Words for the Young* (1 Nov. 1869): 38–9.

Rudd, David. 'Where It Was, There Shall Five Children Be: Staging Desire in *Five Children and It*', in Raymond E. Jones (ed.), *E. Nesbit's Psammead Trilogy: A Children's Classic at 100* (Lanham, Toronto, Oxford: Children's Literature Association and the Scarecrow Press, 2006), pp. 135–49.

Rudwick, Martin J. S. 'The Emergence of a Visual Language for Geological Science, 1760–1840', *History of Science*, 14 (1976): 149–95.

SELECT BIBLIOGRAPHY   297

Rudwick, Martin J. S. *Scenes from Deep Time: Early Pictorial Representations of the Prehistoric World* (Chicago and London: University of Chicago Press, [1992] 1995).

Rupke, Nicolaas A. *Richard Owen: Victorian Naturalist* (New Haven and London: Yale University Press, 1994).

Ruskin, John. *The Ethics of the Dust: Ten Lectures to Little Housewives on the Elements of Crystallisation* (London: Smith, Elder & Co., [1865] 1866).

S., M. B. 'Children on the Shore', *Aunt Judy's May-Day Volume for Young People* (London: Bell and Daldy, 1867), pp. 379–80.

Sandham, Elizabeth. *The Adventures of a Poor Puss* (London: John Harris, 1809).

Saunders, Julia. '"The Mouse's Petition": Anna Laetitia Barbauld and the Scientific Revolution', *The Review of English Studies*, 53.212 (Nov. 2002): 500–16.

Secord, James A. 'Monsters at the Crystal Palace', in Soraya De Chadarevian and Nick Hopwood (eds), *Models: The Third Dimension of Science* (Stanford: Stanford University Press, 2004), pp. 138–69.

Secord, James A. 'Newton in the Nursery: Tom Telescope and the Philosophy of Tops and Balls, 1761–1838', *Journal of History of Science*, 23 (1985): 127–51, reprinted in Laurence Talairach-Vielmas (ed.), *Science in the Nursery: The Popularisation of Science in Children's Literature in Britain and France, 1761–1901* (Newcastle: Cambridge Scholars Publishing, 2011), pp. 34–68.

Selby, Prideaux John. *Illustrations of British Ornithology* (Edinburgh, s.n., 1821–34).

Seymour, Mary. *Little Arthur at the Zoo and the Animals He Saw There* (London, Edinburgh and New York: Thomas Nelson and Sons, 1892).

Seymour, Mary. *Little Arthur at the Zoo and the Birds He Saw There* (London, Edinburgh and New York: Thomas Nelson and Sons, 1892).

Shapin, William. *A Social History of Truth: Civility and Science in Seventeenth-Century England* (Chicago: University of Chicago Press, 1994).

Shteir, Ann B. *Cultivating Women, Cultivating Science: Flora's Daughters and Botany in England 1760–1860* (Baltimore and London: The Johns Hopkins University Press, 1996).

Silver, Carole G. *Strange and Secret Peoples: Fairies and Victorian Consciousness* (Oxford: Oxford University Press, 1999).

Simons, John. *Rossetti's Wombat: Pre-Raphaelites and Australian Animals in Victorian London* (s.l: Middlesex University Press, 2008).

Simons, John. *The Tiger That Swallowed the Boy: Exotic Animals in Victorian England* (Faringdon: Libri Publishing, 2012).

Shuttleworth, Sally. *The Mind of the Child: Child Development in Literature, Science, and Medicine, 1840–1900* (Oxford: Oxford University Press, [2010] 2012).

Smith, Jonathan. *Charles Darwin and Victorian Visual Culture* (Cambridge: Cambridge University Press, 2006).

298 SELECT BIBLIOGRAPHY

Smith, Julie A. 'Representing Animal Minds in Early Animal Autobiography: Charlotte Tucker's *The Rambles of a Rat* and Nineteenth-Century Natural History', *Victorian Literature and Culture*, 43 (2015): 725–44.

Spencer, Jane. 'Behn's Beasts: Aesop's *Fables* and Surinam's Wildlife in *Oroonoko*', in Jane Spencer, Derek Ryan and Karen L. Edwards (eds), *Reading Literary Animals: Medieval to Modern Perspectives on the Non-Human in Literature and Culture* (New York: Routledge, 2019), pp. 46–65.

Spencer, Jane. 'Creating Animal Experience in Late Eighteenth-Century Narrative', *Journal for Eighteenth-Century Studies*, 33.4 (2010): 469–86.

Spencer, Jane, Derek Ryan and Karen L. Edwards (eds), *Reading Literary Animals: Medieval to Modern Perspectives on the Non-Human in Literature and Culture* (New York: Routledge, 2019).

Starr, Laura B. 'Found in Uncle Sam's Mails. [From photographs taken expressly for George Newnes, Ltd.]', *The Strand Magazine*, 16.92 (Sept. 1898): 148–52.

Stevens, Frank. *Adventures in Pondland* (London: Hutchinson and Co., 1905).

Stevens, J. C. *Catalogue of the Menagerie and Aviary at Knowsley* (Liverpool: Joshua Walmsley, 1851).

Stewart, Susan. *On Longing: Narratives of the Miniature, the Gigantic, the Souvenir, the Collection* (Durham and London: Duke University Press, 1993).

Stott, Rebecca. *Theatres of Glass: The Woman who Brought the Sea to the City* (London: Short Books, 2003).

Straley, Jessica. *Evolution and Imagination in Victorian Children's Literature* (Cambridge: Cambridge University Press, 2016).

Susina, Jan. *The Place of Lewis Carroll in Children's Literature* (New York: Routledge, 2010).

Tague, Ingrid H. *Animal Companions: Pets and Social Change in Eighteenth-Century Britain* (University Park, Pennsylvania: The Pennsylvania State University Press, 2015).

Talairach-Vielmas, Laurence. *Fairy Tales, Natural History and Victorian Culture* (Basingstoke: Palgrave Macmillan, 2014).

Talairach-Vielmas, Laurence. 'Shaping the Beast: The Nineteenth-Century Poetics of Palaeontology', in Maria Freddi, Barbara Korte, Joseph Schmied (eds), *The Rhetoric of Science, EJES (European Journal of English Studies)*, 17.3 (2013): 269–82.

Topman, Jonathan. 'Science, Natural Theology, and the Practices of Christian Piety in Early Nineteenth-Century Religious Magazines', in Geoffrey Canto and Sally Shuttleworth (eds), *Science Serialized: Representation of the Sciences in Nineteenth-Century Periodicals* (Cambridge, M.A.: MIT Press, 2003), pp. 33–66.

Travis, Peter W. 'Aesop's Symposium of Animal Tongues', *Postmedieval*, 2 (2011): 33–49.

SELECT BIBLIOGRAPHY 299

Trimmer, Sarah. *Fabulous Histories: Designed for the Instruction of Children Respecting Their Treatment of Animals, in two volumes comprised in one* [1786], 10th edn (London: John Sharpe, 1815).

Tristram, Henry Baker. 'About a Caterpillar', *Good Words for the Young* (1 Nov. 1869): 45–7.

Tristram, Henry Baker. 'About a Fly', *Good Words for the Young* (1 May 1870): 347–54.

Tristram, Henry Baker. 'Ants and Ant-Hills', *Good Words for the Young* (1 March 1869): 242–52.

Tristram, Henry Baker. 'Bees and Beehives', *Good Words for the Young* (1 Jan. 1870): 161–8.

Tristram, Henry Baker. 'Rooks and their relations', *Good Words for the Young* (1 June 1869): 391–6.

Tristram, Henry Baker. 'The Spider and its Webs', *Good Words for the Young* (1 Feb. 1869): 171–6.

Turner, J. *Reckoning with the Beast: Animals, Pain and Humanity in the Victorian Mind* (Baltimore and London: Johns Hopkins University Press, 1980).

Uglow, Jenny. *Nature's Engraver: A Life of Thomas Bewick* (New York: Farrar, Straus and Giroux, 2006).

Valentine, Laura. *Aunt Louisa's Zoological Gardens* (London: Frederick Warne and Co., c. 1876).

Van Dyke, Carolynn. 'Entities in the World: Intertextuality in Medieval Bestiaries and Fables', in Jane Spencer, Derek Ryan and Karen L. Edwards (eds), *Reading Literary Animals: Medieval to Modern Perspectives on the Non-Human in Literature and Culture* (New York: Routledge, 2019), pp. 13–28.

Vila, Anne C. *Enlightenment and Pathology* (Baltimore and London: Johns Hopkins University Press, 1998).

Wakefield, Priscilla. *An Introduction to Botany, in a Series of Familiar Letters* (London: s.n., 1796).

Waterton, Charles. *Wanderings in South America, the North-West of the United States and the Antilles, in the Years 1812, 1816, 1820, & 1824, with Original Instructions for the Perfect Preservation of Birds, & c. for Cabinets of Natural History*, 4th edn (London: B. Fellowes, [1825] 1839).

Watson, Jeanie. '"The Raven: A Christmas Poem". Coleridge and the Fairy Tale Controversy', in James Holt McGavran (ed.), *Romanticism and Children's Literature in Nineteenth-Century England* (Athens and London: The University of Georgia Press, [1991] 2009), pp. 14–33.

Watts, Isaac. *Divine Songs, Attempted in Easie Language for the Use of Children* [1715] (Tamworth: B. Shelton, 1794).

Wyatt, Mary. *Algae Danmonienses; or Dried Specimens of Marine Plants, principally collected in Devonshire; carefully named according to Dr. Hooker's British Flora* (Torquay: Cockrem, 1833).

300 SELECT BIBLIOGRAPHY

Weiser, Elizabeth. *Museum Rhetoric* (University Park: Penn State University Press, 2017).

Wharncliffe, Lord. 'A Day's Elephant Hunting in Ceylon', *Aunt Judy's May-Day Volume for Young People* (London: Bell and Daldy, 1867), pp. 210–15.

White, Daniel E. 'The "Joinerina": Anna Barbauld, the Aikin Family Circle, and the Dissenting Public Sphere', *Eighteenth-Century Studies*, 32.4 (Summer 1999): 511–33.

White, Laura. *The* Alice *Books and the Contested Ground of the Natural World* (Abingdon, New York: Routledge, 2017).

White, R. S. *Natural Rights and the Birth of Romanticism in the 1790s* (Basingstoke: Palgrave Macmillan, 2005).

Wilson, Dudley. *Signs and Portents: Monstrous Births from the Middle Ages to the Enlightenment* (London: Routledge, 1993).

Wollstonecraft, Mary. *Original Stories from Real Life; with Conversations Calculated to Regulate the Affections, and Form the Mind to Truth and Goodness* (London: s.n., 1788).

Wood, John George. *The Boy's Own Book of Natural History* (London: George Routledge & Sons, [1861] 1867).

Wood, John George. *Common Objects of the Sea-Shore; including hints for an aquarium...* (London: George Routledge & Co., 1857).

Wood, John George. *The Fresh and Salt-Water Aquarium* (London: George Routledge & Co., 1868).

Wood, John George. *The Illustrated Natural History* (London: George Routledge & Co., 1853).

Wood, John George. 'On Killing, Setting, and Preserving Insects. I–Killing', *Boy's Own Paper*, 1 (1879): 431–2.

Wood, Theodore. *The Zoo (fourth series)* (London: Society for Promoting Christian Knowledge, 1895).

Wyatt, M. Digby. *Views of the Crystal Palace and Park Sydenham* (Crystal Palace: Day and Son, 1854).

Zimmerman, Virginia 'The Curating Child: Runaways and Museums in Children's Fiction', *The Lion and the Unicorn*, 39.1 (2015): 42–62.

Zimmern, Helen. 'What the Green Lizard Told Me', *Good Words for the Young* (1 March 1870): 265–7.

INDEX[1]

A

Abecedaria, 12
Abecedaria (ABC books; alphabet books), 82–88
Acland, Henry, 186, 186n64
Aesop, 33–35, 38, 74, 126, 173n8, 181, 185, 190, 206
Agasse, Jacques-Laurent, 26
Aikin, John, 49n83, 50, 50n84, 51, 53, 53n89, 59n104, 65, 126
  See also Barbauld, Anna Laetitia, *Evenings at Home; or, the juvenile budget opened*
Allman, George James, 141–143, 141n51, 142n52, 142n54, 143n55, 143n59
A. L. O. E., 56
  The Rambles of a Rat, 56
Andersen, Hans Christian, 89
Animal welfare legislation, 47, 47n77, 65

Audubon, John, 199, 199n113
Aunt Judy's Cot, 147n69
  See also Great Ormond Street Hospital for Sick Children
Aunt Judy's Magazine, 13–15, 88–90, 89n57, 91n59, 92n62, 93n68, 94, 95, 110, 121, 122n154, 126, 127, 129, 130, 138, 139, 141, 142n52, 143, 146n64, 147n69, 152, 154, 162–165, 167, 187, 190n80
Austen, Jane, 23, 24
Auzoux, Louis, 220n2

B

Bagnold, Eliza Sophia Helen, 91, 91n61, 93–95, 93n69, 119, 120, 120n150
Barbauld, Anna Laetitia, 41, 42, 49–53, 49n80, 49–50n83,

---

[1] Note: Page numbers followed by 'n' refer to notes.

© The Author(s), under exclusive license to Springer Nature Switzerland AG 2021
L. Talairach, *Animals, Museum Culture and Children's Literature in Nineteenth-Century Britain*, Palgrave Studies in Animals and Literature, https://doi.org/10.1007/978-3-030-72527-3

301

302 INDEX

Barbauld, Anna Laetitia (*cont.*)
  50n84, 52n89, 57,
  59n104, 65, 126
  *Evenings at Home; or, the juvenile
    budget opened*, 49, 49n83, 50,
    50n84, 53n89, 126
  Lessons for Children, 50, 51
  'Petition for a Mouse,' 49, 52n89
Barker, Mary Anne, 114–119,
  116n140, 128, 146
Barrie, J. M., 242n63
Baudin, Nicolas, 11
Beardsley, Aubrey, 232
Beche, Henry de la, 231n33, 231n34
Becker, Lydia, 91n59
Beeton, Samuel, 88, 89
Bell, George, 137, 137n34
Bennett, Edward Turner, 31n23
Bestiaries, 15, 32–34, 34n36, 83,
  173, 212
Bewick, Thomas, 4n10, 11, 34, 42,
  73–82, 177n23, 194n96,
  195, 195n101
Bishop, James, 75, 76n18, 92n66, 94
Bligh, William, 11
Boreman, Thomas, 32, 34, 39n57
Boswell, James, 25n5
Boys of the Empire, 14n41, 121
Boy's Own Magazine, The, 75n17, 88,
  89, 91n59
Boy's Own Paper, The, 14n41, 88,
  88n56, 89, 91n59, 112n118,
  121, 164
Brewer, Ebenezer Cobham, 91n59
Brightwen, Elizabeth, 127, 164–166,
  164n132, 164n133, 167n143
British Museum, 74, 114, 114n125,
  133n28, 137, 137n38, 169, 170,
  177n23, 183, 207n135, 211,
  211n154, 216, 227, 233, 234,
  253, 267, 267n123,
  267n125, 269–271

Brough, John Cargill, 238, 240
Bryan, Margaret, 41
Buchan, John, 112
Buckland, Francis Trevelyan (Frank
  Buckland), 235, 236
Buckland, William, 224, 236, 249n79
Budge, Ernest Wallis, 267n123
Buffon, Georges-Louis Leclerc, 21,
  21n61, 34
Burroughs, John, 96, 96n74
Burton, Richard Francis, 207
Busk, George, 135, 141, 141n50,
  142n52, 143
Butler, Jacob, 211n155
Byron, Lord, 26

C
Caldecott, Randolph, 89
Caldwell, William Hay, 11n27
Camden, Charles, 99, 99n87, 100,
  100n88, 100n89, 103, 106, 107,
  116, 120
  See also Rowe, Richard
Carey, M. R., 111
Carrington, George, 92,
  92n62, 92n64
Carroll, Lewis, 10, 16–20, 16n50,
  53n92, 89, 131, 167, 176,
  186–217, 188n69, 188n71,
  188n72, 189n73, 202n122,
  203n128, 232, 235, 242n63,
  251, 256
  Alice's Adventures in Wonderland,
    16, 18, 53n92, 131, 176,
    187–192, 188n69, 188n71,
    192n86, 194, 198, 200,
    208–210, 209n147, 212n162,
    235, 251
  The Hunting of the Snark, 19
  Sylvie and Bruno, 190n80, 193n93
  Sylvie and Bruno Concluded, 190n80

INDEX 303

Through the Looking-Glass, and What Alice Found There, 18, 202, 202n122, 203, 203n128, 217n170
Chambers, Robert, 184
Child's Pictorial, The, 75n17
Chunee (Chuny), 68–70, 72, 72n9, 93, 94
Clark, Thomas, 25
   See also Exeter Exchange menagerie (Exeter Change)
Cockle, Mary, 61n112, 62, 63n115
Cook, James, 129, 129n15
Coombe, Fanny Jane Dolly, 178
Cooper, Mary, 39n57
Cornish, Charles John, 75, 75n17
Coughtrey, Millen, 142n52
Craik, Dinah Maria Mulock, 89
Crockford, C., 114n127
Cross, Edward, 26, 69–71, 70n8, 105, 105n92, 115
   See also Exeter Exchange menagerie (Exeter Change)
Cruikshank, George, 89
Crystal Palace aquarium (Sydenham), 122, 132n24
Crystal Palace Gardens (Crystal Palace dinosaur park), 221–229, 222n8, 235, 237, 238, 241, 242, 245, 247, 249, 254, 266
   See also Hawkins, Benjamin Waterhouse
Crystal Palace (Great Exhibition), 132, 195n99
Crystal Palace (Sydenham), 20, 220–222, 226, 227, 229, 230, 232, 244–247, 254, 269, 270
   See also Crystal Palace Gardens; Hawkins, Benjamin Waterhouse
Cupples, George, 96
Cuvier, Georges, 81, 175n14, 249n79

D
Darton, William, 57, 58
Darwin, Charles, 3, 19, 21, 21n61, 82n29, 107n99, 133–134n28, 174, 181n37, 184, 188, 190n80, 194, 196, 198, 198n108, 200, 200n115, 207, 208, 210, 223, 223n10, 234, 237, 240, 241
Dawes, Richard, 219
Dean, Thomas, 75, 76n18, 83
Dew-Smith, Alice, 164, 167n143, 255
Dickens, Charles, 68–70, 69n4, 69n6, 72, 72n10, 134n28, 221, 222
Doré, Gustave, 232
Dorset, Catherine, 10, 59–61, 60n108
   Peacock 'At Home,' 59–62
Drayson, Alfred Wilks, 108–110
Dyer, Gertrude P., 188n71

E
Eastlake, Elizabeth, 222, 223
Edgeworth, Maria, 58, 58n103, 59n104
Edinburgh Museum, 74, 87n50, 177n23
Egyptian Hall, 231n34
1851 Great Exhibition, 7
Ellis, John, 133
Evans, Edmund, 83n36
Evolutionary theory, 88, 91n59, 237, 237n51, 257
   See also Darwin, Charles
Ewing, Alexander, 146n64, 147
Ewing, Juliana Horatia, 14, 89, 110n112, 122, 122n155, 122n156, 132n24, 139–141, 144, 145n62, 146n64, 147, 162
Exeter Exchange menagerie (Exeter Change), 23, 28, 75

F

Fables, 7, 12, 17, 32–39, 34n36, 35n39, 40n62, 41, 51, 58, 59, 74, 78, 83
  See also Aesop
Faraday, Michael, 41
Fielding, Henry, 49n82
Forbes, Edward, 133n28
Ford, Henry Justice, 232
Furniss, Harry, 193n93

G

Gatty, Horatia Katherine Frances, 14, 89, 122n155, 129–131, 134, 135, 141–143, 145, 147, 162
Gatty, Margaret, 13, 14, 89–91, 89n57, 90–91n59, 93n68, 94, 121, 129, 135–141, 147, 151–154, 157, 158, 162–164, 163n127, 187, 187n67, 190n80, 203, 203n127
  See also Aunt Judy's Magazine
Gay, John, 35–39, 37n48, 58, 59
  See also Fables
Gifford, Isabella, 133
Gilbert, W. S., 89
Gilbert, William, 96
Girl's Own Paper, The, 164
Goldsmith, Oliver, 25n5, 26, 34
Good Words, 75n17, 103
Good Words for the Young, 13, 15, 88–90, 91n59, 93n70, 95, 97n81, 99n87, 100n88, 106, 107, 110, 114n127, 115, 121
Gosse, Philip Henry, 133, 187n67, 244n71
Gould, Charles, 174, 205, 206
Gould, John, 179
Grandville, Jean-Jacques, 193
Gravelot, Henri, 35n39
Gray, Asa, 137n38, 141

Gray, John, 133n28
Great Ormond Street Hospital for Sick Children, 147n69
  See also Aunt Judy's Cot
Gresswell, Albert, 216n169
Gresswell, George, 216n169
Greville, Robert Kaye, 133
Griset, Ernest, 84, 84n39
Gwynfryn, 94
  See also Jones, Dorothea

H

Harris, John, 10, 59–61, 63, 194
Harvey, William Henry, 133, 137–139, 137n38
Hawkins, Benjamin Waterhouse, 20, 177n25, 221, 223, 226, 227, 229–235, 237, 238, 241, 245–247, 249, 252, 254
  See also Crystal Palace Gardens
Hawkins, Thomas, 231n33
Henslow, George, 91n59
Hinde, Harry, 179, 200
History of Little Goody Two-Shoes, The, 40, 40n62
Hobson, John Atkinson, 268n132
Holiday, Henry, 19
Home, Everard, 11
Hood, Thomas, 70–72, 70n8
Hood, Tom, 170–173, 170n2, 176, 194–197, 200, 209n147, 216, 240, 241, 249, 253
  From Nowhere to the North Pole. A Noah's Ark-Æcological Narrative, 170, 170n2, 172, 209n147, 240
Hooker, Joseph, 137n38, 156
Hopley, Catherine C., 91, 91n60, 112–114, 112n119, 113n120, 114n125

Horton, E., 91, 153, 154n94
Houghton, William, 91n59
Howe, Cupples, 96
Howe, Edward, 97–99
　See also Rowe, Richard
Hughes, Arthur, 89
Hunter, John, 212n158
Hutchinson, Henry Neville, 242n63
Huxley, Thomas Henry, 141n50,
　　182n41, 196, 224, 235–237,
　　237n50, 240n56

I
Illustrated London News, The,
　228, 242n63

J
Jacobs, Joseph, 263
James, Thomas Rev., 183n42, 192
Jamrach, Charles, 105, 112, 112n118,
　164, 191
Jardine, William, 178
Johns, B. J., 107
Johns, Charles Alexander, 91n59
Johnstone, George, 137
Jones, Dorothea, 94
　See also Gwynfryn

K
Kendall, Edward Augustus,
　50n84, 51n84
Kent, William, 35n39
Kew Gardens, 2, 114n127, 269–270
Kew's Herbarium (Hooker
　Herbarium), 137
Kilner, Dorothy, 51, 53,
　53–54n92, 55–57
　Life and Perambulation of a Mouse,
　53, 54n92, 56, 57

Kingsley, Charles, 16, 18, 39, 61,
　63n115, 89, 91n59, 106,
　107n99, 131n20, 133, 134n28,
　177n22, 186–203, 187n67,
　195n98, 216, 235, 236, 242n63
　*The Water-Babies, or a Fairy-Tale for
　a Land Baby,* 16, 18, 61,
　63n115, 177n22, 194,
　195n98, 203, 235
Kingsley, Henry, 209n147, 235,
　236n45, 237, 237n51, 239,
　256, 257
Kingston, W. H. G., 88, 89
Kipling, Rudyard, 14n41
Knowsley Park (menagerie), 177, 230
Koch, Albert, 231n34

L
Lamb, Charles, 63, 64
Lamb, Mary, 63, 72
　'The Beasts in the Tower,' 63
Landsborough, David, 133, 138
Landseer, Edwin, 26, 26n8
Lang, Andrew, 232
Lear, Edward, 16, 18–20, 167,
　176–186, 200, 216, 230, 249
　'The History of the Seven Families
　of the Lake Pipple-Popple,'
　181, 183, 216–217
　'The Story of the Four Little
　Children Who Went Round the
　World,' 183
Leverian Museum, 211
Lewes, George Henry, 133
Linnaeus, Carl, 3, 81, 208
Linnean Society, 141, 142n54, 143,
　143n55, 177
Lloyd, W. Alford, 132
Locke, John, 33, 33n34, 39, 42
London Natural History Museum,
　141n50, 165, 247n75

306 INDEX

London Zoological Gardens (London Zoo), 12, 17, 23, 26, 56, 62, 65, 67, 75, 75n16, 77, 79, 82, 84–87, 91n60, 93, 104, 104n91, 105, 105n92, 109, 115, 122, 132, 190n80, 209n147, 216, 217n172, 252
London Zoological Society, 12
Loudon, Jane, 79, 80n23, 81, 91n59, 116
   The Young Naturalist's Journey; or, the Travels of Agnes Merton and Her Mamma, 79, 116

M

MacDonald, George, 13, 89
   See also Good Words for the Young
Macleod, Norman, 13, 108, 108n101
   See also Good Words for the Young
Macmillan, Hugh, 107
Magazine for Boys, 88, 89
Magazine of Natural History, 79
Mantell, Gideon Algernon, 231n34, 232n38, 238
Marcet, Jane, 41
Marine aquarium mania, 131
Martin, John, 230, 230–231n33, 232
Martin, Sarah Catherine, 10, 59
   The Comic Adventures of Old Mother Hubbard and Her Dog, 59, 60
Martineau, Harriet, 221, 224, 225, 230, 246–247n74
Measom, George Samuel, 222n8
Meckel, Johann Friedrich, 212n158
Mermaids, 174, 175, 205
Mervyn, Ruth, 91
Millar, H. R., 247n75, 249, 249n79
Milner, Thomas, 231n34
Milton, John, 232
Molesworth, Mary Louisa (Mrs Molesworth), 242n63
Monthly Packet, The, 94

Moore, Hannah, 42
Moore-Park, Carton, 84, 85n41
Morris, Francis Orpen, 91n59, 122n155, 154n94
Museum of Ravensworth Castle, 74, 177n23

N

Nesbit, Anthony, 241, 256, 257, 259, 260, 261n99, 262, 263, 265–268, 271, 272
Nesbit, Edith, 2, 14n41, 20, 190n79, 209n147, 217, 217n172, 241–255, 264n111, 267n123
   Book of Dragons, The, 242
   Enchanted Castle, The, 242, 243, 243n67, 245
   Five Children and It, 241, 244n71, 256, 260, 260n97, 260n98, 263, 268, 268n130
   Magic City, The, 242, 245, 247, 253
   Phoenix and the Carpet, The, 241, 245, 256, 268n130
   Story of the Amulet, The, 2, 14n41, 242, 245, 256, 264n111, 267, 267n123, 268, 270
Newbery, Francis, 1, 2, 173
Newbery, John, 39–41, 59, 83, 83n35
Newcastle Museum, 74, 177n23

O

Obaysh (hippopotamus), 191
Opper, Frederick Burr, 88
O'Reilly, Robert Mrs, 110
Owen, Richard, 68, 133n28, 137n38, 196, 204, 212n158, 212n162, 220, 223n10, 227, 228, 234, 236–238, 240n56, 249, 249n79
Oxford University Museum of Natural History, 189, 189n73

P

Park, Mungo, 188n70

Payne, E. W., 82, 82n32, 206,
206n133, 207n134, 233, 234

Pedley, Ethel C., 257, 259, 260
Dot and the Kangaroo, 257, 257n95

Pepys, Samuel, 25n6

Percival, Thomas, 52,
52–53n89, 53, 65
A Father's Instruction;
Consisting of Moral Tales,
Fables, and Reflections
Designed to Promote the
Love of Virtue, a Taste for
Knowledge, and an Early
Acquaintance with the
Works of Nature, 52, 52n89

Pestalozzi, Johann Heinrich,
31n25, 226

Pidcock, Gilbert, 25, 27, 29
See also Exeter Exchange menagerie
(Exeter Change)

Pilkington, Mary, 43

Ploucquet, Hermann, 7, 170, 171,
171n4, 173, 174, 184,
208, 216–217

Polito, Stephan, 25
See also Exeter Exchange menagerie
(Exeter Change)

Potter, Walter, 7, 171, 171n4

Pratt, Anne, 133

R

Ray, John, 3

Reinhardt, John Theodor, 207n135

Religious Tract Society, 164, 233

Richardson, George, 231n34

Richardson, Samuel, 25n5, 49n82

Roscoe, William, 10, 59, 60,
60n107, 62

The Butterfly's Ball, and the
Grasshopper's Feast, 59, 60,
60n107, 62

Rossetti, Christina, 17

Rossetti, Dante Gabriel, 17, 18,
191, 192n86

Routledge, George, 223n10,
226, 228

Rowe, Richard, 97, 99, 99n87, 106,
116, 117, 120
See also Camden, Charles

Royal College of Surgeons, 237n50

Ruskin, John, 39

S

Sadler, S. W., 96

Saint, Thomas, 34

Sandham, Elizabeth, 49n82

Sedgwick, Adam, 224

Selby, Prideaux John, 74, 177, 177n23

Shaw, George, 11

Shelley, Mary, 207

Sherwood, Mary, 42–43

Smollett, Tobias, 26n8

Society for Promoting Christian
Knowledge, 75n17, 87n50

Society for the Prevention of Cruelty
to Animals, 12, 47n77,
121, 121n154

Stanley, Edward Lord, 177, 178

Stanley, Henry Morton, 207

Stevens, Frank, 188n71

Strahan, Alexander, 89, 121
See also Good Words for the Young

Sunday Magazine, The, 75n17

Surrey Zoological Gardens, 26, 62,
105n92, 193, 201

Swift, Jonathan, 166

Sydenham Crystal Palace
Aquarium, 132n24

308 INDEX

T

Taxidermy, 114, 258
 See also Ploucquet, Herman; Potter,
  Walter; Waterton, Charles
Taylor, Tom, 188n72
Tegetmeier, William Bernhardy,
  235, 237
Tennant, James, 134, 135
Tenniel, John, 192, 192n89, 193,
  193n92, 193n93, 193–194n96,
  199n113, 214n164, 232
Thackeray, William Makepeace, 26n8
Topsell, Edward, 33
Tower of London, 1, 2, 270
Tower of London (Royal menagerie),
  25, 26, 65, 75
Tradescant, John, 273, 276
Trimmer, Joshua, 231n34
Trimmer, Sarah, 41–45, 44n68,
  44n69, 47, 48, 50n84, 57, 65,
  126, 127, 144, 144n61, 199
 *Fabulous Histories: Designed for the
  Instruction of Children
  Respecting Their Treatment of
  Animals,* 44n68, 44n69, 45,
  48, 57, 65, 199 (*see also*
  Trimmer, Sarah, *History of the
  Robins, The*)
 *Family Magazine; or a repository of
  religious instruction and
  rational amusement,* 42
 Guardian of Education, 42
 History of the Robins, The, 43,
  144n61 (*see also* Trimmer,
  Sarah, *Fabulous Histories:
  Designed for the Instruction of
  Children Respecting Their
  Treatment of Animals*)
Tristram, Henry Baker, 107, 107n99
Tucker, Charlotte Maria,
  55n93, 56, 57
 See also A. L. O. E.

Tussaud's, Madame, 2, 270
Twining, Elizabeth, 91n59
Tyndall, John, 224

U

Unger, Franz, 231n34

V

Victoria, 220, 246
Victoria, Queen, 7
Volliner, W. F., 231n34
 See also Zimmerman, W. F. A.

W

Wakefield, Priscilla Bell, 41
Warne, Frederick, 83–85
Waterton, Charles, 171,
  171n4, 173
Watts, Isaac, 58, 192, 216
Webb, Thomas W., 91n59
Whewell, William, 224
White, Gilbert, 90
Wilton, R., 154n94
Wollstonecraft, Mary, 44n69, 59n104
Wood, John George, 75n17,
  91n59, 131n21, 133,
  160, 187n67,
  194n96, 199
Wood, Theodore, 75n17, 87n50
Woodward, Alice
  Bolingbroke, 242n63
Woodward, Benjamin, 186n64
 See also Oxford University Museum
  of Natural History
Wootton, John, 35n39
Wordsworth, William, 24, 26
World Museum Liverpool, 107n99
Wortley, Edward Stuart, 92
Wright, Anne, 91n59

Wright, Bryce McMurdo, 134, 135
Wyatt, Mary, 133
Wyatt, Matthew Digby, 226
Wycliffe Museum, 74, 177n23

Y
Yonge, Charlotte, 94

Z
Zimmerman, W. F. A., 231n34, 270
 See also Volliner, W. F.
Zoological Society Museum,
 177n23, 179
Zoological Society of London, 132,
 132n23, 200
Zornlin, R. M., 91n59